AF361707

Animal Welfare Assessment: Novel Approaches and Technologies

Animal Welfare Assessment: Novel Approaches and Technologies

Editors

Melissa Hempstead
Danila Marini

MDPI • Basel • Beijing • Wuhan • Barcelona • Belgrade • Manchester • Tokyo • Cluj • Tianjin

Editors

Melissa Hempstead
Parasitology
AgResearch
Palmerston North
New Zealand

Danila Marini
School of Animal and
Veterinary Science
The University of Adelaide
Adelaide
Australia

Editorial Office
MDPI
St. Alban-Anlage 66
4052 Basel, Switzerland

This is a reprint of articles from the Special Issue published online in the open access journal *Animals* (ISSN 2076-2615) (available at: www.mdpi.com/journal/animals/special_issues/animal_welfare_assessment).

For citation purposes, cite each article independently as indicated on the article page online and as indicated below:

LastName, A.A.; LastName, B.B.; LastName, C.C. Article Title. *Journal Name* **Year**, *Volume Number*, Page Range.

ISBN 978-3-0365-5562-1 (Hbk)
ISBN 978-3-0365-5561-4 (PDF)

Cover image courtesy of Melissa Hempstead

Contents

About the Editors

Melissa Hempstead

Dr Hempstead received her Ph.D. in biological sciences from the University of Waikato (New Zealand) in partnership with AgResearch. Her thesis investigated improvements to disbudding protocols for dairy goat kids to improve animal welfare. Prior to this, she completed a Bachelor of Science with honours from the University of Waikato. Dr Hempstead was then recruited to a postdoctoral research associate position at Iowa State University College of Veterinary Medicine, where she managed a multi-year research project evaluating dairy goat welfare on commercial farms in the Midwest. Prior to her career in research, Dr Hempstead was a secondary school teacher of science, biology, and mathematics. Currently she has a postdoctoral scientist position with AgResearch investigating whether anthelmintic resistance is a health and welfare issue for sheep with internal parasites.

Danila Marini

Danila Marini is a researcher working at the University Adelaide, Australia. They currently work on a variety of projects, including work on providing pain relief to livestock and ensuring good welfare is maintained during livestock management. Danila obtained a Bachelor of Science (Animal Science) with Honours from The University of Adelaide and completed a Ph.D. and Postdoc at the University of New England and CSIRO. Danila has a keen interest in the learning capabilities of livestock and ensuring animal welfare is not compromised in a changing world.

Preface to "Animal Welfare Assessment: Novel Approaches and Technologies"

Over the last decade, public interest in how animals are raised and managed has grown substantially, and, as a result, there is an increased need for valid and reliable methods of assessing animal welfare. Welfare assessment is an important aspect in the continual improvement of animal welfare, with effective assessment leading to better welfare outcomes. There are, however, many challenges associated with animal welfare assessment, such as the validity and consistency of measures over time and across individuals and the utilization of multiple welfare indicators associated with different species and management styles. Research in this developing field is vital for improving the lives of animals in natural or managed environments.

This Special Issue is focused on recent research and reviews that investigate many areas of welfare assessment, such as novel approaches and technologies used to evaluate the welfare of farmed, captive, or wild animals. Key approaches included welfare assessment, the use of sensors or wearable technologies, non-invasive measures, novel biomarkers of stress, environmental effects on welfare, and discussions of natural behaviour in farm animal welfare.

This Special Issue will be useful for scientists, researchers, animal managers, and policy makers.

Melissa Hempstead and Danila Marini
Editors

animals

Article

Evaluating Potential Cetacean Welfare Indicators from Video of Live Stranded Long-Finned Pilot Whales (*Globicephala melas edwardii*)

Rebecca M. Boys [1,*], Ngaio J. Beausoleil [2], Matthew D. M. Pawley [3], Emma L. Betty [1] and Karen A. Stockin [1,2,*]

[1] Cetacean Ecology Research Group, School of Natural Sciences, College of Sciences, Massey University, Private Bag 102-904, Auckland 1142, New Zealand; e.l.betty@massey.ac.nz
[2] Animal Welfare Science and Bioethics Centre, School of Veterinary Science, College of Sciences, Massey University, Private Bag 11-222, Palmerston North 4442, New Zealand; n.j.beausoleil@massey.ac.nz
[3] School of Mathematical and Computational Sciences, College of Sciences, Massey University, Private Bag 102-904, Auckland 1142, New Zealand; m.pawley@massey.ac.nz
* Correspondence: r.boys@massey.ac.nz (R.M.B.); k.a.stockin@massey.ac.nz (K.A.S.)

Simple Summary: Cetacean strandings occur globally and can impact the welfare as well as the survival of the animals involved. Understanding the welfare status of stranded cetaceans is important to inform appropriate human intervention. However, there is a lack of knowledge on how to assess animal welfare in this context. Here, we used video footage of live stranded animals of four odontocete species to explore which proposed welfare indicators can be assessed at live stranding events. We identified and evaluated potential indicators that could be non-invasively assessed, including 10 non-behavioural and 2 composite behavioural indicators (category of many behaviours). The first data on the fine-scale behaviour of stranded odontocetes and associated human intervention during stranding responses are presented. Our findings suggest that remote assessments of stranded cetacean's welfare states are feasible. These data provide the foundation to develop a systematic, structured welfare assessment framework specific to stranded cetaceans that can inform conservation and management decisions.

Citation: Boys, R.M.; Beausoleil, N.J.; Pawley, M.D.M.; Betty, E.L.; Stockin, K.A. Evaluating Potential Cetacean Welfare Indicators from Video of Live Stranded Long-Finned Pilot Whales (*Globicephala melas edwardii*). *Animals* **2022**, *12*, 1861. https://doi.org/10.3390/ani12141861

Academic Editor: Francesco Filiciotto

Received: 21 June 2022
Accepted: 18 July 2022
Published: 21 July 2022

Publisher's Note: MDPI stays neutral with regard to jurisdictional claims in published maps and institutional affiliations.

Abstract: Despite the known benefit of considering welfare within wildlife conservation and management, there remains a lack of data to inform such evaluations. To assess animal welfare, relevant information must be captured scientifically and systematically. A key first step is identifying potential indicators of welfare and the practicality of their measurement. We assessed the feasibility of evaluating potential welfare indicators from opportunistically gathered video footage of four stranded odontocete species ($n = 53$) at 14 stranding events around New Zealand. The first stranded cetacean ethogram was compiled, including 30 different behaviours, 20 of which were observed in all four species. Additionally, thirteen types of human intervention were classified. A subset of 49 live stranded long-finned pilot whales (*Globicephala melas edwardii*) were assessed to determine indicator prevalence and to quantify behaviours. Four 'welfare status' and six 'welfare alerting' non-behavioural indicators could be consistently evaluated from the footage. Additionally, two composite behavioural indicators were feasible. Three human intervention types (present, watering, and touching) and five animal behaviours (tail flutter, dorsal fin flutter, head lift, tail lift, and head side-to-side) were prevalent (>40% of individuals). Our study highlights the potential for non-invasive, remote assessments via video footage and represents an initial step towards developing a systematic, holistic welfare assessment framework for stranded cetaceans.

Keywords: animal welfare assessment; behaviour; human intervention; marine mammal; cetacean; management; stranding; wildlife

1. Introduction

The welfare of free-ranging animals is increasingly recognised as important to conservation [1–3]. In addition, there is a growing acknowledgment that human activities may directly and indirectly compromise the welfare of wild animals [4]. However, the conservation of wildlife populations is often a focus of government regulations, policies, and biodiversity plans, and the welfare of the individual animals comprising such populations is often overlooked. This is despite the fact that animal survival, and thus conservation success, is inextricably linked to welfare [1,5].

Although the need to assess wild animal welfare has been highlighted [6–9], there are limited systematic, scientific protocols for such assessments [10]. Furthermore, detailed behavioural and physiological data from species in the wild are often lacking [11], hindering the development of welfare assessments for wild populations [12]. Thus, a first step to progressing systematic and holistic welfare assessment for free-living wild animals is developing methods to capture relevant data. Such data need to be species/taxon- and context-specific and should address known or suspected welfare concerns [12,13]. Furthermore, to provide information about the welfare state of the animal, science-based indicators that can be observed and/or measured must be identified [12,14–16].

In the context of free-swimming cetaceans, data on stress hormones [17–19], body condition [20–22], skin disease [23–25], and the impacts of anthropogenic activities on behaviour [26–28] have been collected. However, few studies interpret their findings in terms of welfare or discuss possible welfare implications [9]. During live strandings, cetaceans are subject to both natural [29,30] and anthropogenic stressors [31] that may affect their welfare and survival likelihood [32]. Unfortunately, thus far, such data concerning behavioural and physiological indicators have not been gathered.

Major knowledge gaps concerning the welfare of stranded cetaceans were identified by international experts to be interpreting behavioural and physiological parameters, diagnosing internal injuries, and making end-of-life decisions [32]. Furthermore, these experts stated that major barriers to assessing the welfare of stranded cetaceans related to the limited relevant data collection at strandings and the lack of experts available onsite to interpret parameters and assist in decision making [32]. Notably, the characterisation of stranded cetacean welfare by those experts aligned with contemporary animal welfare science, which interprets interrelated aspects of health, biological function, and behaviour in terms of their impact on the animal's mental state [33,34]. Welfare assessments guided by such characterisations are often facilitated via the use of the Five Domains Model framework for assessing animal welfare [35]. In such a framework, the indicators in domains 1–4 are observed and/or measured, and their cumulative impacts are used to cautiously infer the animal's potential affective state (mental state) in the fifth domain [5,36,37].

Subsequently, these same experts proposed a range of potential welfare indicators for stranded cetaceans [38]. The potential indicators could be grouped into three physical/functional (nutrition, physical environment, and health) and one situation-related domain (behavioural interactions) of the Five Domains Model [35]. The proposed indicators included animal-based parameters, reflecting some aspect of the physical (e.g., body condition), physiological (e.g., respiration rate), or behavioural state of the animal (e.g., vocalisation). Other indicators were resource-/management-based parameters, reflecting aspects of the stranded cetacean's environment (e.g., substrate or duration stranded) or management (e.g., human interaction) that may influence its welfare [39,40]. Resource-/management-based indicators provide welfare-relevant information but do not provide direct evidence of the welfare state and are thus characterised as 'welfare alerting' [12]. Only animal-based indicators can provide direct information about the animal's 'welfare status' and are often preferred in welfare assessments. However, some animal-based indicators may only be 'welfare alerting' in that they can indicate a predisposition for welfare impacts that relate to the animal itself rather than its environment, for example, an animal that is neonatal or unweaned. Welfare alerting indicators are generally more feasible and reliable to assess across time and different observers and are often non-invasive. They are therefore commonly applied in welfare assessments.

To successfully apply indicators in a welfare assessment framework, the feasibility of measuring the indicators, methods of measurement, and validity for inferring welfare states (i.e., mental states) from observable indicators must be evaluated [10,41,42]. In this study we examined the feasibility of assessing animal-based and resource-/management-based indicators proposed by experts at live cetacean stranding events [38]. Furthermore, since experts highlighted the need for assessments to be undertaken by remote, skilled personnel [32], we evaluated which indicators can be observed and/or measured using video footage gathered at live strandings.

Specifically, this study evaluates the use of video footage to (1) identify potential animal and resource-/management-based welfare indicators that could be feasibly measured, (2) examine why certain proposed welfare indicators cannot be identified and/or feasibly measured, and (3) assess whether indicators observed from pilot whales can be quantitatively evaluated via video. There are currently no ethograms available for stranded cetaceans, and there is limited detail on the types of human intervention employed at stranding events. Therefore, we sought to identify and characterise all stranded cetacean behaviours (animal-based indicators) displayed and to provide the first description of the types of human intervention (resource-/management-based indicators) that occurred at these same stranding events. Additionally, we examined whether the features of the stranding circumstances affected the prevalence, frequency, or duration of behaviours displayed by stranded pilot whales.

2. Materials and Methods

2.1. In-field Data Collection

Due to the stochastic nature of strandings, video footage was collected opportunistically at 14 live stranding events between August 2010 and March 2022 around New Zealand (Table S1). Filming occurred with 53 live stranded cetaceans involving four species of odontocete: long-finned pilot whale (*Globicephala melas edwardii*), pygmy killer whale (*Feresa attenuata*), Cuvier's beaked whale (*Ziphus cavirostris*), and Gray's beaked whale (*Mesoplodon grayi*). Most stranding events (11 events, 49 individuals) involved long-finned pilot whales, herein referred to as pilot whales. Accordingly, the analyses presented here focus only on pilot whales. Additional data from the other examined species are included in the initial ground-truthing only to identify and characterise all animal behavioural and human intervention indicators.

The camera set-up varied based on the opportunistic nature of events and equipment availability (Table S1). When the researchers could attend a stranding, two GoPro Hero 7 Black video cameras (GoPro Inc., San Mateo, CA, USA) were mounted on wooden stakes anchored into the ground at 1–2 m from the focal animal. Each camera was mounted at a height of 50–100 cm, positioned cranio-laterally, and angled caudally (0–45°) towards the tail flukes for each focal individual. These recordings were made at 720 p and 60 fps with a wide-angle view, allowing for the focal animal's entire body to be observed bilaterally. Where researchers were unable to access the animals prior to re-floatation, footage was acquired from stranding personnel, including Department of Conservation (DOC) rangers (the government agency responsible for the management of stranding events), marine mammal medics, and the public using camera phones, GoPro cameras, or other similar video cameras. In such circumstances, the videographer stood 1–2 m from the animal and, when possible, positioned themselves cranio-laterally to the focal individual's head, angling the camera caudally towards the tail flukes. The videographer alternated positions around the animal to enable the entire body to be observed bilaterally. In other cases, individuals were filmed from a lateral position, capturing the entire body on one side. The filming duration was dependent upon battery availability, time of day, and the stranding response procedures in progress.

2.2. Selection of Potential Welfare Indicators

Based on the opinions of an international panel of experts in cetacean biology, veterinary medicine, and/or animal welfare [38], we developed a list of theoretically observable/measurable parameters that could be used as potential welfare indicators for stranded cetaceans. The list also included parameters that were deemed observable/measurable from an initial viewing of the video footage collected during the stranding events (Table S1). The indicators and composite behavioural parameters (category including many different behaviours, each of which would be considered an indicator) were organised into the three physical/functional domains (nutrition, physical environment, and health) and the situation-related domain (behavioural interactions) of the Five Domains Model for welfare assessment [35] (Table 1). Within each domain, indicators were further split into animal-based indicators that may directly reflect animal state ('welfare status'), and 'welfare alerting' indicators (both animal- and resource-/management-based), which provide relevant information about the animal or its environment that may affect its state [12] (Table 1).

Table 1. Proposed animal welfare indicators, or composite indicators *, organised into the three physical/functional domains and one situation-related domain of the Five Domains Model for welfare assessment [35]. Within each domain, indicators are organised according to the type of information they provide about the animal's state. See text for details of each indicator and how it was measured or scored.

Domain	Indicators	
	Welfare Status	**Welfare Alerting**
1: Nutrition	Body condition	Animal age class
2: Physical environment	Skin condition/blistering	Initial strand vs. re-strand Dry strand vs. in-water strand Availability of equipment Substrate type * Weather, sea, and tidal conditions
3: Health	Signs of trauma, injuries Signs of skin illness and disease Respiration rate and character/effort Heart rate Bleeding/fluids/mucus from orifices	
4: Behavioural interactions	Body posture * Movements and behaviours Animal vocalisation	Presence and status of pod members * Type and duration of human interaction

2.3. Data Scoring

Each video file, for all species, was examined manually at 0.8× speed by the lead author (RMB) at least twice to identify all observable indicators for each focal individual. A subset of videos was examined by two independent observers to ensure consistency in indicator classification. For each animal, information was collated about which indicators could be assessed (Table 1) and, for those indicators that can vary bilaterally, whether they could only be assessed on the left, right, or both sides. The reasons that particular indicators could not be observed and/or measured for each individual cetacean were also noted. Since 92.5% (*n* = 49) of individual focal stranded cetaceans were pilot whales (Table S1), only data related to that species were subsequently analysed and are presented here. To be considered feasible, an indicator had to be fully assessable (across the whole body) and prevalent, being observed in at least 40% of the pilot whales.

Domain 1: Nutrition

Body condition was assessed visually based on the concavity of the epaxial musculature and nuchal crest following Joblon et al. [43]. The four-point body condition score was assessed as (1) emaciated: severe concavity of the epaxial musculature, visibility of the ribs, and deep depression of the nuchal crest; (2) thin: mild to moderate concavity of the epaxial musculature, no visible ribs, and moderate depression of the nuchal crest; (3) normal: no concavity of the epaxial musculature, no visible ribs, and mild to no depression of the nuchal crest; and (4) robust: convexity of the epaxial musculature and a slight convexity of the nuchal crest (Figure 1).

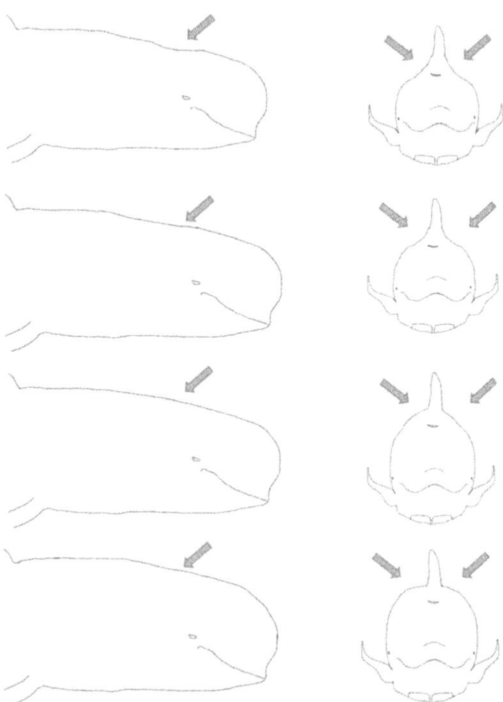

Figure 1. Four-point visual body condition scoring system developed for long-finned pilot whales (*Globicephala melas edwardii*) in this study.

The age class of the animal was qualitatively assessed based on the approximate length relative to the known adult length for the species [44]. Animals were assigned to one of three categories: adult, juvenile, or calf. As the sex of all animals could not be assessed, we assigned adults to be those animals of more than ~432 cm [44]. Juveniles were estimated to be over one third of the length of an adult, while calves were determined to be less than one third of the adult length and/or with foetal folds still visible.

Domain 2: Physical environment

The severity of any skin blistering was qualitatively scored following Groch et al. [45] based on the presence of superficial dermal necrosis (level 1), developed cutaneous bullae (level 2), or developed dermo-epidermal clefting with ulceration (level 3; Figure 2).

Figure 2. Level of skin blistering observed in individual focal animals. (**1**) Dermal necrosis and (**2**) bullae development on two individuals (top), (**2**) bullae development and (**3**) recent dermo-epidermal clefting with ulceration (middle), (**3**) dermo-epidermal clefting with ulceration two days after initial stranding (bottom). Photos credits: Kyle Mulinder (top and middle) and Project Jonah NZ (bottom).

When available, information was collected from stranding response forms about the focal animal's stranding circumstance, specifically, if this was an initial stranding or whether the animal had previously been re-floated and then subsequently re-stranded. Whether the animal was dry-stranded (i.e., on sand only, with no water around the whole body) or in-water-stranded (i.e., whole body surrounded by shallow water but not floating) was determined from the video footage. For animals that were filmed over a prolonged period, the animals were classified as dry-stranded or in-water-stranded based on the conditions present for the longest period during the filming.

The availability of basic stranding response equipment was assessed based on what was in view on the video footage of the focal individual. This included sheets for covering the animals, buckets for pouring water over the animal, spades for digging, and re-floatation mats. The substrate type was assessed from the video footage based on whether the focal animal was stranded on (1) mud flats, (2) sandy beach, (3) pebble beach, or (4) rocky shore. The substrate information was used to provide additional context to potential welfare concerns such as external injuries.

Weather and sea swell were assessed based on what could be viewed on the video. Weather conditions were categorised as (1) sunny, (2) overcast, or (3) precipitation. For animals filmed in prolonged stranding events, the weather conditions were classed as those most prevalent during filming. Due to the potential impact of swell height on the ability to attempt re-floatation, sea conditions were qualitatively assessed based on the approximate swell height as (1) minimal to small swell, ankle- to waist-high waves; (2) medium swell, waist- to shoulder-high waves, or (3) large swell, head-high and larger waves. The tidal conditions were assessed based on whether the tide was low or high and flooding or ebbing, based on tidal charts [46] for the specific stranding date, time, and location.

Domain 3: Health

Externally visible injuries were qualitatively assessed as being superficial or penetrating wounds and were classified by the location on the body. Skin illness/disease was scored based on the perceived appearance of characterised cutaneous manifestations known to occur due to specific infections/diseases [23,24,47–49], including "tattoo", "rounded cutaneous", "whitish velvety", and "whitish to slightly pink verrucous" skin lesions, following Van Bressem et al. [47]. Skin illness/disease lesions were assessed based on being present/absent and the area of the body involved [47].

The respiration rate was assessed based on the visible opening and closing of the blowhole and the audible sound of the focal animal exhaling. The respiration rate was quantitatively assessed, with each audible and visible open/close of the blowhole considered to be a single respiration [50]. The respiration character/effort was qualitatively assessed by examining whether the inhalation and closure of the blowhole occurred immediately following exhalation or there was a period (measured in seconds) between exhalation and inhalation [51,52]. Additionally, unusual respiration was noted through a qualitative assessment of the blowhole opening and closing and audible exhalation, such as whether the animal exhaled twice before inhaling or displayed chuffing [51,53].

The heart rate was quantified, when possible, as a recorded count of the rhythmic movement of the skin [54] on the ventrum, medial to the left pectoral fin. However, the heart rate was only observable in animals in lateral recumbency, as movement in the area close to the ventral surface of the left pectoral fin must be visible. Each of the animal's observable orifices were examined throughout the video duration to assess for any blood, mucus, or other fluids being expelled. Any such excretions were noted qualitatively based on the frequency and the orifice of origin.

Domain 4: Behavioural interactions

Body posture was assessed based on the animal's recumbency position: (1) ventral (lying on the ventrum), (2) lateral (lying on one side of the body), or (3) dorsal (lying on the dorsal surface of the body). Animals could be scored in multiple positions during a video; for example, they may have been moved from a lateral to a ventral position as part of the standard stranding response procedures [31]. Additionally, body posture was assessed based on whether the animal exhibited spinal curvature, most often observed in the peduncle. This feature was assessed based on continuous presence/absence throughout the observation period and was categorised as (1) lateral curvature: body or peduncle is curved laterally to the left or right (Figure 3), (2) dorsal curvature: peduncle is curved dorsally, or (3) ventral curvature: peduncle is curved ventrally.

Figure 3. Example of spinal curvature: left lateral curvature of the peduncle in a stranded long-finned pilot whale. Photo credit: Kyle Mulinder.

Animal movements were assessed based on the type of behaviour and its prevalence, frequency, and relative duration (see Section 2.3.1). Additionally, audible vocalisations were assessed based on the presence/absence and duration as part of the behavioural analysis (see Section 2.3.1). Since it was not possible to confirm whether audible vocalisations were from the focal animal or another animal in the immediate vicinity during mass strandings, the social circumstances of the focal animal was noted when vocalisations were recorded.

Some video footage provided observation of additional stranded animals, and further information on the stranding event was gathered from DOC stranding reports. This enabled the evaluation of whether the pod members of the focal stranded animal were present/absent and, when present, whether pod members were (1) alive or dead and (2) stranded or floating. Human intervention was considered to occur when a human interacted with a focal animal (see Section 2.3.2). The intervention was assessed based on the presence/absence and type of interaction occurring.

2.3.1. Development of Ethogram

Video footage was examined using the program BORIS v7.9.6 [55] to develop a comprehensive ethogram that represents all behaviours observed for the four species of stranded cetaceans, including unusual and rare occurrences [56]. The preliminary ethogram was based on five behaviours detailed in the literature relating to decision-making on the re-floatation of stranded odontocetes: body posture, arching, thrashing, trembling, and vocalisation [31,57–59]. However, these specific behaviours have not previously been described in detail, and their occurrence was not quantified in those studies. Thus, to begin, those behaviours were identified and defined for the two stranded pygmy killer whales due to the length of the video footage available (5 h; Table S1). The footage was then re-examined to characterise all other behaviours expressed by the pygmy killer whales until no new behaviours were noted. This updated ethogram ($n = 20$ behaviours) was then applied to the footage of the other species and stranding events, with new behaviours identified and characterised if there was no prior observation. Additionally, two physiological parameters (respiratory rate and heart rate) were included in the ethogram, as their frequency and duration could be calculated from video footage.

2.3.2. Human Intervention

Video footage was also examined to identify and characterise the types of human intervention for inclusion in the ethogram. The footage for each focal cetacean was examined manually at $0.8\times$ speed by the lead author (RMB) at least twice to identify and ensure intra-observer reliability of characterisation. Additionally, the same two independent observers examined a subset of videos to ensure consistency in the characterisation of intervention types.

Human intervention occurring at live stranding events includes up-righting animals, covering them in wet sheets, and pouring water over the body to reduce the risk of hyperthermia and sunburn [31]. However, previous studies have not provided a detailed characterisation of the types of human intervention occurring with live stranded cetaceans. In this study, a human intervention was considered to occur when a human was observed on the video footage within 1–2 metres of the focal cetacean. Again, the video footage of the pygmy killer whales was examined to characterise all types of human intervention until no new interventions were observed. This ethogram of human intervention was then applied to all other stranding events. New intervention types were identified and characterised if there was no prior observation.

2.4. Analysis of Pilot Whale Data

All behavioural and physiological parameters and human interventions identified in the ethogram for the 49 pilot whales were characterised and coded per second using BORIS v7.9.6 [55]. The prevalence, frequency, and relative duration of the behavioural and physiological parameters and human interventions were calculated from the quantitative scores and standardised by each video's duration to remove any time bias. The behavioural parameters and human interventions were classified as point/event behaviours when they had very short and non-variable durations or as behavioural states when their durations varied. The prevalence of each behaviour and type of human intervention was determined as the percentage of individual pilot whales displaying the parameter or being exposed to the intervention at least once during the observation period. The frequency of point/event behaviours was calculated as the mean rate per minute, including only the individuals displaying that particular behaviour, and the variability was calculated as the standard error of the mean (SEM). The average relative duration of each state behaviour or human intervention was calculated as a percentage of the observation period, including only those individuals that displayed the behaviour or were exposed to the intervention, with variability presented as the range of relative durations.

We further examined whether the features of the stranding circumstances of the individual pilot whales affected the prevalence, frequency, or duration of expression of prevalent behavioural and physiological indicators. We examined the effect of stranding number (initial vs. re-strand) and circumstance (dry vs. in-water) on the prevalence of behaviours and physiological parameters using a Z-test for proportions and on the frequency of point/event behaviours and physiological parameters and the relative durations of state behaviours using a Mann–Whitney U test. To ensure valid statistical inferences, only prevalent parameters (observed in >40% individuals) were included in the analyses. The effects of these features on animals' durations in different postural positions were not evaluated, as they were likely affected by human intervention rather than varying according to the focal animal's state.

3. Results

A total of 427.2 min (7.1 h) of video footage was collected from 11 mass and 3 single stranding events, with observations of 53 focal individuals of four species (Table S1). The duration of focal individual observations ranged from 10 s to 212.9 min (3.6 h) (mean: 483.6 s; 8.1 min).

3.1. Feasibility of Welfare Indicators for Stranded Pilot Whales

There were 49 video clips of individual pilot whales, for a total of 93.5 min (1.6 h), with a mean length of 114.5 s (1.9 min). A total of 16 pilot whales (32.7%) were observable on both sides of the body, while 16 (32.7%) were observable on the left side only, and 17 (34.7%) were observable on the right side only. Table 2 shows the associated results of the 17 non-behavioural welfare indicators that were assessed. Of these, four welfare status indicators were feasible to assess from video footage of more than 40% of the stranded pilot whales. The welfare status indicators that could not be consistently assessed were heart rate, skin blistering, trauma/injuries, and skin disease. A further six welfare alerting indicators were also feasible to assess in at least 40% of pilot whales via video footage, while the other three indicators required data to be gathered from DOC stranding response forms.

Table 2. Non-behavioural welfare indicators assessed for 49 live stranded long-finned pilot whales across 11 stranding events between August 2010 and March 2022 on the New Zealand coast. The number of animals for which the indicator was feasible to assess across the whole body and the percentage of animals for each parameter.

Domain	Welfare Status Indicator (no. Feasible)	Percentage of 49 Individuals (n)	Welfare Alerting Indicator (no. Feasible)	Percentage of 49 Individuals (n)
1: Nutrition				
	Body condition (29)		**Animal age class (49)**	
	Thin	14.3 (7)	Adult	79.6 (39)
	Normal	85.7 (42)	Juvenile	10.2 (5)
			Calf	10.2 (5)
2: Physical environment				
	Skin blistering (8)		**Substrate type (49)**	
	Superficial dermal necrosis	18.4 (9)	Sand beach	100 (49)
	Cutaneous bullae	18.4 (9)	**Stranding circumstance (49)**	
	Dermo-epidermal clefting/ulceration	20.4 (10)	Initial strand	40.8 (20)
			Re-strand	59.2 (29)
			Dry strand	63.3 (31)
			In-water strand	36.7 (18)
			Equipment (49)	
			Sheets covering animal	40.8 (20)
			Buckets pouring water	65.3 (32)
			Spades	20.4 (10)
			Weather (49)	
			Sun	34.7 (17)
			Overcast	65.3 (32)
			Sea condition (24)	
			Minimal/small swell	44.9 (22)
			Medium swell	2.0 (1)
			Tide (49)	
			High	20.4 (10)
			Low	55 (27)
			Incoming	22.4 (11)
			Receding	2.0 (1)

Table 2. *Cont.*

Domain	Welfare Status Indicator (no. Feasible)	Percentage of 49 Individuals (n)	Welfare Alerting Indicator (no. Feasible)	Percentage of 49 Individuals (n)
3: Health				
	Trauma/injuries (8)			
	Superficial wounds	8.2 (4)		
	Penetrating wounds	2.0 (1)		
	Skin illness/disease (8)			
	Present	2.0 (1)		
	Respiration (33)			
	Unusual respiratory character	8.2 (4)		
	Heart rate (3)			
	Bleeding/fluid/mucus from orifice (47)			
	Mucus from mouth	4.1 (2)		
	Mucus from blowhole	4.1 (2)		
	Dark-green fluid from anus	6.1 (3)		
4: Behavioural interactions				
	Curvature of peduncle (49)		**Pod members (49)**	
	Left	12.2 (6)	Present	95.9 (47)
	Right	8.2 (4)	**Status pod members (49)**	
			All alive	4.1 (2)
			All dead	2.0 (1)
			Alive and dead	89.8 (44)
			Stranded	93.9 (46)

Each indicator assessed per domain is boldened.

Domain 1: Nutrition

Body condition was feasible to fully assess in 29 (59.1%) stranded pilot whales. For the remaining animals, sheets covered the epaxial musculature ($n = 20$). Thus, the main visual assessment was based on the concavity of the nuchal crest. Most individuals (85.7%, $n = 42$) were in normal body condition and were adults (79.6%, $n = 39$).

Domain 2: Physical environment

Due to animals being covered in sheets, just eight (16.3%) animals could be assessed across all body regions bilaterally. A further 21 (42.9%) could be assessed across all body regions on one side (10 on the left and 11 on the right). Of these, skin blistering was observed in 72.4% ($n = 21$). Skin blistering around the cranial region (including the mandibles, melon, and blowhole; see Figure 2) could be assessed in all animals—an additional seven animals had blistering present in this cranial region. The level of blistering varied among the 28 affected pilot whales (Table 2).

The stranding circumstance of being in-water- or dry-stranded was feasible to assess in 100% of cases (Table 2). Further information gathered from stranding reports indicated that, at the time of filming, more than half the animals had re-stranded.

The availability of basic stranding response equipment could be assessed in all cases, with variable equipment available (Table 2). The substrate at the stranding location was identified to be sandy in 100% of cases, though in three cases shells were present.

The weather was feasible to assess in all videos. For most animals (65.3%, $n = 32$), the weather was overcast, while for the remainder, it was sunny. Over half of the pilot whales were observed at low tide, with the tidal conditions varying for the rest. The distant low tide mark meant that the sea condition could not be assessed for 26 animals, while most of the remainder were observed with minimal swell (Table 2).

Domain 3: Health

Injuries and wounds across the head and flukes were feasible to assess in all animals, while eight (16.3%) could be assessed across all body regions bilaterally, and 21 (42.9%) could be assessed on one side. Injuries and wounds were rare and mainly involved superficial lacerations (Table 2); these injuries in two animals were likely related to the substrate containing shells. Similarly, skin lesions indicative of illness/disease were not feasible to assess for the 20 animals covered by sheets, and a further 21 could only be assessed on one side. Of the 29 animals that were assessed, one was observed to have tattoo-like lesions on the cranial region (Figure 4; Table 2).

Figure 4. (Left): Observation of mucus from the blowhole and mouth of two live stranded long-finned pilot whales. Area considered the cranial region is defined within the white pentagon. (**Right**): Tattoo-like lesions (within white oval) observed on the cranial region of one individual. Photo credits: Kyle Mulinder (**Left**) and Project Jonah NZ (**Right**).

All respiratory events would have been observable via the video footage if they had occurred. However, due to the short length of some videos, respiration was only observed in 67.3% ($n = 33$) of the animals. In four of these animals, an unusual respiratory character was noted; one animal displayed double chuffing, with short forceful exhalations occurring twice prior to inhalation for almost every respiratory event. Three animals displayed an extended time between exhalation and inhalation. Indeed, in one animal the blowhole remained open for 6 s post-exhalation and prior to inhalation. The heart rate was only feasible to assess in three animals since the other individuals in the necessary position (lateral or dorsal recumbency) were in water ($n = 9$) or were filmed at an angle not conducive to observing the ventrum ($n = 1$).

Bleeding/fluid/mucus from orifices was readily assessable in the case of the blowhole and mouth (95.9%, $n = 47$) of the animals. Mucus excretion was observed to occur in four animals, two from the mouth and two from the blowhole (Figure 4). The genital and anal orifices were less observable due to most animals being in ventral recumbency (71.4%, $n = 35$). However, three animals were observed to defecate dark-green liquid (Figure S1).

Domain 4: Behavioural interactions

Body posture was feasible to assess in 100% of the pilot whales, with most animals only in ventral or lateral recumbency throughout filming (Table 3). Nine (18.4%) were observed in both ventral and lateral recumbency, with an additional animal observed in ventral, lateral, and dorsal recumbency over 1.5 min. Spinal curvature (Figure 3) was feasible to assess in all cetaceans and was noted in ten individuals (20.4%; Table 2). Notably, four pilot whales had their pectoral fins oriented laterally and superior to the dorsal plane (Figure S2), and all were undergoing human intervention when filmed. Behavioural events were observed in 100% of individuals; detailed results are presented in Section 3.2. Audible vocalisations from animals were only detected at mass strandings; these were identified from video of five focal animals (10.2%; Table 3), three of which were identified as calves and two were adults in the presence of calves.

Nearly all (95.9%, $n = 47$) of the focal individuals formed part of mass strandings and, therefore, stranded conspecifics were also present (Table 2). Human interactions with focal animals were observed in 100% of events and included non-invasive (presence only) and invasive interactions (e.g., up-righting animals). Detailed information on the observed human interactions is provided in Section 3.3.

3.2. The Stranded Odontocete Ethogram

Thirty behaviours were identified and described for the four odontocete species when stranded. These included 6 point and 24 state behaviours (Table S2). Aside from the recumbency posture, behavioural parameters were not mutually exclusive in that multiple behaviours could be displayed by an individual simultaneously.

3.2.1. Quantifying Behavioural Observations: Pilot Whales

Table 3 shows how feasible each of the behavioural indicators were and provides the prevalence, frequency, and duration of the assessed behavioural and physiological parameters. Notably, almost all the behavioural indicators (93.3%, $n = 28$) would have been feasible to assess if they had occurred. Eye open left and right were not consistently feasible due to light conditions and pec joint movement was not feasible in covered animals.

Most pilot whales (71.4%, $n = 35$) were observed in ventral recumbency throughout filming. When this recumbency was noted, it lasted for an average of 88.5% of the observation period. A further 10 individuals were moved into ventral recumbency as part of the human intervention during filming. The remaining individuals were filmed in lateral recumbency (28.6%, $n = 14$), which on average, lasted for 60.7% of the observation.

Five behaviours were prevalent, being displayed by over 40% of the pilot whales: tail flutter (69.4%, $n = 34$), dorsal fin flutter (55.1%, $n = 27$), head lift (51%, $n = 25$), tail lift (46.9%, $n = 23$), and head side-to-side (42.9%, $n = 21$). The only behaviour observed in other species but not recorded in pilot whales was head arch. In contrast, nine behaviours were recorded only in pilot whales (Table S2), all with low prevalence (Table 3).

When observed, individuals spent, on average, more than half the monitored time displaying right pectoral fin flutter (57.7%) and tail flutter (54.6%). The mean percentage of the observation period spent displaying dorsal fin flutter in those that did was 41.4%. Although prevalent, head lifting occurred, on average, for only 12.3% of the observation period, whereas tail lift and head side-to-side, also both prevalent, occurred for nearly a quarter of the observation period. All point behaviours had low prevalence and rates of occurrence.

Respiration was recorded at a mean rate of 4.4 breaths/min (SEM $\pm$ 0.4). Notably, inspiration occurred simultaneously with head lifting in nearly 45% of occurrences. The mean recorded heart rate was 48.8 beats/min (SEM $\pm$ 11.6).

Table 3. Observed prevalence (% of individuals displaying or for which the indicator was feasible), mean frequency (rate/minute), or mean relative duration (% of observation period and range) for only long-finned pilot whales that displayed the behaviour, from a total of 49 individuals across 11 stranding events between August 2010 and March 2022. See Table S2 for descriptions of behaviours.

Behaviour	Prevalence	Frequency	Relative Duration
State behaviours			
Ventral recumbency	91.8		88.5 (7.7–100.0)
Lateral recumbency	28.6		60.7 (4.4–99.8)
Dorsal recumbency	2.0		3.4 (3.4–3.4)
Tail flutter	69.4		54.6 (4.6–99.9)
Dorsal fin flutter	55.1		41.4 (6.1–97.3)
Head lift	51.0		12.3 (2.3–32.6)
Tail lift	46.9		23.9 (0.6–72.9)
Head side-to-side	42.9		22.2 (1.1–81.5)
Pec fin flutter R	24.5		57.7 (3.8–98.6)
Pec fin flutter L	22.4		34.8 (4.3–34.75)
Pec joint moves	20.4		22.5 (0.5–78.3)
Tail hover	18.4		22.0 (0.2–55.2)
Tail side-to-side	16.3		15.8 (0.3–46.3)
Body tremble	12.2		24.3 (0.2–84.2)
Vocalisation	10.2		20.7 (5.1–60.2)
Body rocking	10.2		10.4 (5.2–23.2)
Eye open L	10.2		35.4 (2.4–73.7)
Eye open R	8.2		22.1 (2.5–63.8)
Body tenses	6.1		11.2 (5.1–20)
Tail arch	4.1		15.8 (12.4–19.1)
Tail fluke slapping	4.1		28.3 (19.9–36.7)
Whole body arching/thrashing	4.1		5.4 (2.4–8.4)
Mouth open	2.0		17.2 (17.2–17.2)
Point behaviours			
Blowhole twitch	22.4	4.7 ± 1.7	
Nuchal pad twitch	10.2	6.2 ± 2.9	
Open and close blowhole	6.1	3.0 ± 1.6	
Water from blowhole	4.1	2.1 ± 1.4	
Head–pec fin jerk/flinch	2.0	2.6 ± 0.0	
Movement in lower jaw	2.0	6.9 ± 0.0	
Physiological parameters			
Respiration	67.3	4.4 ± 0.4	
Heartbeat	6.1	48.8 ± 11.6	

Differences in Stranding Circumstances: Initial vs. Re-Strand

Of the 49 pilot whales observed, 29 (59.2%) were filmed during a re-stranding, while the remainder (40.8%, $n = 20$) were filmed during their initial stranding event. Body tremble, mouth open, and movement in the lower jaw were only displayed by animals that were stranded for the first time, while the head–pec fin jerk was only observed in re-stranded animals (Table 4).

One prevalent behaviour, dorsal fin flutter, was displayed by a significantly ($z = 2.33$, $p = 0.03$) greater proportion of initially stranded animals than re-stranded animals (Table 4). No evidence of differences in the duration of prevalent behaviours or the rate of respiration was detected ($z = -0.83$, $p = 0.41$; Table 4).

Table 4. Observed prevalence (% of individuals displaying behaviour), mean frequency (rate/minute) ± SEM of point behaviours, and mean relative duration (% of monitored time and range) of state behaviours for only long-finned pilot whales that showed the behaviour, from a total of 20 initial-stranded and 29 re-stranded individuals across 11 stranding events on the New Zealand coast between 2010 and March 2022. [†] Only prevalent indicators could be tested for statistical differences; [*] significant difference ($\alpha = 0.05$) in prevalence between stranding circumstances.

Behaviour	Prevalence		Frequency		Relative Duration	
	Initial Strand	Re-Strand	Initial Strand	Re-Strand	Initial Strand	Re-Strand
State behaviours						
Ventral recumbency	90.0	93.1			85.9 (17.5–100.0)	90.3 (7.7–99.9)
Lateral recumbency	40.0	20.7			54.9 (4.4–99.8)	68.4 (7.2–97.4)
Dorsal recumbency	0.0	3.4			0.0 (0.0–0.0)	3.4 (3.4–3.4)
Tail flutter [†]	75.0	65.5			49.3 (4.6–99.9)	58.8 (12.9–98.1)
Dorsal fin flutter[†]	75.0 [*]	41.4 [*]			38.0 (6.1–95.6)	45.6 (13.0–97.3)
Head lift [†]	65.0	41.4			12.2 (2.6–32.4)	12.4 (2.3–32.6)
Tail lift [†]	45.0	48.3			27.4 (0.6–72.9)	21.6 (2.3–54.1)
Head side-to-side [†]	60.0	31.0			22.6 (5.5–81.5)	21.7 (1.1–47.4)
Pec fin flutter R	35.0	17.2			55.2 (3.8–98.6)	61.2 (30.0–93.7)
Pec fin flutter L	30.0	17.2			22.7 (4.3–69.2)	49.2 (12.2–87.0)
Pec joint moves	25.0	17.2			10.6 (0.5–30.0)	34.5 (11.3–78.3)
Tail hover	20.0	17.2			20.0 (0.2–55.2)	23.6 (2.2–55.1)
Tail side-to-side	15.0	17.2			31.1 (11.7–46.3)	6.7 (0.3–11.8)
Body tremble	30.0	0.0			24.3 (0.2–84.2)	0.0 (0.0–0.0)
Vocalisation	10.0	10.3			33.5 (6.7–60.2)	12.2 (5.1–24.4)
Body rocking	15.0	6.9			11.7 (5.2–23.2)	8.6 (5.9–11.3)
Eye open L	15.0	6.9			37.5 (14.2–73.7)	32.2 (2.4–61.9)
Eye open R	10.0	6.9			33.2 (2.5–63.8)	11.1 (2.9–19.3)
Body tenses	10.0	3.4			14.2 (8.4–20.0)	5.1 (5.1–5.1)
Tail arch	5.0	3.4			19.1 (19.1–19.1)	12.4 (12.4–12.4)
Tail fluke slapping	5.0	3.4			19.9 (19.9–19.9)	36.7 (36.7–36.7)
Whole body arching/thrashing	5.0	3.4			8.4 (8.4–8.4)	2.4 (2.4–2.4)
Mouth open	5.0	0.0			17.2 (17.2–17.2)	0.0 (0.0–0.0)
Point behaviours						
Blowhole twitch	25.0	20.7	6.2 ± 3.0	5.3 ± 1.9		
Nuchal pad twitch	10.0	10.3	1.9 ± 0.7	9.6 ± 4.0		
Open and close blowhole	10.0	3.4	2.8 ± 1.2	5.9 ± 0.0		
Water from blowhole	5.0	3.4	3.6 ± 0.0	0.7 ± 0.0		
Head–pec fin jerk/flinch	0.0	3.4	0.0 ± 0.0	2.6 ± 0.0		
Movement in lower jaw	5.0	0.0	6.9 ± 0.0	0.0 ± 0.0		
Physiological parameters						
Respiration rate [†]	65.0	69.0	3.8 ± 0.7	4.6 ± 0.6		
Heartbeat	10.0	3.4	33.8 and 71.5	41.1		

Differences in Circumstance: Dry vs. In-Water Strandings

Eighteen pilot whales (36.7%) were observed stranded in water, while 31 (63.3%) were recorded as dry-stranded. Body rocking and tail fluke slapping were only observed in individuals that were dry-stranded, while tail side-to-side, tail arch, whole body arching/thrashing, mouth open, head–pec fin jerk, and movement in the lower jaw were only displayed by animals stranded in water (Table 5).

Four prevalent behaviours were displayed in a significantly greater proportion of animals stranded in-water than dry-stranded animals: dorsal fin flutter ($z = -3.03$, $p = 0.00$), head lift ($z = -2.26$, $p = 0.03$), tail lift ($z = -3.29$, $p = 0.00$), and head side-to-side ($z = -2.57$, $p = 0.02$; Table 5). However, no evidence of differences was observed in the duration of prevalent behaviours or in the rate of respiration ($z = -0.97$, $p = 0.33$; Table 5).

Table 5. Observed prevalence (% of individuals displaying behaviour), mean frequency (rate/minute) ± SEM of point behaviours, and mean relative duration (% of monitored time and range) of state behaviours for only long-finned pilot whales that showed the behaviour, from a total of 31 dry- and 18 in-water-stranded individuals across 11 stranding events on the New Zealand coast between 2010 and March 2022. [†] Only prevalent indicators could be tested for statistical differences; * significant difference ($\alpha = 0.05$) in prevalence between stranding circumstances.

Behaviour	Prevalence		Frequency		Relative Duration	
	Dry	In-Water	Dry	In-Water	Dry	In-Water
State behaviours						
Ventral recumbency	90.3	94.4			97.7 (81.9–100.0)	73.3 (7.7–99.9)
Lateral recumbency	16.1	50.0			63.7 (4.4–99.8)	59.1 (7.2–97.3)
Dorsal recumbency	0.0	5.6			0.0 (0.0–0.0)	3.4 (3.4–3.4)
Tail flutter [†]	67.7	72.2			58.1 (4.6–99.9)	49.0 (13.0–92.4)
Dorsal fin flutter [†]	38.7 *	83.3 *			36.6 (6.1–87.6)	45.2 (12.3–97.3)
Head lift [†]	38.7 *	72.2 *			12.7 (2.6–32.6)	11.9 (2.3–32.4)
Tail lift [†]	29.0 *	77.8 *			22.2 (0.6–60.3)	24.9 (2.3–72.9)
Head side-to-side [†]	29.0 *	66.7 *			25.7 (5.5–81.5)	19.5 (1.1–47.4)
Pec fin flutter R	25.8	22.2			61.2 (3.8–98.6)	50.8 (30.0–93.7)
Pec fin flutter L	12.9	38.9			29.4 (4.3–69.2)	37.8 (7.0–87.0)
Pec joint moves	19.4	22.2			18.3 (1.2–46.2)	28.9 (0.5–78.3)
Tail hover	16.1	22.2			26.1 (2.2–55.2)	16.9 (0.2–38.7)
Tail side-to-side	0.0	44.4			0.0 (0.0–0.0)	15.8 (0.3–46.3)
Body tremble	9.7	16.7			28.3 (0.2–84.2)	20.3 (8.6–27.6)
Vocalisation	3.2	22.2			7.0 (7.0–7.0)	24.1 (5.1–60.2)
Body rocking	16.1	0.0			10.4 (5.2–23.2)	0.0 (0.0–0.0)
Eye open L	6.5	16.7			44.0 (14.2–73.7)	29.7 (2.4–61.9)
Eye open R	6.5	11.1			10.9 (2.5–19.3)	33.4 (2.9–63.8)
Body tenses	3.2	11.1			8.4 (8.4–8.4)	12.5 (5.1–20.0)
Tail arch	0.0	11.1			0.0 (0.0–0.0)	15.8 (12.4–19.1)
Tail fluke slapping	6.5	0.0			28.3 (19.9–36.7)	0.0 (0.0–0.0)
Whole body arching/thrashing	0.0	11.1			0.0 (0.0–0.0)	5.4 (2.4–8.4)
Mouth open	0.0	5.6			0.0 (0.0–0.0)	17.2 (17.2–17.2)
Point behaviours						
Blowhole twitch	16.1	38.9	5.6 ± 2.9	6.1 ± 2.3		
Nuchal pad twitch	6.5	11.1	9.5 ± 7.6	5.8 ± 2.1		
Open and close blowhole	3.2	16.7	0.4 ± 0.0	3.1 ± 1.6		
Water from blowhole	3.2	5.6	3.6 ± 0.0	0.7 ± 0.0		
Head–pec fin jerk/flinch	0.0	16.7	0.0 ± 0.0	1.4 ± 0.7		
Movement in lower jaw	0.0	5.6	0.0 ± 0.0	6.9 ± 0.0		
Physiological parameters						
Respiration rate [†]	67.7	61.1	4.1 ± 0.6	4.9 ± 0.8		
Heartbeat	9.7	0.0	48.8 ± 11.6			

3.3. Human Intervention with Stranded Odontocetes

From video footage of all stranded odontocetes, a total of 1061 events were coded from 13 different human interventions (Table S3). The types of human intervention were not mutually exclusive. Indeed, some types of human intervention always occurred simultaneously (e.g., human rolling an individual also required direct contact with the stranded animal).

Quantifying Human Intervention with Stranded Pilot Whales

All types of human intervention would have been feasible to assess if they occurred with the stranded pilot whales. Humans were present at all pilot whale stranding events that were observed, and on average, a human was within 2 metres of the focal animal (present) for 97% of the observed time (Table 6). Aside from human presence, the interventions that were most prevalent, occurring with over half of the stranded pilot whales, were human watering (65%) and human touching (59%). The interactions with the longest average duration per individual focal animal, aside from human presence, were human places sand by sides (96.8%), human touching (61.1%), and human noise (61.2%; Table 6).

Table 6. Types of human intervention that occurred with individual focal stranded pilot whales. Prevalence (% of individual focal stranded cetaceans that the intervention occurred with) and relative duration (% and range) of human intervention with individual focal stranded pilot whales (*n* = 49) calculated for those individuals undergoing the intervention type across 11 stranding events between 2010 and March 2022. See Table S3 for descriptions of intervention types.

Intervention	Prevalence	Relative Duration of Individual Monitoring
Present	100.0	97.4 (35.5–100)
Watering	65.3	36.0 (0.5–86.8)
Touching	59.2	61.1 (3.3–100)
Digging	36.7	51.3 (4.6–99.8)
Rolling	24.5	33.8 (0.4–93.5)
Noise	8.2	61.2 (16.4–98.8)
Holds dorsal fin	6.1	35.0 (2.9–97.6)
Places sand by sides	2.0	96.8 (96.8–96.8)
Rubbing	2.0	21.6 (21.6–21.6)

4. Discussion

A range of potential animal- and resource-/management-based welfare indicators were able to be non-invasively observed and/or measured in stranded cetaceans. We systematically characterised, for the first time, the ethology of stranded odontocetes with 30 different behaviours described. We quantitatively assessed these welfare indicators, including fine-scale behaviour and human intervention, from 49 live stranded pilot whales. Previous studies have highlighted the need for systematic assessment of wild cetacean welfare but have also emphasised challenges due to limited behavioural and physiological data [60,61]. Our study contributes pivotal baseline data that can be used to develop a feasible welfare assessment framework specific to cetacean strandings.

4.1. Holistic Welfare Assessments Are Feasible at Cetacean Stranding Events

A range of indicators related to different aspects of welfare were feasibly evaluated via video footage captured at cetacean strandings. Not only is this useful to enable remote experts to undertake animal assessments [32,38] but the non-invasive measurability of these indicators minimises further welfare compromise for cetaceans that are experiencing physiological stress [29,30]. Although invasive measures (e.g., blood sampling to evaluate haematological parameters) are informative for assessing the health of wild cetaceans [62–64], the use of non-invasive methods for welfare assessments is preferable. Further focus should be to validate the scoring of these indicators from video against live observations and among various indicators that reflect the health and welfare status as well as with known survivorship data.

From the 18 proposed indicators and composite behavioural parameters (Table 1), 10 non-behavioural, 5 animal behaviour, and 3 human intervention indicators were prevalent and thus were feasible to assess from video footage. Importantly, the identified feasible indicators were representative of three physical/functional domains (nutrition, physical environment, and health) and one situation-related domain (behavioural interactions) of the Five Domains Model [35], suggesting that holistic welfare assessments of stranded cetaceans could be achievable using these indicators. Of these, nine welfare status indicators represented three domains. The most feasible to assess were body condition (D1: nutrition), respiration and bleeding/fluid/mucus from orifices (D3: health), and body posture and composite behavioural indicators (D4: behavioural interactions). Potential welfare alerting indicators that could be consistently assessed were the age class (D1), substrate type (D2), dry vs. in-water stranding (D2), the availability of equipment and weather conditions (D2), the presence of other pod members (D4), and the composite behavioural indicator related to the amount and type of human intervention (D4).

Some potential indicators could not be consistently assessed from the video footage. Heart rate could not be evaluated in most animals, as this required a postural position of lateral or dorsal recumbency. However, we do not recommend that stranded cetaceans be placed into lateral recumbency to facilitate the assessment of heart rate, as this may cause pulmonary compression [31]. Thus, heart rate is unlikely to be feasible as a remotely assessed indicator of the welfare state of stranded cetaceans, though in-field assessments via palpation may be possible with trained personnel.

Trauma/injuries, skin blistering, and skin disease could not be assessed across all body regions bilaterally in about 40% of pilot whales, as they were covered to reduce hyperthermia and sunburn risk [31]. Furthermore, in more than two thirds of cases, bilateral observation of an animal's body was not possible due to camera positioning. These factors likely negatively biased the prevalence of observed blistering and injuries. However, if systematic assessment frameworks were implemented to guide evaluations at strandings, video and/or photographs of all body regions could be rapidly captured before interventions occur, allowing for a subsequent assessment of these indicators. This would require minimal time involvement and thus would be unlikely to cause additional welfare compromise. The application of such a framework would ensure consideration of all relevant welfare information and facilitate holistic, multidimensional assessments [5,65].

Additionally, although respiratory events were feasible to assess in all video footage if they occurred, the short duration of some videos utilized in this study compromised our ability to assess the respiratory rate for every individual. Importantly, cetacean species have extended breath holds [66]. Thus, video footage should be collected for at least 5 min to enable assessment of the respiratory rate.

Our results suggest a similar behavioural repertoire among stranded odontocete species. Only one behaviour was not displayed by pilot whales (head arch), and this was only exhibited by two animals, one pygmy killer whale and one Cuvier's beaked whale, possibly indicating severe physiological stress [57,67]. In contrast, nine behaviours were only displayed by pilot whales, likely due to the small sample size of the other species (*n* = 4). Therefore, our findings contribute valuable baseline ethological data from which other studies can assess stranded odontocete behaviour, though future efforts should further examine species-specific differences.

Information on environmental conditions is important to provide context when interpreting welfare status indicators, such as behaviours, and can influence management decisions [42]. In our study, the substrate, whether animals were dry- or in-water-stranded, and the weather conditions could be easily assessed from video footage. However, other alerting indicators required additional information, for example, determining whether individuals were re-stranded required access to stranding reports. Multiple stranding events can cause compounding damage and sustained stress [30], which likely compromise both welfare and survival likelihood [30,32].

Interestingly, almost 60% of the pilot whales had stranded more than once when observed, suggesting that re-floated animals often do not remain at sea, despite re-floatation being considered a 'success' [68]. We examined whether stranding circumstances (re-stranded vs. initially stranded and dry- vs. in-water-stranded) affected the prevalent behaviours displayed by pilot whales, with some differences found (See Section 4.2 for further discussion). However, further data collection is required to enable correlations among resource-/management-based indicators and animal-based indicators to better understand the welfare risk they reflect [69].

4.2. Preliminary Welfare Assessment of Stranded Pilot Whales

Most pilot whales observed were mass stranded and were assessed as adults in normal body condition based on an external visual assessment of the epaxial musculature and the concavity of the nuchal crest [43]. This outwardly healthy appearance has been reported previously at mass strandings [70,71] and generally suggests that hunger or sickness likely have minimal impacts upon these individuals. In contrast, two single stranded animals were in poor (thin) body condition, suggesting they were likely experiencing welfare compromise in the form of hunger and thirst prior to stranding. Indeed, one of these individuals was a neonate that likely stranded due to maternal–filial separation [72,73], suggesting that the welfare of this animal was significantly compromised at stranding. Such animals are also suggested to have low survival likelihood, and end-of-life decisions or long-term captivity are generally indicated [74,75].

Few injuries were observed, with those noted considered to be superficial. These may have occurred due to the stranding event itself and were likely minimal due to the sandy substrate. External injuries are less frequently observed in mass stranded animals, whereas single strandings can be related to some form of trauma [72,76]. Likewise, fluid or mucus discharge from the mouth or blowhole was rare and, when present, was mild. Additionally, faeces were evident from only three animals involved in the same mass stranding. The presence of vomiting and/or faecal discharges can be indicative of underlying health conditions [77] as well as indicating that animals are stressed [57,58]. Prolonged vomiting or diarrhoea can lead to dehydration and therefore should be considered welfare-relevant and included in evaluations.

Notably, despite widespread human interventions, such as covering, and overcast weather conditions, nearly 60% of animals had skin blistering, with serious blistering developed on more than a third. Both the number of affected animals and the severity of skin blistering were likely underestimated since most individuals were covered in sheets and/or had only one side of the body visible in the videos. The common occurrence corroborates the opinions of experts who indicated sunburn as a major welfare concern [32] and suggested it as an indicator for assessing stranded cetacean welfare [38]. Severe forms involving dermo-epidermal clefting with ulceration (observed in 20.4% of pilot whales) are likely to cause pain [38] and critical fluid loss [78], leading to dehydration, hypovolemic shock [45,79], and potential infection. Our results suggest there is considerable cause for welfare concern for many 'managed' live stranded pilot whales based on this indicator alone. Additional assessment of weather conditions will be useful to predict any further skin damage that may occur. Future studies should assess the extent of fluid, protein, and electrolyte loss that may occur when bullae ulcerate and rupture, as this will likely impact both the welfare and survivorship of stranded cetaceans. We suggest that such indicator data will also be important to inform decision making around re-floatation versus euthanasia.

Lateral curvature of the caudal peduncle was noted in 20% of animals in all stranding circumstances. This posture has been reported in stranded cetaceans during rehabilitation and is proposed to predict reduced swimming ability and muscular myopathy [80,81]. Additionally, four animals were observed with their pectoral fins oriented laterally and superior to the dorsal plane, which may indicate damage to joints, such as dislocations. Such postural abnormalities and/or underlying muscle or joint damage are likely to cause pain and, in the longer term, may detrimentally affect swimming and foraging ability [82]. Thus, such individuals may be deemed non-releasable [57,75,81,83]. Postural abnormalities should be correlated with other behavioural, physiological, and/or pathological indicators to better understand their welfare significance [42] and inform the use of this indicator in welfare assessments [13].

Almost all animals were observed in ventral recumbency for most of the video footage. This is likely due to the fact that human intervention occurred at all stranding events, and righting stranded cetaceans onto their ventrum is part of standard stranding response procedures [31,84]. This recumbency position is thought to reduce pulmonary compression compared to lateral recumbency [31] and should minimise the discomfort associated with breathing. Therefore, recumbency position should be considered in welfare assessments.

Interestingly, vocalisation during filming was rare and was only heard where focal animals were calves or adults in the presence of a calf, suggesting a possible maternal–filial connection. Previous studies suggest vocalisations are linked to cetacean welfare state in captive situations [85–87] and may affect epimeletic behaviour provided to wild distressed conspecifics [88]. Accordingly, we recommend additional data collection at stranding events to further assess the validity of vocalisations as a welfare indicator and to compile a comprehensive vocal repertoire for strandings.

All point behaviours had low prevalence and low rates of occurrence, meaning they will not be useful parameters for detecting any effects of environmental conditions or human interventions on cetacean welfare. In contrast, five state behaviours were prevalent, being displayed by more than 40% of the pilot whales (tail flutter, dorsal fin flutter, head lift, tail lift, and head side-to-side). When expressed, tail flutter and dorsal fin flutter were displayed, on average, for more than 40% of the observation time. Additionally, though less prevalent, right pectoral fin flutter occurred for more than 50% of observation time when expressed. Fin fluttering behaviours may be forms of muscle fasciculations or tremors. These fasciculations have previously been suggested as clinical signs of capture myopathy [30,81] and underlying health conditions [89]. Therefore, they are important to consider in welfare assessments.

Notably, dorsal fin flutter was observed in a significantly higher proportion of initial-stranded animals than re-stranded animals and in a greater proportion of in-water-strandings than dry-strandings. In the case of initial versus re-stranded animals, it may be that re-stranded animals become too fatigued to display dorsal fin fluttering. However, in the case of in-water versus dry-stranding, the expression of the behaviour appears to be context-specific and thus may represent the animal's response to its situation. Therefore, the use of such a behaviour as a welfare indicator must consider the animal's conditions and must be interpreted in the specific context of the stranding. Such behaviours may also be affected by human interventions and thus could be used to evaluate the effects on potential welfare state [90]. Future work should correlate these behaviours with physiological and/or pathological indicators to validate their reflection of welfare states [13] and inform their use in decision making.

Although prevalent within the study population, head and tail lift were displayed on average for only 12% and 24% of the observation time, respectively. Notably, both behaviours occurred in a significantly larger proportion of animals that were in-water-stranded than dry-stranded, suggesting that their expression may be context-specific. However, these behaviours may be precursors to arching, which was not observed in pilot whales but is proposed to be a sign of severe physiological stress in cetaceans [57,67]. Further data collection on these behaviours and their correlations with the specific stranding contexts should be undertaken to better understand the welfare state they may reflect and inform their use in welfare assessments for decision-making.

Many of the head lifting events occurred simultaneously with respiration. This is likely due to compression of the thoracic cavity when the animal is not supported by water (the case for all pilot whales in this study), which can cause breathing difficulties [51,67]. Furthermore, three pilot whales from the same mass stranding displayed delayed inhalation following exhalation for up to 6 seconds. Such respiratory delays are suggested to be indicative of shock and typically imply an end-of-life decision [51]. Further observation of head lifting during respiration events and delayed inhalation and the correlation of these behaviours with pathology will be important to assess, as this could provide data to infer the unpleasant experience of breathlessness [91,92]. These indicators should be considered important aspects to include in welfare assessments [69,93] and to inform decision-making around re-floatation versus euthanasia.

There were negligible differences in the frequency and duration of prevalent behaviours between initial-stranded animals and those observed during re-stranding and between dry- and in-water-stranded animals. This may be due to the inherent physiologically stressful situation of stranding, whereby behavioural differences caused by stranding circumstances are likely minimal. However, it is also possible that the lack of statistically significant effects is due to the sample size being too small to detect biologically relevant differences in behavioural expression. These common behaviours should be further correlated with physiological and/or pathological indicators to better understand their welfare significance [13]. They can then be considered for investigating the effects of various human interventions or stranding situations on animal welfare [42,90,93].

Human presence occurred nearly constantly for almost all observed pilot whales. Watering, touching, and digging out occurred with more than a third of the pilot whales and, when occurring, lasted for more than a third of the observation period. These high levels of interventions may negatively affect the welfare state of stranded cetaceans since humans may be perceived as threatening [35], particularly when encountered in an inherently physiologically stressful situation. However, appropriate, minimal intervention may also reduce other welfare concerns. For example, the provision of sheets and cooling water over the body should reduce the risk of hyperthermia and sunburn [31], which may otherwise cause pain and discomfort [32,38]. Future research should examine differences in stranded cetacean behavioural and physiological parameters with and without human intervention to investigate the effects of differing interventions on animal welfare [13].

4.3. Study Considerations

Due to the stochastic nature of stranding events, opportunistic filming by the public was an important data source in our study. Despite many videos being short in duration, we were able to identify and evaluate physical and environmental indicators and characterise behaviour. Similar video lengths have been used elsewhere [10,94]. However, these data are unlikely to provide accurate estimates of the behavioural time budgets and respiratory rates of stranded cetaceans. Furthermore, welfare compromise is expected to worsen throughout a stranding [30,32], and time stranded is considered a major concern for survival likelihood [32]. Accordingly, we recommend the application of standardized methods for data collection as a routine part of cetacean stranding response. This should include video recording from cameras mounted on poles and a longer filming duration, ideally from the onset of stranding to re-floatation or euthanasia, in order to fully evaluate the severity, duration, and progression of welfare impacts [95,96]. Standardized and continuous automated data collection will facilitate further investigation of indicators and the effects of human activities without hindering timely intervention to improve animal welfare and survival likelihood.

The experts consulted in Boys et al. [38], considered animal responsiveness via reflex testing to be a valuable and practical indicator. However, this was not tested at the stranding events presented here, despite it featuring in the New Zealand Standard Operating Procedures for cetacean strandings [75]. Nonetheless, it is likely that responsiveness could be evaluated via video footage with correct camera positioning. Thus, its feasibility should be assessed at future stranding events. Other valuable measures, such as body temperature, may also be taken in-field to augment the remote evaluation from video, though this may be limited by equipment and the availability of appropriately trained/skilled personnel. Finally, future studies should aim to collect data from both single and mass strandings to enable the statistical evaluation of the effects of stranding type on the presented indicators. The evaluation of these additional data will ensure comprehensive welfare assessments at future stranding events to inform decision making.

5. Conclusions

Video data provided valuable welfare-relevant information and highlighted the potential for experts to undertake assessments remotely. Importantly, our findings present an initial proof of concept concerning the feasibility of non-invasive welfare indicators, including behaviour, relevant to stranded odontocetes. However, additional data are required to explore the value of such indicators for predicting stranding outcomes, such as remaining at sea following re-floatation and longer-term survival, and to understand the effects of environmental conditions and human interventions on welfare and survivorship. Such information will better support decision-making concerning re-floatation versus euthanasia. Our study highlights the value of applying the Five Domains Model to facilitate holistic welfare assessments, allowing for more rapid informed prognoses of individual cetaceans. Including indicators that are practical to measure and validated in welfare assessment protocols will allow for more holistic, transparent, and justifiable evaluations of stranded cetacean welfare states. This will facilitate appropriate management interventions, leading to the best animal welfare and conservation outcomes from stranding events.

Supplementary Materials: The following supporting information can be downloaded at: https: //www.mdpi.com/article/10.3390/ani12141861/s1. **Figure S1**: Observation of dark green liquid (within black ovals) defecated from live-stranded long-finned pilot whale. Photo credit: Rob Leenheer; Figure S2: Observation of pectoral fin oriented laterally and superior to dorsal plane (within black ovals) in live stranded long-finned pilot whale. Photo credit: Kyle Mulinder; Table S1: Stranding events (n = 14) and details of video footage collected of individual live cetaceans (n = 53) of four odontocete species between August 2010 and March 2022, New Zealand. Only pilot whale data were used in analyses, with data from other species providing ground-truthing to identified behavioural indicators. In the case of mass strandings, footage may have included multiple individuals, however the video length noted included only the focal animal. *Note three animals were filmed both cranio-laterally and laterally; Table S2: Ethogram of stranded odontocete behaviour derived from video observations of 53 focal individuals (4 species, 14 stranding events) on the New Zealand coast between August 2010 and March 2022. Two physiological parameters are included. Note: behaviours displayed only by pilot whales** vs those not displayed by pilot whales*; Table S3: Types of human intervention that occurred with individual focal stranded cetaceans (n = 53; 4 species across 14 stranding events) on the New Zealand coast between August 2010 and March 2022.

Author Contributions: Conceptualization, R.M.B. and K.A.S.; methodology, R.M.B., N.J.B. and K.A.S.; validation, R.M.B., N.J.B. and K.A.S.; formal analysis, R.M.B. and M.D.M.P.; investigation, R.M.B. and M.D.M.P.; resources, R.M.B. and K.A.S.; data curation, R.M.B.; writing—original draft preparation, R.M.B.; writing—review and editing, N.J.B., M.D.M.P., E.L.B. and K.A.S.; supervision, N.J.B., M.D.M.P., E.L.B. and K.A.S.; project administration, K.A.S.; funding acquisition, R.M.B., E.L.B. and K.A.S. All authors have read and agreed to the published version of the manuscript.

Funding: This manuscript is part of Ph.D. research completed by Rebecca M. Boys. Rebecca M. Boys was supported by an Association of Commonwealth Universities Doctoral Scholarship, and Karen A. Stockin was supported by a New Zealand Royal Society Te Apārangi Rutherford Discovery Fellowship (2019–2024). The research was additionally supported by an Animal Ethics Inc. Research Grant USA, GoPro Inc. USA, the Wildbase Research Trust Fund New Zealand, the New Zealand Veterinary Association Marion Cunningham Memorial Fund Grant, and the Animal Behavior Society Amy R. Samuels Grant USA. The funders had no involvement in the study design, data collection, analysis, interpretation, or the writing of the article.

Institutional Review Board Statement: This project was evaluated by the Massey University Animal Ethics Committee and was approved under notification number 20/10. Research permits from the New Zealand Department of Conservation were also approved following indigenous Māori consultation under permit number 86298-MAR.

Informed Consent Statement: Not applicable.

Data Availability Statement: Restrictions apply to the availability of these data. Data were obtained from the Department of Conservation and members of the public and are available from the authors with the permission of mana whenua, the Department of Conservation, and the public that provided the video data.

Acknowledgments: We would like to thank the mana whenua (representing the Indigenous people of Aotearoa New Zealand) and staff at the Department of Conservation Te Papa Atawhai New Zealand. We would also like to thank Project Jonah and all the individuals who provided video footage of stranded cetaceans. Furthermore, we also thank IFAW for their continued advice, support, and collaboration on all the stranding work being undertaken. We would also like to thank the three anonymous reviewers for their comments, which greatly improved our manuscript.

Conflicts of Interest: The authors declare no conflict of interest.

References

1. Paquet, P.; Darimon, C. Wildlife Conservation and Animal Welfare Two Sides of the Same Coin. *Anim. Welf.* **2010**, *19*, 177–190.
2. Butterworth, A. *Marine Mammal Welfare: Human Induced Change in the Marine Environment and Its Impacts on Marine Mammal Welfare*; Animal Welfare; Butterworth, A., Ed.; Springer International Publishing: New York, NY, USA, 2017.
3. Scholtz, W. Injecting Compassion into International Wildlife Law: From Conservation to Protection? *Transnatl. Environ. Law* **2017**, *6*, 463–483. [CrossRef]
4. Fraser, D.; MacRae, A.M. Four Types of Activities That Affect Animals: Implications for Animal Welfare Science and Animal Ethics Philosophy. *Anim. Welf.* **2011**, *20*, 581–590.
5. Beausoleil, N.J.; Mellor, D.J.; Baker, L.; Baker, S.E.; Bellio, M.; Clarke, A.S.; Dale, A.; Garlick, S.; Jones, B.; Harvey, A.; et al. "Feelings and Fitness" Not "Feelings or Fitness"-The Raison d'etre of Conservation Welfare, Which Aligns Conservation and Animal Welfare Objectives. *Front. Vet. Sci.* **2018**, *5*, 296. [CrossRef]
6. Fraser, D. Toward a Synthesis of Conservation and Animal Welfare Science. *Anim. Welf.* **2010**, *19*, 121–124.
7. Menkhorst, P.; Clemann, N.; Sumner, J. Fauna-Rescue Programs Highlight Unresolved Scientific, Ethical and Animal Welfare Issues. *Pac. Conserv. Biol.* **2016**, *22*, 301–303. [CrossRef]
8. Hampton, J.O.; Hyndman, T.H. Underaddressed Animal-Welfare Issues in Conservation. *Conserv. Biol.* **2019**, *33*, 803–811. [CrossRef] [PubMed]
9. Clegg, I.L.K.; Boys, R.M.; Stockin, K.A. Increasing the Awareness of Animal Welfare Science in Marine Mammal Conservation: Addressing Language, Translation and Reception Issues. *Animals* **2021**, *11*, 1596. [CrossRef]
10. Harvey, A.M.; Morton, J.M.; Mellor, D.J.; Russell, V.; Chapple, R.S.; Ramp, D. Use of Remote Camera Traps to Evaluate Animal-Based Welfare Indicators in Individual Free-Roaming Wild Horses. *Animals* **2021**, *11*, 2101. [CrossRef]
11. Hill, S.P.; Broom, D.M. Measuring Zoo Animal Welfare: Theory and Practice. *Zoo Biol.* **2009**, *28*, 531–544. [CrossRef]
12. Harvey, A.; Beausoleil, N.J.; Ramp, D.; Mellor, D.J. A Ten-Stage Protocol for Assessing the Welfare of Individual Non-Captive Wild Animals: Free-Roaming Horses (Equus Ferus Caballus) as an Example. *Animals* **2020**, *10*, 148. [CrossRef]
13. Watters, J.V.; Krebs, B.L.; Eschmann, C.L. Assessing Animal Welfare with Behavior: Onward with Caution. *J. Zool. Bot. Gard.* **2021**, *2*, 6. [CrossRef]
14. Richmond, S.E.; Wemelsfelder, F.; de Heredia, I.B.; Ruiz, R.; Canali, E.; Dwyer, C.M. Evaluation of Animal-Based Indicators to Be Used in a Welfare Assessment Protocol for Sheep. *Front. Vet. Sci.* **2017**, *4*, 210. [CrossRef] [PubMed]
15. Harley, J.J.; Stack, J.D.; Braid, H.; McLennan, K.M.; Stanley, C.R. Evaluation of the Feasibility, Reliability, and Repeatability of Welfare Indicators in Free-Roaming Horses: A Pilot Study. *Animals* **2021**, *11*, 1981. [CrossRef]
16. Whittaker, A.L.; Golder-Dewar, B.; Triggs, J.L.; Sherwen, S.L.; McLelland, D.J. Identification of Animal-Based Welfare Indicators in Captive Reptiles: A Delphi Consultation Survey. *Animals* **2021**, *11*, 2010. [CrossRef]
17. Siebert, U. Investigations of Thyroid and Stress Hormones in Free-Ranging and Captive Harbor Porpoises (Phocoena Phocoena): A Pilot Study. *Aquat. Mamm.* **2011**, *37*, 443–453. [CrossRef]
18. Burgess, E.A.; Hunt, K.E.; Kraus, S.D.; Rolland, R.M. Quantifying Hormones in Exhaled Breath for Physiological Assessment of Large Whales at Sea. *Sci. Rep.* **2018**, *8*, 10031. [CrossRef] [PubMed]
19. Teerlink, S.; Horstmann, L.; Witteveen, B. Humpback Whale (Megaptera Novaeangliae) Blubber Steroid Hormone Concentration to Evaluate Chronic Stress Response from Whale-Watching Vessels. *Aquat. Mamm.* **2018**, *44*, 411–425. [CrossRef]
20. Bradford, A.L.; Weller, D.W.; Punt, A.E.; Ivashchenko, Y.V.; Burdin, A.M.; VanBlaricom, G.R.; Brownell, R.L. Leaner Leviathans: Body Condition Variation in a Critically Endangered Whale Population. *J. Mammal.* **2012**, *93*, 251–266. [CrossRef]
21. Hart, L.B.; Wischusen, K.; Wells, R.S. Rapid Assessment of Bottlenose Dolphin (Tursiops Truncatus) Body Condition: There's an App for That. *Aquat. Mamm.* **2017**, *43*, 635–644. [CrossRef]
22. Christiansen, F.; Dawson, S.M.; Durban, J.W.; Fearnbach, H.; Miller, C.A.; Bejder, L.; Uhart, M.; Sironi, M.; Corkeron, P.; Miller, C. Population Comparison of Right Whale Body Condition Reveals Poor State of the North Atlantic Right Whale. *Mar. Ecol. Prog. Ser.* **2020**, *640*, 1–16. [CrossRef]
23. Van Bressem, M.; Waerebeek, K.; Aznar, F.; Raga, J.; Jepson, P.; Duignan, P.J.; Deaville, R.; Flach, L.; Viddi, F.; Baker, J.; et al. Epidemiological Pattern of Tattoo Skin Disease: A Potential General Health Indicator for Cetaceans. *Dis. Aquat. Org.* **2009**, *85*, 225–237. [CrossRef]
24. Hart, L.B.; Rotstein, D.S.; Wells, R.S.; Allen, J.; Barleycorn, A.; Balmer, B.C.; Lane, S.M.; Speakman, T.; Zolman, E.S.; Stolen, M.; et al. Skin Lesions on Common Bottlenose Dolphins (Tursiops Truncatus) from Three Sites in the Northwest Atlantic, USA. *PLoS ONE* **2012**, *7*, e33081. [CrossRef] [PubMed]

25. Clegg, I.L.K.; Borger-Turner, J.L.; Eskelinen, H.C. C-Well: The Development of a Welfare Assessment Index for Captive Bottlenose Dolphins (Tursiops Truncatus). *Anim. Welf.* **2015**, *24*, 267–282. [CrossRef]
26. Stockin, K.; Lusseau, D.; Binedell, V.; Wiseman, N.; Orams, M.B. Tourism Affects the Behavioural Budget of the Common Dolphin Delphinus Sp. in the Hauraki Gulf, New Zealand. *Mar. Ecol. Prog. Ser.* **2008**, *355*, 287–295. [CrossRef]
27. Meissner, A.M.; Christiansen, F.; Martinez, E.; Pawley, M.D.M.; Orams, M.B.; Stockin, K.A. Behavioural Effects of Tourism on Oceanic Common Dolphins, Delphinus Sp., in New Zealand: The Effects of Markov Analysis Variations and Current Tour Operator Compliance with Regulations. *PLoS ONE* **2015**, *10*, e0116962. [CrossRef]
28. Bas, A.A.; Christiansen, F.; Öztürk, B.; Öztürk, A.A.; Erdoğan, M.A.; Watson, L.J. Marine Vessels Alter the Behaviour of Bottlenose Dolphins Tursiops Truncatus in the Istanbul Strait, Turkey. *Endanger. Species Res.* **2017**, *34*, 1–14. [CrossRef]
29. Cowan, D.F.; Curry, B.E. Histopathology of the Alarm Reaction in Small Odontocetes. *J. Comp. Pathol.* **2008**, *139*, 24–33. [CrossRef] [PubMed]
30. Fernandez, A.; Bernaldo de Quiros, Y.; Sacchini, S.; Sierra, E. Pathology of Marine Mammals: What It Can Tell Us About Environment and Welfare. In *Marine Mammal Welfare*; Butterworth, A., Ed.; Springer International Publishing: Cham, Switzerland, 2017; pp. 585–608. ISBN 978-3-319-46993-5.
31. Geraci, J.R.; Lounsbury, V. *Marine Mammals Ashore: A Field Guide for Strandings*, 2nd ed.; National Aquarium in Baltimore: Baltimore, MD, USA, 2005.
32. Boys, R.M.; Beausoleil, N.J.; Pawley, M.D.M.; Littlewood, K.E.; Betty, E.L.; Stockin, K.A. Fundamental Concepts, Knowledge Gaps and Key Concerns Relating to Welfare and Survival of Stranded Cetaceans. *Diversity* **2022**, *14*, 338. [CrossRef]
33. Boissy, A.; Manteuffel, G.; Jensen, M.B.; Moe, R.O.; Spruijt, B.; Keeling, L.J.; Winckler, C.; Forkman, B.; Dimitrov, I.; Langbein, J.; et al. Assessment of Positive Emotions in Animals to Improve Their Welfare. *Physiol. Behav.* **2007**, *92*, 375–397. [CrossRef]
34. Mellor, D.J. Updating Animal Welfare Thinking: Moving beyond the "Five Freedoms" towards "A Life Worth Living". *Animals* **2016**, *6*, 21. [CrossRef] [PubMed]
35. Mellor, D.J.; Beausoleil, N.J.; Littlewood, K.; McLean, A.; McGreevy, P.; Jones, B.; Wilkins, C. The 2020 Five Domains Model: Including Human–Animal Interactions in Assessments of Animal Welfare. *Animals* **2020**, *10*, 1870. [CrossRef]
36. Beausoleil, N.J.; Mellor, D.J. Validating Indicators of Sheep Welfare. In *Achieving Sustainable Production of Sheep*; Greyling, J., Ed.; Burleigh Dodds Science Publishing: Cambridge, UK, 2017; ISBN 978-1-78676-084-5.
37. Mellor, D.J.; Beausoleil, N.J. Extending the "Five Domains" Model for Animal Welfare Assessment to Incorporate Positive Welfare States. *Anim. Welf.* **2015**, *24*, 241–253. [CrossRef]
38. Boys, R.M.; Beausoleil, N.J.; Pawley, M.D.M.; Littlewood, K.E.; Betty, E.L.; Stockin, K.A. Identification of Potential Welfare and Survival Indicators for Stranded Cetaceans through International, Interdisciplinary Expert Opinion. *R. Soc. Open Sci.* in review.
39. Fraser, D. *Understanding Animal Welfare: The Science in Its Cultural Context*; Wiley-Blackwell: Ames, IA, USA, 2008; ISBN 978-1-4051-3695-2.
40. Hemsworth, P.; Mellor, D.J.; Cronin, G.; Tilbrook, A. Scientific Assessment of Animal Welfare. *New Zealand Vet. J.* **2015**, *63*, 24–30. [CrossRef] [PubMed]
41. Kaurivi, Y.B.; Laven, R.; Hickson, R.; Stafford, K.; Parkinson, T. Identification of Suitable Animal Welfare Assessment Measures for Extensive Beef Systems in New Zealand. *Agriculture* **2019**, *9*, 66. [CrossRef]
42. Lesimple, C. Indicators of Horse Welfare: State-of-the-Art. *Animals* **2020**, *10*, 294. [CrossRef] [PubMed]
43. Joblon, M.; Pokras, M.; Morse, B.; Harry, C.; Rose, K.; Sharp, S.M.; Niemeyer, M.E.; Patchett, K.M.; Sharp, W.; Moore, M. Body Condition Scoring System for Delphinids Based on Short-Beaked Common Dolphins (Delphinus Delphis). *J. Mar. Anim. Ecol.* **2014**, *7*, 5–13.
44. Betty, E.L.; Stockin, K.A.; Hinton, B.; Bollard, B.A.; Smith, A.N.H.; Orams, M.B.; Murphy, S. Age, Growth, and Sexual Dimorphism of the Southern Hemisphere Long-Finned Pilot Whale (Globicephala Melas Edwardii). *J. Mammal.* **2022**, *103*, 560–575. [CrossRef]
45. Groch, K.R.; Diaz-Delgado, J.; Marcondes, M.C.C.; Colosio, A.C.; Santos-Neto, E.B.; Carvalho, V.L.; Boos, G.S.; Oliveira de Meirelles, A.C.; Ramos, H.; Guimaraes, J.P.; et al. Pathology and Causes of Death in Stranded Humpback Whales (Megaptera Novaeangliae) from Brazil. *PLoS ONE* **2018**, *13*, e0194872. [CrossRef]
46. NIWA Tidal Forecaster. Available online: https://tides.niwa.co.nz/ (accessed on 31 March 2022).
47. Van Bressem, M.F.; Reyes, J.; Felix, F.; Echegaray, M.; Siciliano, S.; Di Beneditto, A.; Flach, L.; Viddi, F.; Avila, I.; Herrera, J.; et al. A Preliminary Overview of Skin and Skeletal Diseases and Traumata in Small Cetaceans from South American Waters. *Lat. Am. J. Aquat. Mamm.* **2007**, *6*, 7–42. [CrossRef]
48. Van Bressem, M.-F.; Van Waerebeek, K.; Flach, L.; Reyes, J.; César, M.; Santos, O.; Siciliano, S.; Echegaray Skontorp, M.; Viddi, F.; Félix, F.; et al. *Skin Diseases in Cetaceans*; International Whaling Commission: Santiago, Chile, 2008.
49. Kautek, G.; Van Bressem, M.-F.; Ritter, F. External Body Conditions in Cetaceans from La Gomera, Canary Islands, Spain. *J. Mar. Anim. Ecol.* **2019**, *11*, 4–17.
50. Kremers, D.; Célérier, A.; Schaal, B.; Campagna, S.; Trabalon, M.; Böye, M.; Hausberger, M.; Lemasson, A. Sensory Perception in Cetaceans: Part II—Promising Experimental Approaches to Study Chemoreception in Dolphins. *Front. Ecol. Evol.* **2016**, *4*, 50. [CrossRef]
51. Mazzariol, S.; Cozzi, B.; Centelleghe, C. *Handbook for Cetaceans' Strandings*; NETCET; Massimo Valdina: Padua Italy, 2015.

52. Martins, M.C.I.; Miller, C.; Hamilton, P.; Robbins, J.; Zitterbart, D.P.; Moore, M. Respiration Cycle Duration and Seawater Flux through Open Blowholes of Humpback (Megaptera Novaeangliae) and North Atlantic Right (Eubalaena Glacialis) Whales. *Mar. Mammal Sci.* **2020**, *36*, 1160–1179. [CrossRef]

53. Fire, S.E.; Miller, G.A.; Wells, R.S. Explosive Exhalations by Common Bottlenose Dolphins during Karenia Brevis Red Tides. *Heliyon* **2020**, *6*, e03525. [CrossRef]

54. Al-Naji, A.; Tao, Y.; Smith, I.; Chahl, J. A Pilot Study for Estimating the Cardiopulmonary Signals of Diverse Exotic Animals Using a Digital Camera. *Sensors* **2019**, *19*, 5445. [CrossRef]

55. Friard, O.; Gamba, M. BORIS: A Free, Versatile Open-Source Event-Logging Software for Video/Audio Coding and Live Observations. *Methods Ecol. Evol.* **2016**, *7*, 1325–1330. [CrossRef]

56. Martin, P.; Bateson, P. *Measuring Behaviour. An Introductory Guide*, 3rd ed.; Cambridge University Press: Cambridge, UK, 2009.

57. Townsend, F.I. Medical Management of Stranded Small Cetaceans. In *Zoo and Wild Animal Medicine*; Fowler, M., Ed.; WB Saunders: Philadelphia, PA, USA, 1999; pp. 485–493.

58. Sampson, K.; Merigo, C.; Lagueux, K.; Rice, J.; Cooper, R.; Weber Iii, E.S.; Kass, P.; Mandelman, J.; Innis, C. Clinical Assessment and Postrelease Monitoring of 11 Mass Stranded Dolphins on Cape Cod, Massachusetts. *Mar. Mammal Sci.* **2012**, *28*, E404–E425. [CrossRef]

59. Sharp, S.M.; Knoll, J.S.; Moore, M.J.; Moore, K.M.; Harry, C.T.; Hoppe, J.M.; Niemeyer, M.E.; Robinson, I.; Rose, K.S.; Brian Sharp, W.; et al. Hematological, Biochemical, and Morphological Parameters as Prognostic Indicators for Stranded Common Dolphins (Delphinus Delphis) from Cape Cod, Massachusetts, U.S.A. *Mar. Mammal Sci.* **2014**, *30*, 864–887. [CrossRef]

60. Nicol, C.; Bejder, L.; Green, L.; Johnson, C.; Keeling, L.; Noren, D.; Van der Hoop, J.; Simmonds, M. Anthropogenic Threats to Wild Cetacean Welfare and a Tool to Inform Policy in This Area. *Front. Vet. Sci.* **2020**, *7*, 57. [CrossRef] [PubMed]

61. King, K.; Joblon, M.; McNally, K.; Clayton, L.; Pettis, H.; Corkeron, P.; Nutter, F. Assessing North Atlantic Right Whale (Eubalaena Glacialis) Welfare. *J. Zool. Bot. Gard.* **2021**, *2*, 52. [CrossRef]

62. Wells, R.S.; Rhinehart, H.L.; Hansen, L.J.; Sweeney, J.C.; Townsend, F.I.; Stone, R.; Casper, D.R.; Scott, M.D.; Hohn, A.A.; Rowles, T.K. Bottlenose Dolphins as Marine Ecosystem Sentinels: Developing a Health Monitoring System. *EcoHealth* **2004**, *1*, 246–254. [CrossRef]

63. Schwacke, L.H.; Smith, C.R.; Townsend, F.I.; Wells, R.S.; Hart, L.B.; Balmer, B.C.; Collier, T.K.; De Guise, S.; Fry, M.M.; Guillette, L.J., Jr.; et al. Health of Common Bottlenose Dolphins (Tursiops Truncatus) in Barataria Bay, Louisiana, Following the Deepwater Horizon Oil Spill. *Environ. Sci. Technol.* **2014**, *48*, 93–103. [CrossRef] [PubMed]

64. Barratclough, A.; Wells, R.S.; Schwacke, L.H.; Rowles, T.K.; Gomez, F.M.; Fauquier, D.A.; Sweeney, J.C.; Townsend, F.I.; Hansen, L.J.; Zolman, E.S.; et al. Health Assessments of Common Bottlenose Dolphins (Tursiops Truncatus): Past, Present, and Potential Conservation Applications. *Front. Vet. Sci.* **2019**, *6*, 444. [CrossRef]

65. Fraser, D.; Weary, D.M.; Pajor, E.A.; Milligan, B.N. A Scientific Conception of Animal Welfare That Reflects Ethical Concerns. *Anim. Welf.* **1997**, *6*, 187–205.

66. Berta, A.; Sumich, J.L.; Kovacs, K.M. Chapter 10—Respiration and Diving Physiology. In *Marine Mammals*, 3rd ed.; Berta, A., Sumich, J.L., Kovacs, K.M., Eds.; Academic Press: San Diego, CA, USA, 2015; pp. 299–343. ISBN 978-0-12-397002-2.

67. Townsend, F.I.; Smith, C.R.; Rowles, T.K. Health Assessment of Bottlenose Dolphins in Capture–Release Studies. In *CRC Handbook of Marine Mammal Medicine*; Gulland, F.M.D., Dierauf, L.A., Whitman, K.L., Eds.; CRC Press: Boca Raton, FL, USA, 2018; pp. 823–834. ISBN 978-1-315-14493-1.

68. Wiley, D.; Early, G.; Mayo, C.; Moore, M. Rescue and Release of Mass Stranded Cetaceans from Beaches on Cape Cod, Massachusetts, USA; 1990–1999 a Review of Some Response Actions. *Aquat. Mamm.* **2001**, *27*, 162–171.

69. Boulton, A.C.; Kells, N.J.; Cogger, N.; Johnson, C.B.; O'Connor, C.; Webster, J.; Palmer, A.; Beausoleil, N.J. Risk Factors for Bobby Calf Mortality across the New Zealand Dairy Supply Chain. *Prev. Vet. Med.* **2020**, *174*, 104836. [CrossRef]

70. Gales, N.J. Mass Stranding of Striped Dolphin, Stenella Coeruleoalba, at Augusta, Western Australia: Notes on Clinical Pathology and General Observations. *J. Wildl. Dis.* **1992**, *28*, 651–655. [CrossRef]

71. Bogomolni, A.L.; Pugliares, K.R.; Sharp, S.M.; Patchett, K.; Harry, C.T.; LaRocque, J.M.; Touhey, K.M.; Moore, M. Mortality Trends of Stranded Marine Mammals on Cape Cod and Southeastern Massachusetts, USA, 2000 to 2006. *Dis. Aquat. Org.* **2010**, *88*, 143–155. [CrossRef]

72. Diaz-Delgado, J.; Fernandez, A.; Sierra, E.; Sacchini, S.; Andrada, M.; Vela, A.I.; Quesada-Canales, O.; Paz, Y.; Zucca, D.; Groch, K.; et al. Pathologic Findings and Causes of Death of Stranded Cetaceans in the Canary Islands (2006–2012). *PLoS ONE* **2018**, *13*, e0204444. [CrossRef]

73. Câmara, N.; Sierra, E.; Fernández, A.; Suárez-Santana, C.M.; Puig-Lozano, R.; Arbelo, M.; Herráez, P. Skeletal and Cardiac Rhabdomyolysis in a Live-Stranded Neonatal Bryde's Whale With Fetal Distress. *Front. Vet. Sci.* **2019**, *6*, 476. [CrossRef] [PubMed]

74. Whaley, J.; Borkowski, R. *Policies and Best Practices: Marine Mammal Stranding Response, Rehabilitation, and Release: Standards for Release*; NOAA National Marine Fisheries Service: Silverspring, MD, USA, 2009.

75. Boys, R.M.; Beausoleil, N.J.; Betty, E.L.; Stockin, K.A. When and How to Say Goodbye: An Analysis of Standard Operating Procedures That Guide End-of-Life Decision-Making for Stranded Cetaceans in Australasia. *Mar. Policy* **2022**, *138*, 104949. [CrossRef]

76. Arbelo, M.; de Los Monteros, A.E.; Herráez, P.; Andrada, M.; Sierra, E.; Rodríguez, F.; Jepson, P.D.; Fernández, A. Pathology and Causes of Death of Stranded Cetaceans in the Canary Islands (1999-2005). *Dis. Aquat. Org.* **2013**, *103*, 87–99. [CrossRef]

77. Waltzek, T.B.; Cortés-Hinojosa, G.; Wellehan, J.F.X., Jr.; Gray, G.C. Marine Mammal Zoonoses: A Review of Disease Manifestations. *Zoonoses Public Health* **2012**, *59*, 521–535. [CrossRef]

78. Gales, N.; Woods, R.; Vogelnest, L. Marine Mammal Strandings and the Role of the Veterinarian. In *Medicine of Australian Mammals*; Vogelnest, L., Woods, R., Eds.; CSIRO Publishing: Clayton, Australia, 2008.

79. Martinez-Levasseur, L.M.; Gendron, D.; Knell, R.J.; O'Toole, E.A.; Singh, M.; Acevedo-Whitehouse, K. Acute Sun Damage and Photoprotective Responses in Whales. *Proc. Biol. Sci.* **2011**, *278*, 1581–1586. [CrossRef] [PubMed]

80. Gulland, F.M.D.; Dierauf, L.A.; Whitman, K.L. *CRC Handbook of Marine Mammal Medicine*, 3rd ed.; CRC Press: Boca Raton, FL, USA, 2018; ISBN 978-1-4987-9687-3.

81. Câmara, N.; Sierra, E.; Fernández, A.; Arbelo, M.; Bernaldo de Quirós, Y.; Arregui, M.; Consoli, F.; Herráez, P. Capture Myopathy and Stress Cardiomyopathy in a Live-Stranded Risso's Dolphin (Grampus Griseus) in Rehabilitation. *Animals* **2020**, *10*, 220. [CrossRef] [PubMed]

82. Fish, F.E. Balancing Requirements for Stability and Maneuverability in Cetaceans. *Integr. Comp. Biol.* **2002**, *42*, 85–93. [CrossRef] [PubMed]

83. Harms, C.A.; Greer, L.; Whaley, J.; Rowles, T.K. Euthanasia. In *CRC Handbook of Marine Mammal Medicine*; CRC Press: Boca Raton, FL, USA, 2018; pp. 675–691. ISBN 978-1-4987-9687-3.

84. Simeone, C.; Moore, K. Stranding Response. In *CRC Handbook of Marine Mammal Medicine*; Gulland, F., Dierauf, L., Whitman, K., Eds.; CRC Press: Boca Raton, FL, USA, 2018; pp. 3–13. ISBN 978-1-315-14493-1.

85. Castellote, M.; Fossa, F. Measuring Acoustic Activity as a Method to Evaluate Welfare in Captive Beluga Whales (Delphinapterus Leucas). *Aquat. Mamm.* **2006**, *32*, 325–333. [CrossRef]

86. Dibble, D.; Van Alstyne, K.; Ridgway, S. Dolphins Signal Success by Producing a Victory Squeal. *Int. J. Comp. Psychol.* **2016**, *29*, 1–10. [CrossRef]

87. Eskelinen, H.C.; Richardson, J.L.; Tufano, S. Stress, Whistle Rate, and Cortisol. *Mar. Mammal Sci.* **2021**, *38*, 1–13. [CrossRef]

88. Kuczaj, S.A.; Frick, E.E.; Jones, B.L.; Lea, J.S.E.; Beecham, D.; Schnöller, F. Underwater Observations of Dolphin Reactions to a Distressed Conspecific. *Learn. Behav.* **2015**, *43*, 289–300. [CrossRef] [PubMed]

89. van Elk, C.E.; van de Bildt, M.W.G.; Jauniaux, T.; Hiemstra, S.; van Run, P.R.W.A.; Foster, G.; Meerbeek, J.; Osterhaus, A.D.M.E.; Kuiken, T. Is Dolphin Morbillivirus Virulent for White-Beaked Dolphins (Lagenorhynchus Albirostris)? *Vet. Pathol.* **2014**, *51*, 1174–1182. [CrossRef] [PubMed]

90. Palmer, A.L.; Beausoleil, N.J.; Boulton, A.C.; Cogger, N. Prevalence of Potential Indicators of Welfare Status in Young Calves at Meat Processing Premises in New Zealand. *Animals* **2021**, *11*, 2467. [CrossRef] [PubMed]

91. Lansing, R.W.; Gracely, R.H.; Banzett, R.B. The Multiple Dimensions of Dyspnea: Review and Hypotheses. *Respir. Physiol. Neurobiol.* **2009**, *167*, 53–60. [CrossRef]

92. Beausoleil, N.J.; Mellor, D.J. Introducing Breathlessness as a Significant Animal Welfare Issue. *N Z Vet. J.* **2015**, *63*, 44–51. [CrossRef]

93. Yon, L.; Williams, E.; Harvey, N.D.; Asher, L. Development of a Behavioural Welfare Assessment Tool for Routine Use with Captive Elephants. *PLoS ONE* **2019**, *14*, e0210783. [CrossRef]

94. Ghaskadbi, P.; Habib, B.; Qureshi, Q. A Whistle in the Woods: An Ethogram and Activity Budget for the Dhole in Central India. *J. Mammal.* **2016**, *97*, 1745–1752. [CrossRef]

95. Mellor, D.; Patterson-Kane, E.; Stafford, K. *The Sciences of Animal Welfare*; Wiley-Blackwell: Hoboken, NJ, USA, 2009.

96. Mellor, D.J. Operational Details of the Five Domains Model and Its Key Applications to the Assessment and Management of Animal Welfare. *Animals* **2017**, *7*, 60. [CrossRef]

Article

Health and Welfare Survey of 30 Dairy Goat Farms in the Midwestern United States

Melissa N. Hempstead [†], Taylor M. Lindquist, Jan K. Shearer, Leslie C. Shearer and Paul J. Plummer *

Veterinary Diagnostic and Production Animal Medicine, College of Veterinary Medicine, Iowa State University, Ames, IA 50011, USA; melissa.hempstead@agresearch.co.nz (M.N.H.); tml@iastate.edu (T.M.L.); jks@iastate.edu (J.K.S.); lshearer@iastate.edu (L.C.S.)
* Correspondence: pplummer@iastate.edu
† Present address: AgResearch Ltd., Grasslands Research Center, Palmerston North 4410, New Zealand.

Simple Summary: There appears to be a rapid expansion of dairy goat farming in the United States and the information available to producers on health, welfare, and production applicable to those in the Midwestern US is limited. This study intended to survey 30 dairy goat farms in the Midwestern US to provide insight into husbandry practices pertaining to health, welfare, and production, and to identify areas of future research. Pain relief for disbudding and castration, education and training programs, early kid management, and hoof trimming were identified as potential areas of future research. This study provided insight into the husbandry practices carried out on 30 dairy goat farms in the Midwestern US and areas of research to improve health and welfare.

Abstract: Dairy goat production in the Midwestern United States is increasing at a rapid rate and information on dairy goat husbandry practices applicable for producers in this region is limited. The objective of this study was to survey 30 dairy goat farms in the Midwestern US to provide insight into husbandry practices pertaining to health, welfare, and production, and to identify areas of future research. A questionnaire was developed and comprised 163 questions that were organized into categories including information on the producer (e.g., farming experience), staff, and goats (e.g., herd size, breed), housing, feeding and nutrition, milking practices and production, kid management, husbandry practices (e.g., disbudding, castration, hoof trimming), and health. Areas of future research that can improve goat health, production and welfare include pain relief for husbandry practices such as disbudding and castration, early kid management during birth to prevent illness/disease or mortality (e.g., warm and dry areas for kid rearing), eradication programs for common contagious diseases, training programs and education for claw trimming, disbudding, and udder health. In conclusion, this study provided insight into the husbandry practices carried out on 30 dairy goat farms in the Midwestern US and areas of research to improve health and welfare.

Keywords: animal welfare; animal husbandry; welfare assessment; wellbeing; goat; caprine; dairy

Citation: Hempstead, M.N.; Lindquist, T.M.; Shearer, J.K.; Shearer, L.C.; Plummer, P.J. Health and Welfare Survey of 30 Dairy Goat Farms in the Midwestern United States. *Animals* **2021**, *11*, 2007. https://doi.org/10.3390/ani11072007

Academic Editor: Vera Baumans

Received: 12 May 2021
Accepted: 1 July 2021
Published: 5 July 2021

1. Introduction

In the United States (US) there are approximately 2.7 million goats, and of these, 440,000 are dairy goats [1]. The number of milking goats within the study population comprising Minnesota, Iowa, Wisconsin, and Illinois is 14,000, 29,000, 82,000 and 10,000, respectively [1], which is representative of approximately 30% of the total population of dairy goats in the US.

To date, information on dairy goat husbandry practices applicable for the Midwestern US is limited, which may be due to them historically being regarded as a minor species in comparison with dairy cattle [2]. However, demand for goat milk products, such as cheese and yoghurt, is rising and likely related to changes in demographic composition, which has driven the proliferation of dairy goat operations throughout the US [2].

Welfare assessment of dairy goat farms is increasing, with research available in the UK [3], Europe [4–7] and South America [8]. However, to our knowledge, there are few studies of dairy goat welfare assessment in the US, with the exception of our own study [9].

The objective of this study was to survey 30 dairy goat farms in the Midwestern US (enrolled in a welfare assessment study [9]) to provide insight into current husbandry practices pertaining to health, welfare, and production, and to identify potential areas of future research. Based on our knowledge of the Midwestern US dairy industry, we hypothesized that producers would have knowledge gaps in the areas of early kid management, and routine husbandry procedures, such as disbudding and hoof trimming.

2. Materials and Methods

This study was part of a larger study that performed welfare assessment of lactating dairy goats from 30 farms across the Midwestern US and identified the most prevalent welfare issues [9]. The present study was conducted between January and December 2019 and surveyed human participants; therefore, the survey protocol was reviewed by the Iowa State University Institutional Review Board (IRB; No. 18-497) prior to study commencement. Our protocol was deemed exempt from IRB approval. However, we ensured that the rights and privacy of the study participants were protected. After providing informed consent to be involved in this study, participants were given an identification number between 1 and 30, which was used to de-identify the collected data and distinguish between individual responses. The first author had access to the personal details of the participants, which were linked to their unique identification number on a computer that was stored in a lockable filing unit when not in use.

2.1. Survey Development

The questionnaire used to survey producers was developed from a review of available literature on existing surveys of goat producers, goat welfare assessment, producer attitude to goat behavior and welfare, husbandry practices and farm management [5,6,10–14] and the researchers' own experiences with dairy goat farming. The questionnaire was designed using a web-based data collection software (REDCap®, Vanderbilt University, Nashville, TN, USA). Questions were entered into REDCap®, which generated the questionnaire.

The draft questionnaire was initially reviewed by a small committee of veterinary practitioners and an animal scientist, and a social scientist in the Department of Sociology of Iowa State University. The modified draft questionnaire was then reviewed by eight members of the American Dairy Goat Association, a representative of a milk company operating within the Midwestern US, and two commercial dairy goat producers. Based on these reviews, a final version of the questionnaire was produced. The questionnaire comprised 163 questions that were organized into 11 major topics including producer-specific information (e.g., farming experience), staffing, goat-specific information (e.g., herd size, breed), housing and environment, goat behavior, feeding and nutrition, milking practices and milk production, goat kid management, husbandry practices (e.g., disbudding, castration, hoof trimming, euthanasia), cleaning and sanitation, and health and veterinary care. Information on participant demographics were also collected. The questionnaire included a mixture of multiple choice, rating scale/slider, Likert scale, matrix, and open-ended questions. The questionnaire took approximately 60 min to complete based on responses from those that reviewed the questionnaire during the drafting phase.

A blank copy of the questionnaire is available in the Supplementary Materials (Table S1).

2.2. Study Participants

Producers were recruited by distribution of advertising material by a milk company operating in the Midwestern region on our behalf, and visitation of the first author with a feed representative to directly distribute the advertising material over a month-long period. Participation was incentivized by receipt of compensation for the producers' involvement in this study. Those that indicated their interest in this study were provided an introductory letter describing the study objectives and additional information to help them decide whether to commit to enrolling in this study. Producers were enrolled in this study over a 6-month period and the questionnaire was distributed once informed consent was received. Participation was voluntarily and producers were self-selected. Further details of farmer recruitment are described by Hempstead et al. [9].

2.3. Survey Delivery

Questionnaires were sent to producers between February and November and responses were received between February and December 2019. Producers were sent a link to REDCap® (via email), which presented the questionnaire and collected their responses (18 participants). However, not all producers had email capabilities or computer access; therefore, a hardcopy version of the online questionnaire was mailed to these producers (12 participants). When required, producers were sent a reminder until all questionnaires had been received.

2.4. Data Management and Statistical Analysis

Data were exported from REDCap® as a comma-separated values file and used with Excel (Microsoft Corporation, Redmond, WA, USA). The data were presented as (1) percentages, with the actual number of farms in brackets and means with standard error, or (2) mean values with standard deviation (SD) or interquartile range (where appropriate). Due to the self-selected nature of the study participants, no formal hypothesis testing was conducted.

3. Results

3.1. Response/Completion Rate

All 30 producers completed the questionnaire, which included one respondent who completed the online questionnaire in part twice; these data were combined, and any differences in reported values were averaged. No duplicate questionnaires were entered. Eighteen questionnaires were completed online and 12 were completed on hardcopy. The number of days for producers to return the questionnaire ranged from 0 to 175 days, with a mean of 34 days (SD = 46).

3.2. Demographics

Producers had a mean age of 44 (SD = 11; range = 17–62 years old) and 64% (18/28) of producers were male (36% were female [10/28]; age not disclosed for two participants). The mean number of years of experience farming goats was 14 years (SD = 14; range = 3–52 years), and 73% (22/30) of producers had experience on a cow dairy; for these respondents, the mean number of years of experience was 19 years (SD = 10; range = 1–38 years).

3.3. Producer-Specific Information

Producers rated how they felt about animal welfare and whether it was a key priority for how their farm was run along a rating/slider scale (where numbers 0 through 100 were presented along a scale; 0 = strongly disagree, 50 = neither agree nor disagree, 100 = strongly agree). The responses of 29/30 producers (one participant did not respond) ranged from neither agree nor disagree to strongly agree (mean ± SD = 90 ± 11). In addition, producers were asked to rate the level of importance they placed on staff training along a rating/slider scale (where 0 = not at all important, 50 = important, 100 = very important). The responses of 26/30 producers (four participants did not respond) ranged from important to very important (mean ± SD = 77 ± 5). However, three of the producers rated training between not at all important and important (≤37 ± 5). Producers were asked what type of training was provided to staff; 13% (4/30) of producers responded they did not provide any training. The types of training provided by the remaining participants are presented in Table 1. Interestingly, producers that rated training as between not at all important and important reported they provided staff training. Those that provided no staff training did not provide an explanation.

Table 1. Type of staff training provided by dairy goat producers.

Staff Training	Percentage (Number) of Farms
Animal handling	81% (21/26)
Goat behavior	46% (12/26)
Kid rearing practices	88% (23/26)
Identifying sick/injured animals	85% (22/26)
Feeding/nutrition	65% (17/26)
Routine husbandry procedures (e.g., ear tagging, disbudding)	77% (20/26)
Machinery/equipment operation	50% (13/26)
Milking routines	92% (24/26)
Housing	38% (10/26)
Transportation of goats	27% (7/26)
Record keeping	62% (16/26)
Other—genetics and breeding	4% (1/26)

3.4. Farm Characteristics

Based on producer responses to the questionnaire, the median number of lactating does was 166 (IQR = 70), which ranged between 2 and 6500 does (mean $\pm$ SD = 618 $\pm$ 1519). The median herd size (i.e., number of lactating and non-lactating does) was 185 (IQR = 175), which ranged from 55 and 9000 does (mean $\pm$ SD = 788 $\pm$ 1964). The mean lactation length was 314 days (SD = 66; range = 190–600 days), with a mean yearly production of 7 pounds/doe/day (SD = 2; range = 5–11 pounds/doe/day). Ten percent (3/30) of producers responded as being certified organic. Ninety three percent of producers (28/30) farmed Saanen and Alpine breeds. American Lamancha, Anglo-Nubian, Toggenburg, Oberhasli, Sable and Kiko were present on 60% (18/30), 37% (11/30), 33% (10/30), 20% (6/30), 7% (2/30) and 3% (1/30) of farms, respectively.

3.5. Goat Husbandry

3.5.1. Resources

The most common diet included hay and grain/concentrate, which was approximately four-times more popular than fermented forage or other feeds (Table 2). Access to outdoor spaces were provided on 73% (22/30) of farms (Table 2), of which earthen and pasture surface types were most common compared to concrete and rock piles (Table 3).

Table 2. Dairy goat producer responses relating to resource availability.

Resources	Category	Percentage (Number) of Farms
Diet	Hay	90% (27/30)
	Grain/concentrate	90% (27/30)
	Fermented forage	23% (7/30)
	Total mixed ration (fermented and fresh forage)	10% (3/30)
	Fresh cut grass	7% (2/30)
	Corn	3% (1/30)
	Other—banana and watermelon peel	3% (1/30)
Outdoor pen surface	Earthen	86% (19/22)
	Pasture	59% (13/22)
	Concrete	27% (6/22)
	Rock piles	9% (2/22)

Table 3. Dairy goat producer responses relating to claw trimming practices.

Claw Trimming Practice	Category	Percentage (Number) of Farms
Operator	Producer	87% (26/30)
	Staff	20% (6/30)
	Friends or family	40% (12/30)
	Paid contractor(s)	27% (8/30)
	None, self-taught	42% (11/26)
	Trained by friends or family	46% (12/26)
Producer training	Trained by a veterinarian	8% (2/26)
	Trained by a paid contractor	4% (1/26)
	Other—reading material	8% (2/26)
Tool used	Grinder	10% (3/30)
	Hand-powered trimmer or shears	83% (25/30)
	Pneumatic hoof trimmer	13% (4/30)
	Blade	7% (2/30)
Frequency	Every 1–2 months	7% (2/30)
	Every 3–4 months	43% (13/30)
	Every 5–6 months	33% (10/30)
	Every 7–12 months	10% (3/30)
	When needed	7% (2/30)
Goat age at first claw trim	<3 months	3% (1/30)
	3–5 months	10% (3/30)
	6–8 months	33% (10/30)
	9–12 months	40% (12/30)
	Over 12 months	13% (4/30)

3.5.2. Milking Procedures

Machine milking was used on 93% (28/30) of farms. Hand milking was used on 7% (2/30) of farms. Udder preparation prior to milking was conducted on 90% (27/30) of farms; this included cleaning the teats of debris on 89% (24/27) of farms, sanitizing the teats on 44% (12/27) of farms and checking the fore milk on 30% (8/27) of farms. Ten percent (3/30) of producers reported that no udder preparation was carried out. Udder treatment following milking was carried out on 70% (21/30) of farms, and 100% (21/21) of these farms used a teat dip or spray and 23% (5/21) of these farms used teat conditioner; 30% (9/30) of producers reported they did not use any post-milking treatment.

Eighty percent (24/30) of producers said that they actively check for evidence of mastitis. Of these farms, 100% (24/24) assessed milk quality, 79% (19/24) assessed swelling, 75% (18/24) assessed heat and firmness to the touch and 58% assessed udder color (14/24).

3.5.3. Goat Husbandry Practices

All producers reported that claw trimming was carried out on the goats. Responses to questions on corrective claw trimming practice including who performs claw trimming and sources of training, the tool used, frequency and average age of goat at first claw trim are presented in Table 3.

Producer responses relating to health including treatment for gastrointestinal and external parasites, pain relief usage, and veterinary experience with goats are presented in Table 4. Producers rated how often veterinary treatment was sought along a rating/slider scale (where 0 = never, 50 = sometimes, 100 = always). The responses of 27/30 producers (three participants did not respond) ranged from never to sometimes (mean ± SD = 49 ± 22).

Table 4. Dairy goat producer responses relating to routine health practices.

Health	Category	Percentage (Number) of Farms
Frequency of treatment for gastrointestinal parasites	Every 3–5 months	14% (4/29)
	Every 6–8 months	17% (5/29)
	Every 9–12 months	38% (11/29)
	When needed	21% (6/29)
	Not clarified	10% (3/29)
Frequency of treatment for external parasites	Every 3–5 months	17% (5/29)
	Every 6–8 months	7% (2/29)
	Every 9–12 months	28% (8/29)
	When needed	21% (6/29)
	No treatment	7% (2/29)
	Not clarified	21% (6/29)
Pain relief usage	Yes	70% (21/30)
	No	30% (9/30)
Pain relief used for	Disbudding	29% (6/21)
	Castration	5% (1/21)
	Disease	67% (14/21)
	Injury	100% (21/21)
	Other—kidding	5% (1/21)
Factors that affect use of pain relief	Cost	10% (2/21)
	Time taken to administer pain relief	14% (3/21)
	Use of a veterinarian	5% (1/21)
	Benefits for the animal	90% (19/21)
	Benefits for humans (e.g., ease of handling)	10% (2/21)
Veterinarian used has adequate goat experience	Yes	73% (22/30)
	No	27% (8/30)

Producer responses relating to diagnosis of diseases on their farms including caseous lymphadenitis (CL), caprine arthritis encephalitis (CAE) and Johne's disease are presented in Table 5. Approximately two-thirds of producers reported having CL on their farms, with just under half of participants having diagnosed CAE; Johne's disease was less commonly diagnosed on farms (Table 5).

Table 5. Dairy goat producer responses to the question: has caseous lymphadenitis, caprine arthritis encephalitis or Johne's disease ever been diagnosed on your farm?

Disease	Yes	No
Caseous lymphadenitis	64% (18/28)	36% (10/28)
Caprine arthritis encephalitis	48% (14/29)	52% (15/29)
Johne's disease	7% (2/30)	93% (28/30)

3.6. Goat Kid Husbandry

Producer responses relating to goat kid rearing practices including removal from dam after birth (a practice common place in goat farming to reduce the risk of disease transmission such as CAE), rearing method, colostrum management, navel disinfection, housing strategies, and weaning age are presented in Table 6.

Table 6. Dairy goat producer responses relating to common kid rearing practices.

Goat Kid Rearing Practices	Category	Percentage (Number) of Farms
Rearing method	Hand-rearing	97% (29/30)
	Dam-rearing	3% (1/30)
Removal from dam after birth	Immediately	76% (22/29)
	≤4 h of birth	17% (5/29)
	5–12 h	3% (1/29)
	at least 48 h	3% (1/29)
	≤10 ounces	27% (8/30)
	11–20 ounces	50% (15/30)
Amount of colostrum in 24 h after birth	21–30 ounces	3% (1/30)
	≥31 ounces	10% (3/30)
	Free choice	10% (3/30)
	Heat-treated goat colostrum	43% (13/30)
Colostrum type	Raw goat colostrum	20% (6/30)
	Powdered goat colostrum	33% (10/30)
	Powered cow colostrum	20% (6/30)
Navel disinfection	Yes	60% (18/30)
	No	40% (12/30)
	Electric/gas heating	50% (15/30)
Kid housing strategies (during winter months)	Heat lamps	33% (10/30)
	Insulated walls	40% (12/30)
	High pen walls to prevent drafts	37% (11/30)
	Hutches	7% (2/30)
Weaning age	5–6 weeks	10% (3/30)
	7–8 weeks	63% (19/30)
	9–10 weeks	23% (7/30)
	11–12 weeks	3% (1/30)

Producer responses relating to disbudding including method, operator and training received, goat kid age at disbudding, power source, iron application time, horn bud removal and antiseptic application are presented in Table 7. Producers rated how confident they felt in their practice to disbud kids effectively without complication along a rating/slider scale (where 0 = not confident at all, 50 = confident, 100 = very confident). The responses of all (26/26) of the producers that perform disbudding themselves ranged from confident to very confident (mean ± SD = 86 ± 16). Producers were also asked to rate the likelihood of changing their practice if there was a better option available (where 0 = not at all likely, 50 = likely, 100 = very likely). The responses of 25/26 producers (one participant did not answer) ranged from likely to very likely (mean ± SD = 63 ± 24). Although, 3/25 producers rated the likelihood of changing their practice between not at all likely and likely (≤24). Additionally, producers rated how much pain they think disbudding causes (where 0 = no pain, 50 = some pain, 100 = extreme pain). The responses of 24/26 producers (two participants did not answer) ranged from some pain to extreme pain (mean ± SD = 60 ± 16), but 3/25 producers rated the amount of pain caused as being between no pain and some pain (≤41).

Producer responses relating to castration including whether castration is performed and what buck kids are kept for, the method used, goat kid age at castration, the operator and training received are presented in Table 8. Producers rated how confident they felt in their practice to castrate kids effectively without complication along a rating/slider scale (where 0 = not confident at all, 50 = confident, 100 = very confident). The responses of 7/8 producers (one participant did not answer) that perform castration themselves ranged from confident to very confident (mean ± SD = 85 ± 18). Producers were also asked to rate the likelihood of changing their practice if there was a better option available (where 0 = not at all likely, 50 = likely, 100 = very likely). The responses of all (8/8) producers ranged from likely to very likely (mean ± SD = 62 ± 27). Although, 3/8 producers rated the likelihood of changing their practice between not at all likely and likely (≤32). Additionally, producers rated how much pain they think castration causes (where 0 = no pain, 50 = some pain, 100 = extreme pain). The responses of 6/8 producers (two participants did not answer) ranged from no pain to some pain (mean ± SD = 49 ± 23).

Table 7. Dairy goat producer responses relating to disbudding practices.

Disbudding Practices	Category	Percentage (Number) of Farms
Method	Cautery iron	97% (29/30)
	Not clarified	3% (1/30)
Operator	Producer	87% (26/30)
	Staff	13% (4/30)
	Friends or family	27% (8/30)
	Veterinarian	7% (2/30)
	None, self-taught	39% (10/26)
Training	Friends/family	46% (12/26)
	Veterinarian	23% (6/26)
Goat kid age at disbudding	≤7 days	33% (10/30)
	8–14 days	43% (13/30)
	15–21 days	13% (4/30)
	22–28 days	3% (1/30)
	≥29 days	7% (2/30)
Iron power source	Electric	66% (19/29)
	Liquefied petroleum gas (LPG)	10% (3/29)
	Gas canister (butane)	31% (9/29)
	≤4 s	7% (2/29)
Total iron application time	5–7 s	21% (6/29)
(per horn bud)	8–12 s	55% (16/29)
	≥13 s	17% (5/29)
Horn bud removal	Yes	62% (18/29)
	No	38% (11/29)
Antiseptic use	Yes	20% (6/30)
	No	80% (24/30)

Table 8. Dairy goat producer responses relating to castration practices.

Castration Practice	Category	Percentage (Number) of Farms
Castration performed	Yes	27% (8/30)
	No	73% (22/30)
Buck kids kept for	Showing	13% (1/8)
	Meat	88% (7/8)
	Pets	25% (2/8)
	Other—supplying an antigen laboratory	13% (1/8)
Method	Rubber ring/band	100% (8/8)
Goat kid age at castration	<1 week	13% (1/8)
	1–3 weeks	25% (2/8)
	4–6 weeks	13% (1/8)
	7–9 weeks	13% (1/8)
	10–12 weeks	38% (3/8)
Operator	Producer	100% (8/8)
	Staff	13% (1/8)
	Friends/family	13% (1/8)
	No one, self-taught	38% (3/8)
Training received	Friends/family	50% (4/8)
	Veterinarian	13% (1/8)

The key research areas that have be identified from this study are presented in Table 9.

Table 9. Summary of the key areas of research for the Midwestern US dairy goat industry. The summary is arranged based on the percentage of producers (highest to lowest) that responded as *not performing* (major) or *performing* (minor) various husbandry practices.

Major Research Areas	Percentage (Number) of Farms
Effective and practical pain relief for castration	75% (6/8) thought castration did not cause pain
Effective and practical pain relief for disbudding	71% (15/21) did not provide pain relief for disbudding
Early kid management training and education (e.g., exposure to cold temperatures)	50% (15/30) did not use heating in kid rearing areas
Eradication or testing programs for common diseases (e.g., caseous lymphadenitis (CL), caprine arthritis encephalitis (CAE))	64% (18/28) and 48% (14/29) diagnosed CL, and CAE, respectively
Claw trimming training and education	40% (11/26) did not receive training
Standardized disbudding training and education	39% (10/26) did not receive training
Udder health	20% (6/30) did not monitor mastitis
General training programs for Midwestern US dairy goat producers	13% (4/30) did not provide staff training
Minor research areas	
Effect of outdoor space on goat behavior, hoof wear, parasite burden and productivity	73% (22/30) provided outdoor space
Corrective claw trimming frequency and claw growth	43% (13/30) trimmed claws every 3–4 months
Parasite management training and education	38% (11/29) drenched every 9–12 months

4. Discussion

Our study has provided insight into the current husbandry practices of 30 dairy goat producers in the Midwestern US and highlighted potential areas of future research to improve dairy goat health, production, and welfare.

4.1. Producer-Specific Information

The producers involved in the present study largely agreed that good animal welfare was a key priority for their farms; however, whether this correlates with actual positive welfare on-farm is not yet understood [9]. Some research suggests that consumer preference to buy products from animals that experience a high level of animal welfare does not necessarily correlate with those products being purchased, and that consumers may select products based on price and not ethical standards [15].

Training of staff responsible for animal care and management is crucial for providing a high standard of animal health and welfare [16–18]. Handlers that received both hands-on and online training on cornual nerve block application for cautery disbudding of calves had a much higher success rate and better outcomes for the calves undergoing this procedure than those operators that received online training only [18]. A large-scale study of dairy cattle farms in southern and central Italy reported that levels of staff training were inversely related to prevalence failures in almost all areas of welfare assessed; for example, farms that had no parasite control, foot bathing, routine footcare, or vaccination programs also failed to provide animal welfare training for stock people [17]. Combined, this research demonstrates the importance of training specific for animal welfare in order to ensure good welfare outcomes for livestock. Most producers in this study rated highly the importance of staff training; however, a small number of those producers reported they did not provide training (13%). These results may suggest that some producers responded as they thought they should, rather than what they actually thought. Those that did not provide training were generally small family-run farms and responded that they worked with goats from a young age. A recent study from Turkey reported that 73% (67/92) of dairy goat producers did not provide staff training for milking practices, which may not have been necessary as the producers stated that staff had many years of milking experience [19]. Additionally, Gökdai, Sakarya, Contiero and Gottardo [19] reported that producers were aware that training was a necessity, but that time constraints prevented them from providing adequate training to their staff. Training programs specific to dairy goat producers in the Midwestern US setting are required and should be actively encouraged to ensure stockperson uptake.

4.2. Goat Husbandry

Access to outdoor spaces was provided on approximately three-quarters of the farms in this study, and over half provided the goats the opportunity to graze on pasture, with more than one-quarter of producers providing access to concrete or rocks; this may reflect producer preference for goats to have access to environments that encourage performance of natural behaviors, or contain enrichment (e.g., climbing structures, pasture). A survey of 46 dairy goat farms in the UK reported that 17% of farms grazed goats outdoors, which the authors postulated was associated with the difficulties maintaining a high-yielding herd on pasture and the susceptibility of goats to parasitism with gastrointestinal nematodes (GIN) [10]. Reducing parasite load was also cited as an explanation for 12/13 dairy goat farms in New Zealand maintaining their does in intensive production (i.e., housed in barns where forage and crops are transported and fed to goats) [20]. Differences in availability of outdoor spaces between the Midwestern US and other nations may be due to variance in climatic conditions that allow GIN to be problematic (e.g., decrease production). However, these explanations require validation.

Sixty-nine percent of producers reported treating their does for GIN at least once a year. The GIN eggs are passed in the feces, and onto the pasture where the larvae hatch and are ingested during grazing [21]. As stated earlier, many producers in this study provided the goats with access to pasture, justifying regular worming regimens. However, anthelmintic resistance in nematode populations is a major issue in goats [22]; resistance to anthelmintics is associated with multiple factors including the fast metabolism of goats of drugs leading to shorter residence time facilitating nematode resistance and over use of anthelmintics at incorrect dosage [23]. Targeted selective treatment of individuals with high fecal egg counts has been suggested as a useful method of preserving anthelmintic efficacy by ensuring a number of untreated GIN in refugia [24]. Only two producers from the present study responded as using this technique. Another strategy involves the use of tannin-rich plants, which have natural anthelmintic properties to control nematode populations in goats [25]. Future research on tannin-rich plants that consistently show efficacy in vitro or the supplementation of grazing sheep with products containing Duddingtonia fungi to reduce pasture larvae [26] as well as better education programs for producers on careful GIN management to reduce resistance to anthelmintic drugs are required.

More than one-quarter of producers provided concrete in the outdoor area. Hard surfaces such as concrete may increase natural wear of the hooves. Additionally, goats spent more time lying on rubber matting and plastic slats than wood shavings indicating that goats may prefer solid surfaces to lie on than straw or wood chip [27]. Whether concrete and other solid surfaces are preferable over soft bedding materials and natural hoof wear is observed remains to be seen.

Ninety percent of producers reported that udder preparation was carried out prior to milking with physical removal of debris on the teats being most common. However, less than half of the farms sanitized the teats. A survey of 46 dairy goat farms in the UK, reported that 83% of farms used some form of udder and teat preparation [10]. Teat sanitizing dairy cattle prior to milking reduces the presence of *Staphylococcus aureus* (common mastitis pathogen affecting cattle) in milk [28]. The difference in these practices may be associated with the difference in acceptance of somatic cell count (SCC) levels by milk companies (1.5 million somatic cells and 750,000 somatic cells for goats and cows, respectively); therefore, cow dairies must reduce bacteria as much as possible. Multiple factors are associated with increased SCC in goats (compared with dairy cattle), such as increased dry matter intake, lactation number/parity, stage of lactation, and lower mature equivalent milk production [29]. Further, increased SCC is not necessarily correlated with intra-mammary infections [29]. Results of the National Animal Health Monitoring Survey in 2009 reported that 2.8% of lactating does in the US had clinical mastitis and 30.7% of operations had at least one doe with clinical mastitis; the most common form of identification was visual inspection [30]. Interestingly, 20% of producers we surveyed did not check for mastitis, indicating that the previously reported prevalence rate may be underrepresented. More education and training surrounding udder health for dairy goats in the Midwestern US is required.

Corrective claw trimming to remove excess claw growth is an important husbandry practice for goats' claws that are not worn naturally. Severely overgrown claws (and diseases such as laminitis and CAE) may be associated with lameness, which is a major health issue in dairy goats [3,31,32]. Claw trimming was conducted on all farms and most commonly by the producer. Although 51.4% of goats observed on these farms (2325/4520) had overgrown claws [9]. A recent survey of goat producers in the UK reported that all producers perform claw trimming [10] yet an earlier study from the same group identified claw overgrowth as a major issue [3]. In the present study, claws were trimmed most commonly every 3–4 months (43%) and 5–6 months (33%), which is similar to previous surveys of dairy goat farmer in the UK [10] (36% and 33% for every 3–4 months and 5–6 months, respectively). There was, however, a higher percentage of producers that trimmed every 1–2 months in UK [10] compared with the Midwestern US (16% vs. 7%). Together, these results may indicate that more frequent hoof trimming is required to prevent claw overgrowth. Future research should evaluate best practice recommendations for claw trimming frequency to ensure minimal rates of claw overgrowth.

Pain relief was used on 70% of farms in the present study with the most common reasons including injury and disease. Factors that affected the use of pain relief on farms were the associated costs (including veterinary personnel), and time taken to administer, but most producers used pain relief because of the benefits to the animal. The relatively high number of producers that used pain relief on their farms may be associated with the self-selected nature of this study in that producers who were interested in improving animal welfare opted into this study.

Producers reported that they felt the veterinary practitioner they used had adequate goat experience on 73% of farms. In comparison, 83% of 46 producers in the UK felt their local veterinary practitioner had sufficient knowledge and experience with dairy goats [10]. Recent data from a USDA survey suggests that the number of goat producers (including meat, fiber and dairy) in the US consulting with veterinarians has increased, rising from 39.5% in 2009 to 49.7% in 2019 [33].

More than half of the producers reported that CL was diagnosed on their farms, with just under half reporting diagnosis of CAE on their farms. We are unaware of whether the producers in the present study were part of a testing or eradication program or the financial implications associated with these diseases, but this information would be useful to include in future studies. In the UK, producers observed these diseases far less often (CL: 7%, 10/45 farms; CAE: 11%, 5/45 farms) [10]. The relatively high number of goats with these diseases on farms in the present study may be associated with non-selective breeding for animals that are susceptible to these conditions, or management practices that increase the risk of transmission such as a failure to separate or cull infected animals. Johne's disease was the most commonly reported disease on UK dairy goat farms (49%; 22/45), whereas producers in the present study responded as having this disease on their farms less often (7%; 2/30 farms); this may be due farmers may not be actively checking for Johne's disease. With the apparent trend of dairy goat farming expansion in the US, sourcing goats from multiple herds and comingling, likely increases the risk of disease transmission. Therefore, careful checking or testing of new animals brought into the herd prior to comingling, can reduce the risk of disease. A disease eradication program to control CAE, CL and Johne's diseases may be beneficial for the Midwestern US dairy goat industry as has been largely successful on Norwegian dairy goat farms [34].

4.3. Goat Kid Husbandry

Interestingly, only half of the producers used heating in the kid rearing areas, even though the study area ambient air temperatures can reach $-20\,^{\circ}$C (or below) [35] in the winter during kidding season. In our recent study, approximately 65% (17/30) of farms had goats with frostbitten ears (i.e., any amount of pinna is missing; appears straight cut), which was likely caused by extended time spent in extreme low temperatures at birth [9]. It is vital that producers ensure better management of neonatal kids during extreme temperatures by completely drying kids after birth and moving them to warm environments. Goat kids that experience cold air temperatures (-3 to $10\,^{\circ}$C) for at least 5 days after parturition have lower survival rates compared to kids exposed to warmer temperatures [36,37]. However, an earlier study observed no mortalities in kids raised in non-insulated rooms (-10 to $-4\,^{\circ}$C) compared to insulated rooms (9 to $14\,^{\circ}$C) [38]. Adult goat behavior is affected by temperature as lying time was reduced in goats experiencing low (-6 to $-8\,^{\circ}$C) ambient air temperatures compared with moderate (10 to $12\,^{\circ}$C) ambient temperatures [39]. Additionally, goats will spend reduced time in outdoor spaces (compared with indoors) with decreasing temperatures (below $-10\,^{\circ}$C) [40]. More education and training programs are required to increase the use of insulated or heated rooms for newborn kids.

The overwhelming majority of producers in this study used the cautery method to disbud goat kids (only one producer did not state the method used). Cautery disbudding (with the provision of pain relief) appears to be the best option to date for having hornless goats [41–44]. In the present study, 29% (6/21) of producers used pain relief for disbudding, which appears a relatively low number when compared to the UK of which all kids require anesthesia and/or analgesia for disbudding [45]. To improve the welfare of goat kids undergoing disbudding, it is important that some form of pain relief is used. Caustic paste disbudding is the second most common method used in calves [14,46,47], but it can run into the eyes or be rubbed onto other areas of the body or pen mates [48]. Additionally, caustic paste causes more pain in goat kids than cautery disbudding and may not consistently prevent scurs [41,42]. In a previous survey of goat producers and veterinarians in the US and Canada, 97% (39/40) of veterinarians and 95% (218/229) of producers stated they used a cautery iron to disbud their goat kids [49]. The authors stated that caustic paste was not used, but provided no insight into what method was used instead [49].

Disbudding was predominantly carried out by producers in the present study, which was likely to reduce costs associated with veterinary practitioners; only a small number of producers employed veterinary practitioners to perform disbudding. In comparison with the UK, only veterinary practitioners are permitted to disbud goat kids (with a cautery iron and under anesthesia) under UK law [45,50]. Additionally, just over one-fifth of the producers that disbudded kids themselves, were trained by a veterinary practitioner, which is a smaller percentage than those that were trained by family and friends (12/26; 46%), again likely due to reduced costs. Producers received cautery disbudding training for calves in Canada and Czech Republic in 98% and 85% of producers surveyed, respectively [14,51]; however, we found that only 60% of producers received any cautery disbudding training, highlighting the need for a standardized cautery disbudding protocol for goat kids and training programs for producers in the Midwestern US.

It is typically considered good practice to disbud goat kids once the horn buds are palpable and within a week of age [52]. In a recent review of comparisons of disbudding practices between calves and kids, on average kids were disbudded at 10.6 days of age (SD = 5.7) (on average calves were 5.3 weeks of age at disbudding (SD = 2.0)) [53]. The majority of producers (77%) in the present study disbudded kids less than 2 weeks of age, although 23% of producers in the present study disbudded kids at an advanced age (i.e., more than 2 weeks of age). In the UK, kids are commonly disbudded at less than 2 weeks of age (93%; 42/45) [10]. Disbudding goat kids beyond 2 weeks of age increases the difficulty of completely removing the horn bud and may increase the incidence of scurs and prolong healing [53], further highlighting the need for producer education and training in this area.

Seventy percent of producers used the cautery iron for 8 s or more to disbud their kids, which may increase the risk of thermal injury to the brain. In a recent pilot study by our group on the effect of cautery iron application duration on brain injury in goat kids (by a trained and experienced operator), we observed brain injury in at least some goat kids at all duration applications (5, 10, 15, and 20 s); however, longer applications resulted in more severe and consistent brain injury [54]. Application times of 15 and 20 s should be avoided as these durations resulted in severe histopathological changes, including a branching region of edema across multiple gyri, hemorrhage, and microscopic lesions consisting of leptomeningeal and cerebrocortical necrosis [54]. Cautery disbudding training should be included as part of routine training programs for farm staff by a veterinary practitioner or an experienced operator.

In the present study, 27% of producers performed castration on their animals, and of these producers, most buck kids were castrated for reducing odor in the meat or because they were pets. The producers that did not perform castration tended to keep bucks for breeding purposes. Castration was carried out almost exclusively by the producer and with a ring or band, most likely as the practice requires minimal training and is not as technical as surgical castration. Ring castration does, however, cause acute and long-term pain in lambs (reviewed by [55,56]). Use of the Burdizzo method appears to cause similar behavioral and stress responses as ring castration in lambs, which can be reduced by local anesthesia and non-steroidal anti-inflammatory drugs [57]. Producers rated themselves as confident to very confident they could castrate their kids effectively without complication, although they were open to changing their practice if there was a better method available. This may be associated with using another method, which may be more time efficient. Producers rated castration as causing no pain to some pain. Future studies should investigate pain or distress associated with ring castration of goat kids.

This study was not without limitations. Ideally, we would have selected 30 farms at random from the Midwestern region to be involved in this study, but we were unable to access a database of producers in this region. Therefore, we must be cautious when trying to extrapolate our findings and make statements about the wider Midwestern dairy goat industry as there is likely to be some degree of self-selection bias, with consequent overestimation (or under-representation) of certain themes. For example, farmers that chose to be involved with this study likely already had an interest in improving animal welfare, therefore we may not have captured the views of those that had no interest in improving the welfare of their goats. In future, the survey should be made available to all producers within the Midwestern US region as a standalone survey without requiring a farm visit to perform welfare assessment. Further, it would have been useful to include a greater number range within the rating/slider scale to increase the sensitivity about the data collected.

5. Conclusions

In conclusion, this study has provided insight into the husbandry practices carried out on 30 dairy goat farms in the Midwestern US, which can be used by the industry to inform and improve routine husbandry practices.

Our study has highlighted many potential areas of future research to improve the welfare of dairy goats. Major potential areas that showed the highest number of producers that did not perform various husbandry procedures were the use of pain relief for husbandry practices such as castration and disbudding, early kid management during birth to prevent illness/disease or mortality (e.g., warm and dry areas for kid rearing), eradication programs for common contagious diseases such as CAE and CL, training programs and education for claw trimming, disbudding, and udder health. Minor research areas included the highest number of producers that performed various husbandry practices such as the effect of outdoor space on goat behavior, hoof wear, parasite burden and productivity, corrective claw trimming frequency and the effect on claw growth, and parasite management.

Supplementary Materials: The following are available online at https://www.mdpi.com/article/10.3390/ani11072007/s1, Table S1: A blank copy of the questionnaire provided to dairy goat producers in the dairy goat welfare study.

Author Contributions: Conceptualization, M.N.H., T.M.L., J.K.S., L.C.S. and P.J.P.; methodology, M.N.H., T.M.L., J.K.S., L.C.S. and P.J.P.; formal analysis, M.N.H.; investigation, M.N.H.; data curation, M.N.H.; writing—original draft preparation, M.N.H., J.K.S. and P.J.P.; writing—review and editing, M.N.H., T.M.L., J.K.S., L.C.S. and P.J.P.; funding acquisition, P.J.P. and J.K.S. All authors have read and agreed to the published version of the manuscript.

Funding: This research was funded by the United States Department of Agriculture (USDA) through the National Institute for Food and Agriculture (NIFA), grant number 2018-67015-28136.

Institutional Review Board Statement: This study was reviewed by the Institutional Review Board of Iowa State University (protocol code 18-497-00 and 14/12/2018) and received an exemption determination as the research was within exempt categories (i.e., research involved use of educational tests (cognitive, diagnostic, aptitude, achievement), survey procedures, interview procedures, or observations of public behavior, unless (i) information obtained is recorded in such a manner that human subjects can be identified, and (ii) any disclosure of the human subjects' responses outside the research could reasonably place the subject at risk of criminal or civil liability or be damaging to the subjects' financial standing, employability, or reputation).

Data Availability Statement: The data presented in this study are available on request from the corresponding author. The data are not publicly available due to restrictions on privacy.

Acknowledgments: We would like to acknowledge the Iowa State University staff Stephen Sapp and Rebecca Burzette their assistance on survey construction and analysis of responses. We also thank the farm management and staff for their time and effort completing the survey.

Conflicts of Interest: The authors declare no conflict of interest. The funders had no role in the design of the study; in the collection, analyses, or interpretation of data; in the writing of the manuscript, or in the decision to publish the results.

References

1. National Agricultural Statistics Service (NASS). *Sheep and Goats*; United States Department of Agriculture (USDA): Fort Collins, CO, USA. Available online: https://www.nass.usda.gov/Publications/Todays_Reports/reports/shep0120.pdf (accessed on 21 January 2020).
2. Miller, B.A.; Lu, C.D. Current status of global dairy goat production: An overview. *Asian Australas. J. Anim. Sci.* **2019**, *32*, 1219–1232. [CrossRef]
3. Anzuino, K.; Bell, N.J.; Bazeley, K.J.; Nicol, C. Assessment of welfare on 24 commercial UK dairy goat farms based on direct observations. *Vet. Rec.* **2010**, *167*, 774–780. [CrossRef] [PubMed]
4. Battini, M.; Barbieri, S.; Vieira, A.; Stilwell, G.; Mattiello, S. Results of testing the prototype of the AWIN welfare assessment protocol for dairy goats in 30 intensive farms in Northern Italy. *Ital. J. Anim. Sci.* **2016**, *15*, 283–293. [CrossRef]
5. Can, E.; Vieira, A.; Battini, M.; Mattiello, S.; Stilwell, G. On-farm welfare assessment of dairy goat farms using animal-based indicators: The example of 30 commercial farms in Portugal. *Acta Agric. Scand. Sect. A Anim. Sci.* **2016**, *66*, 43–55. [CrossRef]
6. Muri, K.; Stubsjøen, S.M.; Valle, P. Development and testing of an on-farm welfare assessment protocol for dairy goats. *Anim. Welf.* **2013**, *22*, 385–400. [CrossRef]
7. Arsoy, D. Herd management and welfare assessment of dairy goat farms in Northern Cyprus by using breeding, health, reproduction, and biosecurity indicators. *Trop. Anim. Health Prod.* **2019**, *52*, 71–78. [CrossRef]
8. Leite, L.; Stamm, F.; Souza, R.; Filho, J.C.; Garcia, R. On-farm welfare assessment in dairy goats in the Brazilian Northeast. *Arq. Bras. Med. Vet. Zootec.* **2020**, *72*, 2308–2320. [CrossRef]
9. Hempstead, M.N.; Lindquist, T.M.; Shearer, J.K.; Shearer, L.C.; Cave, V.M.; Plummer, P.J. Welfare Assessment of 30 Dairy Goat Farms in the Midwestern United States. *Front. Vet. Sci.* **2021**, *8*. [CrossRef] [PubMed]
10. Anzuino, K.; Knowles, T.G.; Lee, M.; Grogono-Thomas, R. Survey of husbandry and health on UK commercial dairy goat farms. *Vet. Rec.* **2019**, *185*, 267. [CrossRef] [PubMed]
11. Battini, M.; Vieira, A.; Barbieri, S.; Ajuda, I.; Stilwell, G.; Mattiello, S. Invited review: Animal-based indicators for on-farm welfare assessment for dairy goats. *J. Dairy Sci.* **2014**, *97*, 6625–6648. [CrossRef]
12. Muri, K.; Tufte, P.; Skjerve, E.; Valle, P. Human-animal relationships in the Norwegian dairy goat industry: Attitudes and empathy towards goats (Part I). *Anim. Welf.* **2012**, *21*, 535–545. [CrossRef]
13. Muri, K.; Valle, P. Human-animal relationships in the Norwegian dairy goat industry: Assessment of pain and provision of veterinary treatment (Part II). *Anim. Welf.* **2012**, *21*, 547–558. [CrossRef]

14. Winder, C.B.; Leblanc, S.J.; Haley, D.B.; Lissemore, K.D.; Godkin, M.A.; Duffield, T.F. Practices for the disbudding and dehorning of dairy calves by veterinarians and dairy producers in Ontario, Canada. *J. Dairy Sci.* **2016**, *99*, 10161–10173. [CrossRef]
15. Cornish, A.; Jamieson, J.; Raubenheimer, D.; McGreevy, P. Applying the Behavioural Change Wheel to Encourage Higher Welfare Food Choices. *Animals* **2019**, *9*, 524. [CrossRef] [PubMed]
16. Lockworth, C.R.; Craig, S.L.; Liu, J.; Tinkey, P.T. Training Veterinary Care Technicians and Husbandry Staff Improves Animal Care. *J. Am. Assoc. Lab. Anim. Sci.* **2011**, *50*, 84–93. [PubMed]
17. Peli, A.; Pietra, M.; Giacometti, F.; Mazzi, A.; Scacco, G.; Serraino, A.; Scagliarini, L. Survey on animal welfare in 943 Italian dairy farms. *Ital. J. Food Saf.* **2016**, *5*, 5832. [CrossRef]
18. Winder, C.B.; Leblanc, S.J.; Haley, D.B.; Lissemore, K.D.; Godkin, M.A.; Duffield, T.F. Comparison of online, hands-on, and a combined approach for teaching cautery disbudding technique to dairy producers. *J. Dairy Sci.* **2018**, *101*, 840–849. [CrossRef]
19. Gökdai, A.; Sakarya, E.; Contiero, B.; Gottardo, F. Milking characteristics, hygiene and management practices in Saanen goat farms: A case of Canakkale province, Turkey. *Ital. J. Anim. Sci.* **2020**, *19*, 213–221. [CrossRef]
20. Solis-Ramirez, J.; Lopez-Villalobos, N.; Blair, H.T. Dairy goat production systems in Waikato, New Zealand. In Proceedings of the New Zealand Society of Animal Production, Invercargill, New Zealand, 15 January 2011; pp. 86–91.
21. Craig, T.M. Gastrointestinal Nematodes, Diagnosis and Control. *Vet. Clin. N. Am. Food Anim. Pr.* **2018**, *34*, 185–199. [CrossRef] [PubMed]
22. Hoste, H.; Sotiraki, S.; Torres-Acosta, J.F.D.J. Control of Endoparasitic Nematode Infections in Goats. *Vet. Clin. N. Am. Food Anim. Pract.* **2011**, *27*, 163–173. [CrossRef] [PubMed]
23. Hennessy, D. Physiology, pharmacology and parasitology. *Int. J. Parasitol.* **1997**, *27*, 145–152. [CrossRef]
24. Charlier, J.; Morgan, E.; Rinaldi, L.; van Dijk, J.; Demeler, J.; Höglund, J.; Hertzberg, H.; Van Ranst, B.; Hendrickx, G.; Vercruysse, J.; et al. Practices to optimise gastrointestinal nematode control on sheep, goat and cattle farms in Europe using targeted (selective) treatments. *Vet. Rec.* **2014**, *175*, 250–255. [CrossRef] [PubMed]
25. Santos, F.O.; Cerqueira, A.P.M.; Branco, A.; Batatinha, M.J.M.; Botura, M.B. Anthelmintic activity of plants against gastrointestinal nematodes of goats: A review. *Parasitology* **2019**, *146*, 1233–1246. [CrossRef] [PubMed]
26. Healey, K.; Lawlor, C.; Knox, M.R.; Chambers, M.; Lamb, J. Field evaluation of Duddingtonia flagrans IAH 1297 for the reduction of worm burden in grazing animals: Tracer studies in sheep. *Vet. Parasitol.* **2018**, *253*, 48–54. [CrossRef]
27. Sutherland, M.A.; Lowe, G.L.; Watson, T.J.; Ross, C.M.; Rapp, D.; Zobel, G.A. Dairy goats prefer to use different flooring types to perform different behaviours. *Appl. Anim. Behav. Sci.* **2017**, *197*, 24–31. [CrossRef]
28. Da Costa, L.; Rajala-Schultz, P.; Schuenemann, G.; Rajala-Schultz, P. Management practices associated with presence of Staphylococcus aureus in bulk tank milk from Ohio dairy herds. *J. Dairy Sci.* **2016**, *99*, 1364–1373. [CrossRef]
29. Wilson, D.J.; Stewart, K.N.; Sears, P.M. Effects of stage of lactation, production, parity and season on somatic cell counts in infected and uninfected dairy goats. *Small Rumin. Res.* **1995**, *16*, 165–169. [CrossRef]
30. USDA. *Disease and Mortality on U.S. Goat Operations*; USDA: Animal and Plant Health Inspection Service; Veterinary Services: Centers for Epidemiology Animal Health: Fort Collins, CO, USA, 2009.
31. Christodoulopoulos, G. Foot lameness in dairy goats. *Res. Vet. Sci.* **2009**, *86*, 281–284. [CrossRef]
32. Hill, N.P.; Murphy, P.E.; Nelson, A.J.; Mouttotou, N.; Green, L.E.; Morgan, K.L. Lameness and foot lesions in adult British dairy goats. *Vet. Rec.* **1997**, *141*, 412–416. [CrossRef]
33. USDA. *Use of Veterinarians on Goat Operations*; USDA: Animal and Plant Health Inspection Service: Fort Collins, CO, USA, 2020.
34. Muri, K.; Leine, N.; Valle, P. Welfare effects of a disease eradication programme for dairy goats. *Animals* **2016**, *10*, 333–341. [CrossRef]
35. NOAA National Centers for Environmental Information. Climate at a Glance: Statewide Mapping. Available online: https://www.ncdc.noaa.gov/cag/ (accessed on 14 January 2021).
36. Luo, N.; Wang, J.; Hu, Y.; Zhao, Z.; Zhao, Y.; Chen, X. Cold and heat climatic variations reduce indigenous goat birth weight and enhance pre-weaning mortality in subtropical monsoon region of China. *Trop. Anim. Health Prod.* **2019**, *52*, 1385–1394. [CrossRef]
37. Mellado, M.; Vera, T.; Meza-Herrera, C.; Ruíz, F. A note on the effect of air temperature during gestation on birth weight and neonatal mortality of kids. *J. Agric. Sci.* **2000**, *135*, 91–94. [CrossRef]
38. Eik, L. Performance of goat kids raised in a non-insulated barn at low temperatures. *Small Rumin. Res.* **1991**, *4*, 95–100. [CrossRef]
39. Bøe, K.E.; Andersen, I.L.; Buisson, L.; Simensen, E.; Jeksrud, W.K. Flooring preferences in dairy goats at moderate and low ambient temperature. *Appl. Anim. Behav. Sci.* **2007**, *108*, 45–57. [CrossRef]
40. Bøe, K.E.; Ehrlenbruch, R. Thermoregulatory behavior of dairy goats at low temperatures and the use of outdoor yards. *Can. J. Anim. Sci.* **2013**, *93*, 35–41. [CrossRef]
41. Hempstead, M.N.; Waas, J.R.; Stewart, M.; Cave, V.M.; Sutherland, M.A. Evaluation of alternatives to cautery disbudding of dairy goat kids using physiological measures of immediate and longer-term pain. *J. Dairy Sci.* **2018**, *101*, 5374–5387. [CrossRef]
42. Hempstead, M.; Waas, J.R.; Stewart, M.; Cave, V.M.; Sutherland, M.A. Evaluation of alternatives to cautery disbudding of dairy goat kids using behavioural measures of post-treatment pain. *Appl. Anim. Behav. Sci.* **2018**, *206*, 32–38. [CrossRef]
43. Hempstead, M.N.; Waas, J.R.; Stewart, M.; Cave, V.M.; Sutherland, M.A. Can Isoflurane and Meloxicam Mitigate Pain Associated with Cautery Disbudding of 3-Week-Old Goat Kids? *Animals* **2020**, *10*, 878. [CrossRef]

44. Hempstead, M.N.; Waas, J.R.; Stewart, M.; Dowling, S.K.; Cave, V.M.; Lowe, G.L.; Sutherland, M.A. Effect of isoflurane alone or in combination with meloxicam on the behavior and physiology of goat kids following cautery disbudding. *J. Dairy Sci.* **2018**, *101*, 3193–3204. [CrossRef] [PubMed]
45. British Veterinary Association. *BVA and Goat Veterinary Society Policy Position on Goat Kid Disbudding and Analgesia*; BVA: London, UK, 2018.
46. Cozzi, G.; Gottardo, F.; Brscic, M.; Contiero, B.; Irrgang, N.; Knierim, U.; Pentelescu, O.; Windig, J.; Mirabito, L.; Eveillard, F.K.; et al. Dehorning of cattle in the EU Member States: A quantitative survey of the current practices. *Livest. Sci.* **2015**, *179*, 4–11. [CrossRef]
47. USDA. *Health and Management Practices on U.S. Dairy Operations, 2014*; USDA: Fort Collins, CO, USA, 2018; pp. 1–202.
48. Winder, C.; LeBlanc, S.; Haley, D.; Lissemore, K.D.; Godkin, M.A.; Duffield, T.F. Clinical trial of local anesthetic protocols for acute pain associated with caustic paste disbudding in dairy calves. *J. Dairy Sci.* **2017**, *100*, 6429–6441. [CrossRef]
49. Valdmanis, L.; Menzies, P.; Millman, S. A survey of dehorning practices and pain management in goats. In *41st Congress of the International Society for Applied Ethology*; Galindo, F., Alvarez, L., Eds.; ISAE: Mérida, Mexico, 2007.
50. *Veterinary Surgeons Act*; UK Government: London, UK, 1966; Volume 36. Available online: https://www.legislation.gov.uk/ukpga/1966/36 (accessed on 10 February 2021).
51. Staněk, S.; Šárová, R.; Nejedlá, E.; Šlosárková, S.; Doležal, O. Survey of disbudding practice on Czech dairy farms. *J. Dairy Sci.* **2018**, *101*, 830–839. [CrossRef]
52. Smith, M.C.; Sherman, D.M. Dehorning and descenting. In *Goat Medicine*, 2nd ed.; Wiley-Blackwell: Ames, IA, USA, 2009; pp. 723–731. [CrossRef]
53. Hempstead, M.; Waas, J.; Stewart, M.; Sutherland, M. Goat kids are not small calves: Species comparisons in relation to disbudding. *Anim. Welf.* **2020**, *29*, 293–312. [CrossRef]
54. Hempstead, M.N.; Shearer, J.K.; Sutherland, M.A.; Fowler, J.L.; Smith, J.S.; Smith, J.D.; Lindquist, T.M.; Plummer, P.J. Cautery Disbudding Iron Application Time and Brain Injury in Goat Kids: A Pilot Study. *Front. Vet. Sci.* **2021**, *7*. [CrossRef]
55. Mellor, D.; Stafford, K. Acute castration and/or tailing distress and its alleviation in lambs. *N. Z. Vet. J.* **2000**, *48*, 33–43. [CrossRef] [PubMed]
56. Smith, J.S.; Schleining, J.; Plummer, P. Pain Management in Small Ruminants and Camelids. *Vet. Clin. N. Am. Food Anim. Pr.* **2021**, *37*, 17–31. [CrossRef] [PubMed]
57. Molony, V.; Kent, J.; Hosie, B.; Graham, M. Reduction in pain suffered by lambs at castration. *Vet. J.* **1997**, *153*, 205–213. [CrossRef]

Article

Heart Rate and Heart Rate Variability Change with Sleep Stage in Dairy Cows

Laura B. Hunter [1,2,3,*], Marie J. Haskell [2], Fritha M. Langford [2], Cheryl O'Connor [1], James R. Webster [1] and Kevin J. Stafford [3]

[1] Animal Behaviour and Welfare, Ethical Agriculture, AgResearch Ltd. Ruakura Research Centre, Hamilton 3214, New Zealand; Cheryl.OConnor@agresearch.co.nz (C.O.); Jim.Webster@agresearch.co.nz (J.R.W.)

[2] Animal Behaviour and Welfare, Animal and Veterinary Sciences, Scotland's Rural College (SRUC), Edinburgh EH9 3JG, UK; Marie.Haskell@sruc.ac.uk (M.J.H.); Fritha.Langford@sruc.ac.uk (F.M.L.)

[3] School of Agriculture and Environment, Massey University, Palmerston North 4474, New Zealand; K.J.Stafford@massey.ac.nz

* Correspondence: laura.hunter@agresearch.co.nz

Citation: Hunter, L.B.; Haskell, M.J.; Langford, F.M.; O'Connor, C.; Webster, J.R.; Stafford, K.J. Heart Rate and Heart Rate Variability Change with Sleep Stage in Dairy Cows. *Animals* 2021, *11*, 2095. https://doi.org/10.3390/ani11072095

Academic Editor: Frank J. C. M. Van Eerdenburg

Received: 18 June 2021
Accepted: 7 July 2021
Published: 14 July 2021

Publisher's Note: MDPI stays neutral with regard to jurisdictional claims in published maps and institutional affiliations.

Simple Summary: The amount of sleep acquired and changes to patterns of sleep could be a useful tool to assess cow welfare, particularly in response to changes or stressors in their environment. However, the current most accurate method to assess sleep, polysomnography (PSG), is difficult and time consuming. In humans, heart rate (HR) and variability in time between heart beats (HRV) can be used to identify sleep stages, and this could be a useful alternative to investigate sleep in cows. We compared measures of HR and HRV with PSG in two groups of dairy cows in different environments and investigated the effects of lying posture on these measures. We found that HR decreased with deepening sleep stages in both groups of cows, that rapid eye movement sleep (REM) was associated with higher HRV and that HR and HRV also changed with different lying postures. The patterns of differences between sleep stages were similar between the two groups of cows. Our results suggest that HR and HRV change with sleep stages in cows and that these measures could be a useful, and more easily applied, method of assessing sleep stages in dairy cows.

Abstract: Changes to the amount and patterns of sleep stages could be a useful tool to assess the effects of stress or changes to the environment in animal welfare research. However, the gold standard method, polysomnography PSG, is difficult to use with large animals such as dairy cows. Heart rate (HR) and heart rate variability (HRV) can be used to predict sleep stages in humans and could be useful as an easier method to identify sleep stages in cows. We compared the mean HR and HRV and lying posture of dairy cows at pasture and when housed, with sleep stages identified through PSG. HR and HRV were higher when cows were moving their heads or when lying flat on their side. Overall, mean HR decreased with depth of sleep. There was more variability in time between successive heart beats during REM sleep, and more variability in time between heart beats when cows were awake and in REM sleep. These shifts in HR measures between sleep stages followed similar patterns despite differences in mean HR between the groups. Our results show that HR and HRV measures could be a promising alternative method to PSG for assessing sleep in dairy cows.

Keywords: dairy cows; heart rate; sleep; heart rate variability; polysomnography

1. Introduction

Two main stages of sleep exhibited by animals are known as rapid eye movement sleep (REM) and non-REM sleep. Non-REM sleep has been associated with restorative functions in the body and brain, for example, the clearance of potentially harmful toxins produced by normal cellular function [1]. REM sleep has been associated with memory, learning and dreaming [2]. Changes to the amount and patterns of sleep stages could be used to assess animal welfare, as these aspects of sleep are known to be affected by factors such as environmental conditions, stressful occurrences during the day, pain or illness [3]. For example, after moving into an unfamiliar environment, cows were found to spend less time lying in postures associated with sleep than their baseline, which could be an indication of stress [4]. However, in dairy cows, without using neuro-electrophysiological methods, it is difficult to accurately identify sleep from wakefulness, let alone different sleep stages.

As sleep is a homeostatic function originating in the brain, the most accurate way to study it is through polysomnography (PSG), the study of multiple electrophysiological signals, namely brain activity, eye movements and muscle activity [5]. PSG can be successfully used to study sleep in calves [6] and in adult cows [7,8]; however, it is costly, the equipment is fragile, and interpretation of the signals is time-consuming [7]. Being able to identify sleep in dairy cows with other more easily applied or less invasive devices would be beneficial not only for the cow's comfort and welfare, but also for ease of application by researchers, thus facilitating the study of the sleep of cows and opening several new avenues for investigation of the effects of sleep loss or importance of sleep for cows.

During sleep and different sleep stages, changes occur in the regulation of the mammalian autonomic nervous system (ANS) and its subdivisions, the parasympathetic (PNS) and sympathetic nervous systems (SNS), affecting many functions such as heart rate, respiration rate and body temperature [9]. Specifically, during REM sleep, there is variability in ANS activity, leading to more variability in the associated physiological functions, whereas in non-REM sleep stages, there is more activity of the PNS while SNS activity is reduced [9]. Because the ANS affects the heart, measures of heart rate (HR) and heart rate variability (HRV: the measurement of the variability in the time between successive heart beats) can be used as a way to identify activation of the PNS and SNS [10]. In humans, changes in HR and HRV have been used to accurately identify and differentiate between sleep stages [11,12]. HRV can be quantified with different methods. Time domain indices of HRV identify differences in the time between successive heart beats or inter-beat-interval (IBI) while frequency domain indices classify the signal into frequency bands [13].

The study of HRV in cow welfare to date has focussed mainly on the application of HR and HRV to identify and assess stress. HR and HRV were found to be affected by severe lameness which may cause chronic stress in cattle [14]. Calves being disbudded without local anaesthetic showed an increase in frequency domain metrics of HRV [15]. HR and HRV has also been used to identify positive interactions in dairy cows, and social licking between cows was found to reduce heart rate in receivers [16]. Body posture has been found to affect HR and HRV measures. Heart rate was lower and variability in time between heart beats was higher in cows lying down compared to when standing [17]. However, to our knowledge, investigations of HR and HRV during sleep in cows have not been done. In previous work, we have found that sleep occurs when cows are lying down, but specific lying postures could not be used reliably to identify sleep stages compared to PSG in dairy cows. Furthermore, housing conditions have been shown to affect the relationship between lying postures and sleep [18].

It is possible that HR and HRV could be used to identify sleep stages in cows. The equipment required to assess HR and HRV is less invasive and more easily applied than equipment used for PSG. Therefore, the objective of this study was to determine if HR and HRV differ between sleep stages in dairy cows, and to determine if this is repeatable between cows in different areas, housing conditions and lying postures.

2. Materials and Methods

2.1. Ethical Statement

The study was designed in accordance with the relevant guidelines and legislation in both Scotland and New Zealand (NZ) where the studies took place. Ethical approval was obtained from the UK Home Office (Project Licence P204B097E), SRUC Animal Ethics

Committee (Ref. ED AE 03-2018) and Ruakura Animal Ethics committee (AE 14708) prior to the start of animal manipulations.

2.2. Cows and Housing

Twelve cows were recruited for this study from two locations. Six non-lactating and non-pregnant Holstein cows (age 3.9 ± 0.7) from the Acrehead unit of SRUC's Dairy Research Centre (Dumfries, Scotland) and six, three-year-old, non-lactating, pregnant Kiwi-cross (Friesian × Jersey) cows from the DairyNZ Lye Farm (Newstead, NZ) were used. The small sample size was necessary due to the time-intensive methods required to habituate the cows to the recording devices. Non-lactating cows were selected to avoid disruptions to the cow's sleep patterns due to fetching for milking and the risk of damage to recording devices in the milking parlour. The Scottish cows, destined to be culled from the herd due to reduced fertility, were healthy during the trial.

The cows were managed in a large group pen and moved into a smaller adjacent 'test' pen individually for recording sessions (Figure 1). The Scottish cows were held on deep-bedded straw in a barn. The group pen measured 20 m × 5.2 m and test pen 5.2 m × 5.1 m. The cows were fed silage and always had access to water. The NZ cows were managed outdoors in a large paddock. They were able to graze and were provided with silage ad libitum and always had access to water. The group pen measured 44 m × 29 m and was created with live electric fencing. The 10 m × 10 m test pen was created with non-live electric fencing tape, to prevent potential interference with the electrophysiological recordings. The fencing set-up for both group and test pens could be moved around the paddock when ground conditions became wet or muddy. In both locations, a 2 m buffer zone was created between the test pen and group pen, to inhibit contact and reduce damage to recording devices from social interactions.

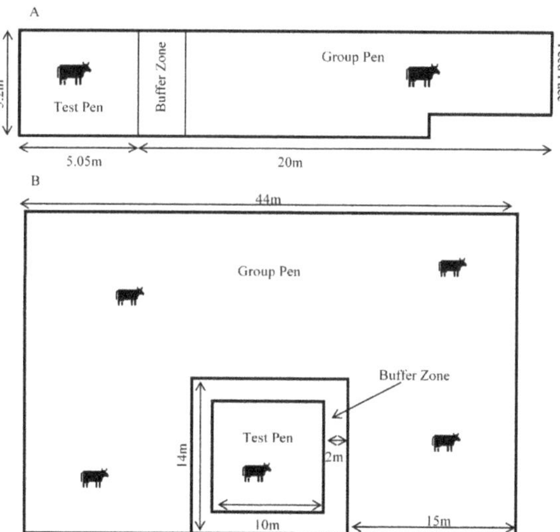

Figure 1. Diagrams of group and test pen design in the Scottish indoor-housed study (**A**) and in the NZ outdoor pasture study (**B**). During recordings, the test cow was moved into the test pen. When not recording, the cow was moved back into to the group pen.

The cows were fitted with the recording devices and moved into the test pens individually for a maximum of 7 days. The devices and recording gear were downloaded, re-charged and re-set twice daily. Devices were removed if the cow showed signs of skin or behavioural irritation or in the case of forecasted heavy rain (NZ outdoor group).

2.3. Heart Rate Recording

HR and HRV were measured using a polar RS800 CX watch and Polar equine monitoring belt (Polar Electro Oy, Kempele, Finland). Patches of hair at the electrode locations were clipped and the electrodes were generously coated with ultrasound gel (Aquasonic 100 gel, Parker Laboratories, Fairfield, NJ, USA) to improve contact with the skin and signal transmission. The belt and watch were checked frequently and adjusted as needed throughout the recording. An elastic surcingle was attached over the belt to keep it tight to the skin. The clasps of the Polar belt and surcingle were padded with felt and wrapped in cohesive bandage to reduce the chance of irritation to the cows and also reduce the chance of the surcingle loosening throughout the recording. The watch was synchronized to the recording computer's time and was programmed to record heart rate and R-R intervals which are used for HRV calculations. R-R intervals are the time (in milliseconds) from the R peak of one heartbeat to the R peak of the next heartbeat.

The data were downloaded and analysed using Polar Pro Trainer (v5.35.160) and artefacts in the R-R data were filtered and corrected using moderate filter power. Only traces containing less than 1% of identified errors were used in the analysis. Filtered data were exported and HR and HRV statistics were calculated in 30 s intervals (epochs) corresponding to the scored PSG data. Only time domain features of the HRV were calculated since frequency domain features of the HRV may not be an accurate representation of the data in such small time periods [19]. Time domain features included mean HR (in beats per minute—BPM), root mean square of successive differences of the R-R signal (RMSSD), and standard deviation of the R-R signal (SDRR) in 30 s epochs.

2.4. PSG Recording and Sleep Scoring

2.4.1. PSG Recording Protocols

PSG were recorded as described by Hunter et al. (2021) [20]). Pre-gelled adhesive snap ECG electrodes (Natus Neurology, Ottawa, ON, Canada) were used to record four EEG, a reference (REF), patient grounding (PGND), and two EOG and two EMG channels from the cows. Lead wires were snapped on, bundled down the neck and plugged into the Embletta MPR PG + St proxy PSG recording device (Embla Systems, Ottawa, ON, Canada). The device was placed in a padded plastic box within a pouch sewn to the elastic surcingle covering the HR monitor belt. The device was programmed, data were downloaded, and traces processed and scored using RemLogic 3.4.3 software (Embla Systems, Ottawa, ON, Canada). Good quality recordings, which had a minimum of one complete EEG, EOG and EMG trace each with good impedance (1–14kHz), minimal artefacts and with good quality corresponding HR traces were used in the analysis. Recordings lasted a maximum of 10 h due to device memory limitations.

2.4.2. Sleep Scoring

The good quality traces were scored according to criteria developed from a combination of previous cow sleep EEG studies [6,7,21] as well as human sleep scoring criteria [22]. Five stages of sleep and wakefulness were scored: Awake (W), REM (R) and Non-REM (which was subdivided into 3 stages, light N1 and N2, and deep N3). Rumination was also scored from the PSG as substantial artefacts due to jaw muscle movements when chewing obscured the actual signals of the traces and it was impossible to tell what stage the cow was in during that time. Intra-observer accuracy was calculated using "irr" package [23] in R (v4.0.5) using Cohen's Kappa with $\kappa = 0.83$ and overall agreement of 89.4% indicating good agreement [24].

2.4.3. Lying Postures

Lying postures were identified from video recordings made from four surveillance cameras equipped with automatic infra-red night vision capability (Geovision monitoring system, Viewlog, GeoVision Inc., Taiwan (Scottish Cows)) and Vivotek ND9541P H.265 NVR (Vivotek Inc., Taiwan) (NZ Cows)). Lying postures, with head held up above the point of the shoulder (UP), head held below the shoulder (HL), head resting on the ground to the front (HF), neck turned with head resting on flank or "tucked" (T), lateral lying or "flat out" (FO) and moving (MV) as well as not scored (NS) (Table 1), were scored instantaneously every 30 s corresponding to the start of the PSG and HR epochs. Intra-observer reliability was conducted in R (v4.0.5) [25] using the Cohen's kappa in the 'irr' package [23] and the kappa statistic was $\kappa = 0.95$ demonstrating a high level of agreement [24].

Table 1. Behavioural ethogram for scoring lying postures in dairy cows, including head positions and photographs from surveillance videos.

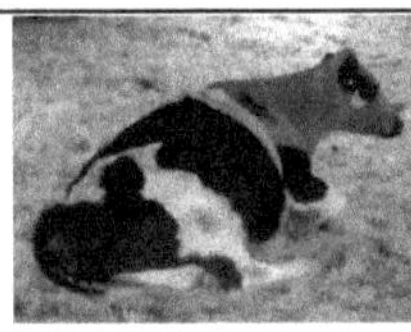

Head Up (UP)
Lying sternally recumbent with head held up

Head Low (HL)
Lying sternally recumbent with head held low

Head Resting Front (HF)
Lying sternally recumbent with the head and or neck resting on the ground

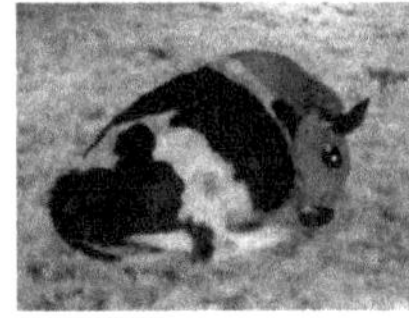

Tucked (T)
Lying sternally recumbent with head turned and resting on the flank

Flat-out (FO)
Lying laterally with legs extended and head and neck resting on the ground

2.5. Data Analysis

Scored sleep stages, lying postures and heart rate data were aligned by time stamps. In cows, sleep occurs when lying down [21], therefore only epochs identified as lying were included in the analysis. Epochs with posture 'not scored' (NS) due to observer inability to accurately observe behaviour or other extraneous circumstances were also removed from the dataset. As the stages of sleep or wakefulness could not be determined while ruminating, these epochs were also removed.

We fitted a mixed-effects model to determine if the cow's HR changed by sleep stage using the 'lme4' [26] and 'lmertest' [27] packages in R (v4.0.5) [25]. The fixed effects were study (country), sleep stage and their interaction. We included recording day nested within individual cow ID as random factors. We then used the same model with each of the remaining variables, and RMSSD and SDRR as the response variables. Using the 'predictmeans' [28] package we calculated the predicted means, standard error of the means (SEM) and least significant differences (LSD).

We then re-ran the same models, now including the cow's lying posture as a fixed effect with interaction with study and calculated predicted means of cow's HR and HRV measures by lying posture and study.

3. Results

Overall, with rumination, standing, and unscored lying behaviour removed, 1968 epochs totaling 16.4 h of good quality data were obtained from 10 cows in 29 recordings days. Data from one Scottish and one NZ cow were removed as they each had only one limited good quality recording that did not contain any lying periods. The data set was skewed towards more time in the awake (W) state, as 629 epochs were scored as W, 315 epochs in N1, 593 epochs in N2, 197 epochs in N3 and finally 234 epochs in REM (Table 2).

Table 2. Count of the total number of epochs of data in each posture by sleep stage (tucked (T), head resting front (HF), head low (HL), head up (UP), moving (MV) and flat out (FO)) and study country.

	NZ							Scot						
	T	**HF**	**HL**	**UP**	**MV**	**FO**	**Total**	**T**	**HF**	**HL**	**UP**	**MV**	**FO**	**Total**
W	24	2	67	116	14	1	224	42	4	133	157	49	20	405
N1	4	2	34	58	3	0	101	20	1	103	52	13	25	214
N2	9	4	61	120	0	0	194	51	13	217	46	6	66	399
N3	8	2	21	47	0	0	78	17	1	72	15	3	11	119
R	90	0	0	1	0	0	91	114	1	4	3	1	20	143
Total by Posture	135	10	183	342	17	1	688	244	20	529	273	72	142	1280

3.1. Lying Posture

The HR and HRV parameters changed with specific body posture while lying. In the Scottish cows, moving and lying flat out postures resulted in significantly higher mean heart rate (MV = 56.43 ± 3.17 bpm, FO = 55.53 ± 3.15 bpm) than all other postures (lying with the head up, or low, or resting on the ground or with the head tucked) (Figure 2). Flat out lying was rare in the NZ data, with only one epoch over all observations. Moving was also associated with a higher mean HR in the NZ group (84.22 ± 3.41 bpm). In the NZ cows, tucked posture was also associated with significantly higher RMSSD values than the head low, head up and moving postures, indicating more variability in the time between successive heart beats. Similar results were found with RMSSD in the Scottish group, who had higher RMSSD in T compared to HL ($p = 0.007$, df = 1936, t = 2.7), UP ($p = 0.0017$, df = 1936, t = 3.15) and lower compared to FO ($p = 0.0035$, df = 1936, t = -2.92). We also found a significant effect of sleep stage and its interaction with study location on the HR and HRV parameters. Table 2 shows the means for the different sleep stages.

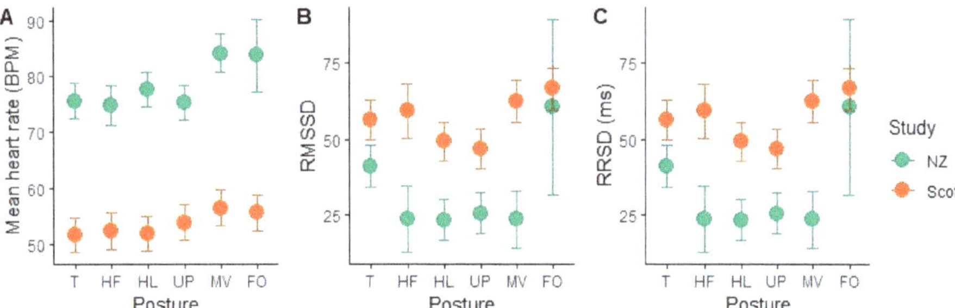

Figure 2. Plots of predicted means with error bars indicating standard error of the means for the mean heart rate in beats per minute (bpm) (**A**), the RMSSD (**B**) and SDRR (**C**) by lying postures (Tucked (T), head resting front (HF), head low (HL), head up (UP), moving (MV) and flat out (FO)) for the New Zealand group (NZ), Scottish group (Scot). Figure produced in R v4.0.5 using ggplot2 package (https://cran.r-project.org/web/packages/ggplot2/index.html, accessed on 3 May 2021).

3.2. Mean HR

After accounting for variation between cows and study days, we found a large effect of study group on mean HR. Mean HR was around 20 BPM lower in the Scottish cows than the NZ cows. After accounting for this variation, significant differences between sleep stages were evident. In both the indoor-housed Scottish group and outdoor-managed NZ cows, mean heart rate was significantly slower in the REM sleep stage compared to awake (W) (NZ: $p < 0.001$, df = 1934, t = 5.51) (Scottish: $p < 0.001$, df = 1934, t = 12.16). In the Scottish group, N2 and N3 stages were not significantly different from one another ($p = 0.09$, df = 1934, t = 1.68), but all others (W, N1, R) were. In the NZ group, W and N1 were not different from each other ($p = 0.46$, df = 1934, t = −0.74), and neither were N3 and REM ($p = 0.89$, df = 1934, t = 0.14). Overall, heart rate declined successively from W to N1 and then to N2, while N3 and REM were significantly lower than the other sleep stages.

3.3. RMSSD-Variability between Successive Heart Beats

As heart rates were significantly different between the study groups, it is unsurprising that they also had a significant effect of study on the RMSSD (Figure 3). As the mean heart rate was lower in the Scottish group, their RMSSD was 15–30 ms higher than the NZ group, indicating longer inter-beat intervals (Table 3). Accounting for the random effects, we found significantly higher RMSSD values during REM sleep epochs, indicating more variability in the time between successive heart beats in this stage. In the NZ group, the RMSSD during REM sleep was significantly higher than W ($p = 0.0061$, df = 1937, t = −2.74), N1 ($p = 0.01$, df = 1937, t = −2.57) and N2 ($p = 0.0056$, df = 1937, t = −2.78) but not significantly different from N3 ($p = 0.18$). In the Scottish group, RMSSD during REM sleep was highly significantly different from all other sleep stages, which did not differ greatly from one another. However, N2 did differ significantly from W ($p = 0.002$, df = 1937, t = −3.09) and N1 ($p = 0.028$, df = 1937, t = −1.554). Overall, the time between successive heart beats during REM sleep was significantly more variable than the other stages. N3 was more variable than W but not compared to the other NREM stages (N1 & N2).

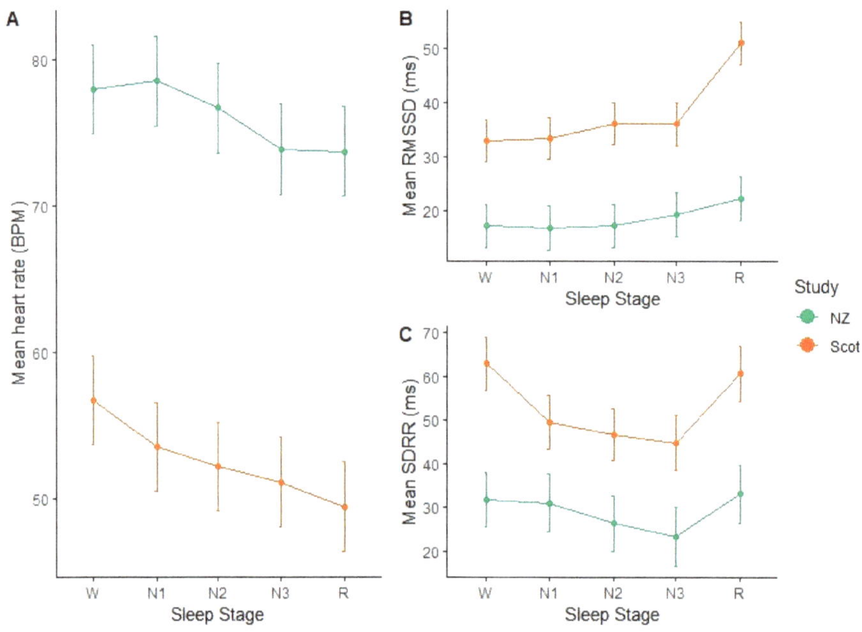

Figure 3. Plots of predicted means with error bars indicating standard error of the mean in each sleep stage for the NZ and Scotland groups for mean HR (**A**), RMSSD (**B**) and SDRR (**C**)). Figure produced in R v4.0.5 using ggplot2 package (https://cran.r-project.org/web/packages/ggplot2/index.html, accessed on 3 May 2021).

Table 3. Table of predicted means ± standard error of the HR mean, RMSSD and SDRR for each of the stages, awake (W), Non-REM: N1–N3 and REM sleep overall data and by study group in NZ and Scotland (SC).

Sleep Stage	W	N1	N2	N3	REM
HR Mean					
Predicted Mean NZ	78.03 [A] ± 3.05	78.57 [A] ± 3.09	76.72 [B] ± 3.06	73.90 [C] ± 3.10	73.78 [C] ± 3.09
Predicted Mean SC	56.77 [D] ± 3.04	53.56 [E] ± 3.05	52.20 [F] ± 3.04	51.14 [F] ± 3.08	49.48 [G] ± 3.07
RMSSD					
Predicted Mean NZ	17.15 [A] ± 3.97	16.80 [A] ± 4.12	17.14 [A] ± 3.99	19.22 [AB] ± 4.18	22.14 [BC] ± 4.14
Predicted Mean SC	32.92 [CD] ± 3.82	33.41 [DE] ± 3.88	36.03 [F] ± 3.83	35.96 [EF] ± 3.98	50.96 [G] ± 3.94
SDRR					
Predicted Mean NZ	31.74 [AB] ± 6.22	30.95 [ABC] ± 6.60	26.30 [AC] ± 6.27	23.32 [C] ± 6.75	33.08 [ABD] ± 6.63
Predicted Mean SC	62.93 [E] ± 6.03	49.61 [D] ± 6.19	46.69 [BD] ± 6.04	44.83 [BD] ± 6.42	60.71 [E] ± 6.34

[A–G] Means within each HR/HRV measure without a common superscripted letter are significantly different at $p < 0.05$.

3.4. SDRR-Total Variability of Time between Heart Beats

There were differences between the groups, but this was not as wide as for the other variables. SDRR was higher for the awake and REM stages compared to the other stages, indicating that there was higher variability in the overall time between heart beats for these stages. In the NZ cows, SDRR during REM sleep was significantly higher than N3 ($p = 0.0288$, df = 1947, t = -2.1882), but not the other stages. N3 was significantly lower than W ($p = 0.0337$, df = 1947, t = 2.1254), however N3 was not significantly different from the other stages. In the Scottish group, SDRR was not significantly different between W and REM ($p = 0.4422$, df = 1947, t = 0.7687), but these stages were significantly higher than N1, N2 and N3 that were not significantly different from one another.

4. Discussion

Our results show that cardiac outputs could be useful in assessing sleep stages in dairy cows. However, we found major differences in mean HR between the two groups of cows, which may be due to different cow characteristics. Understandably, despite replication in data collection methods, there were marked differences in the housing, breed, size, physiological stage, and diet of the cows in each study location. The NZ cows were all in late pregnancy, whereas the Scottish cows were non-pregnant and non-lactating. Late pregnant heifers and cows have been found to have higher mean HR than earlier on in pregnancy [29]. While there was not a particularly large difference in cow age, there was a difference between cow size and breed. The Scottish cows were very large Holstein cows, and the NZ cows were much smaller being Jersey-Holstein crosses (Kiwi-cross). Other studies have found significant differences in HR and HRV measures between different breeds (Brown Swiss and Simmental) when standing, lying and milking [30]. Body size is also known to affect HR and HRV, and a decrease in HR was found with increasing weight in horses and ponies [31]. The Scottish recordings were made indoors in spring/summer months with a daytime temperature average of 15.2 °C (range 8–22 °C) and overnight temperature of 11.2 °C (range 5–22 °C). The NZ recordings were made outdoors over winter with average an average daytime temperature of 10.2 °C (range 2–18 °C) and overnight average of 8.4 °C (range 2–14 °C). Seasonal thermal stress has been found to affect behaviour, stress and immune response in dairy cows [32,33], and increasing temperature humidity index has been associated with decreased HRV measures in sheep and goats [34]. Although the cows in the winter conditions in NZ had higher HR and lower HRV than cows in summer conditions in Scotland, the environmental conditions could have affected the HR and HRV activity in this study.

Importantly, despite these group differences, we found that HR and HRV changes with sleep stages in both groups and clearly, Figure 2 shows that the differences are in the same direction. These results indicate that the patterns of the changes in HR and HRV measures between the sleep stages are stable and as such these measures could be used with all cows, although further research is needed to assess if these patterns are also observed in lactating cows and cows in other stages of pregnancy.

Surprisingly, we found that mean HR during REM sleep was lower than when awake and in the lighter NREM sleep stages (N1 and N2). This is different than results in humans, where HR has been found to decrease with the progression of NREM sleep stages, and speeds up again in REM sleep [35]. However, similar results with an overall lower heart rate during REM sleep were also found in dogs [36]. Despite the lower mean HR, HRV measures of RMSSD and SDRR were higher in REM sleep, indicating more variation between heart beats. This observation is similar to that shown for HRV patterns during human sleep, where HRV tends to be more variable when awake and in REM sleep than during N3 and other NREM sleep stages [37]. Mean HR and RMSSD may be useful to distinguish between awake (W) and REM stages; however, they are not particularly useful to distinguish between NREM stages (N1, N2, N3). SSDRR was useful to identify N3 stages in both groups as it was significantly lower. These patterns of differences in sleep stage could be useful in future applications to predict sleep stage of dairy cows, particularly if prior to recording, a lying awake baseline could be specified. Then sleep stages could be identified or predicted from differences from that baseline.

A previous study has found that body position was associated with difference in HR and HRV measures in cows [17]; however, they did not specify body posture while lying and were unable to identify awareness levels. We found that the specific posture that cows adopted during lying affected their HR and HRV, and in particular that epochs identified as being in the flat out (lateral lying) posture and epochs with the head moving resulted in higher HR and more variability in the HRV. As moving is a physically active behaviour, this activity may have had a carry-over effect on the heart rate for an extended period. Therefore, an epoch in which the cow moved her head at the start may have higher HR across that epoch and into the next. Flat out posture was rarely observed in the NZ group, and only scored once, and even then, was only observed as a transition behaviour between other postures. In the Scottish group, flat out postures were far more commonly observed, and most often occurred while the cow was in N2 sleep as well as awake and in N1. It is unclear if the increased HR in this posture was due to the position of the body which could have facilitated a faster movement of blood, or because most epochs scored as flat out happened to occur in sleep stages that had higher heart rates. In the NZ group almost all REM sleep occurred in the tucked posture. The tucked posture was found to have significantly higher RMSSD; however, since REM sleep was also found to have higher RMSSD it is likely that the effect of the posture on the HRV measure was more likely due to the sleep stage in this case.

The intra-observer reliability for sleep scoring was 89%, which according to inter-scorer reliability in human sleep studies is very good [38,39]. However, there is still some possibility that the 11% uncertainty in sleep scoring was a contributing factor to the variability of the HR and HRV measures within sleep stages. Additionally, we analysed the HR data in 30 s epochs, specifically to correspond to sleep scoring. The 30 s epoch is a standard practice in scoring sleep stages from PSG, as it corresponds well to the structure of human sleep, containing fewer stage shifts than longer epochs which would be more likely to contain many stage shifts [40]. Despite shorter epoch length, some mid-epoch stage shifts could still have occurred. In these instances, although the PSG was scored one way, the HR measures could have reflected another stage, and this could also be another source of variability in the HR and HRV measures within sleep stage.

Similarly, the HR and HRV data may have also been influenced by the short epoch windows. Typically, human HRV measures are conducted in 5 min increments, although ultra-short windows such as 30 s windows have been found to be acceptable for the assessment of HRV at rest in humans [41]. Bouts of cow sleep stages can typically be quite short [7,21] and thus multiple stage shifts would be captured in a longer epoch length of 5 min. This was a major reason for choosing to analyse the HR and HRV in ultra-short windows. However, some have questioned the accuracy of windows shorter than 2-min for the analysis of HRV in human athletes [19]. RMSSD measurements in even shorter 10 s windows were also found to correspond well to standard longer intervals in humans, but SDRR did not [42]. Therefore, the short time window selection could have affected the accuracy of the cow HRV RMSSD and SDRR measurements. HR and HRV may be useful for the assessment of sleep stages in dairy cows, however, further investigation into the validity of ultra-short HRV measures in dairy cows and additional validation with PSG is needed.

5. Conclusions

We have shown that sleep stage is associated with changes in HR and HRV in dairy cows. Mean HR was significantly lower in the indoor-housed, non-pregnant, and non-lactating cows compared to pregnant, dry, outdoor managed cows. We also found that mean HR decreased with sleep depth, SDRR was more variable in awake and REM states, and RMSSD was significantly higher in REM sleep than the other stages. These results indicate that HR and HRV could be a useful measure for the future identification of sleep stages in dairy cows using less invasive devices than PSG, making sleep research for animal welfare more accessible.

Author Contributions: Conceptualization, L.B.H., M.J.H., F.M.L., C.O., J.R.W., K.J.S.; methodology, L.B.H.; software, L.B.H.; formal analysis, L.B.H.; investigation, L.B.H.; data curation, L.B.H.; writing—

original draft preparation, L.B.H.; writing—review and editing, M.J.H., F.M.L., C.O., K.J.S.; visualization, L.B.H.; supervision, M.J.H., F.M.L., C.O., J.R.W., K.J.S.; funding acquisition, M.J.H., F.M.L., C.O., J.R.W., K.J.S. All authors have read and agreed to the published version of the manuscript.

Funding: SRUC receives core funding from the Scottish Government's Rural and Environmental Science and Analytical Services Division (RESAS). AgResearch funding provided by the New Zealand Ministry of Business, Innovation and Employment (A23975).

Institutional Review Board Statement: The study was designed in accordance with the relevant guidelines and legislation in both Scotland and New Zealand where the studies took place. Ethical approval was obtained from the UK Home Office (Project Licence P204B097E), SRUC Animal Ethics Committee (Ref. ED AE 03-2018) and Ruakura Animal Ethics committee (AE 14708) prior to the start of animal manipulations.

Data Availability Statement: The data that support the findings of this study are available from the corresponding author, L.B.H., upon reasonable request.

Acknowledgments: We acknowledge statistical and R programming assistance from Harold Henderson. We also thank all those who have been involved in the on-farm data collection and animal management work in Scotland at Acrehead farm and in New Zealand at AgResearch Ltd. and Dairy NZ Lye farm.

Conflicts of Interest: The authors declare no conflict of interest.

References

1. Xie, L.; Kang, H.; Xu, Q.; Chen, M.J.; Liao, Y.; Thiyagarajan, M.; O'Donnell, J.; Christensen, D.J.; Nicholson, C.; Iliff, J.J.; et al. Sleep Drives Metabolite Clearance from the Adult Brain. *Science* **2013**, *342*, 373–377. [CrossRef]
2. Le Bon, O. Relationships between REM and NREM in the NREM-REM sleep cycle: A review on competing concepts. *Sleep Med.* **2020**, *70*, 6–16. [CrossRef]
3. Langford, F.M.; Cockram, M.S. Is sleep in animals affected by prior waking experiences? *Anim. Welf.* **2010**, *19*, 215–222.
4. Fukasawa, M.; Komatsu, T.; Higashiyama, Y. The change of sleeping and lying posture of Japanese black cows after moving into new environment. *Asian Australas. J. Anim. Sci.* **2018**, *31*, 1828–1832. [CrossRef]
5. Carley, D.W.; Farabi, S.S. Physiology of sleep. *Diabetes Spectr.* **2016**, *29*, 5–9. [CrossRef]
6. Hänninen, L.; Mäkelä, J.P.; Rushen, J.; de Passillé, A.M.; Saloniemi, H. Assessing sleep state in calves through electrophysiological and behavioural recordings: A preliminary study. *Appl. Anim. Behav. Sci.* **2008**, *111*, 235–250. [CrossRef]
7. Ternman, E.; Hänninen, L.; Pastell, M.; Agenäs, S.; Nielsen, P.P. Sleep in dairy cows recorded with a non-invasive EEG technique. *Appl. Anim. Behav. Sci.* **2012**, *140*, 25–32. [CrossRef]
8. Kull, J.A.; Proudfoot, K.L.; Pighetti, G.M.; Bewley, J.M.; O'Hara, B.F.; Donohue, K.D.; Krawczel, P.D. Effects of acute lying and sleep deprivation on the behavior of lactating dairy cows. *PLoS ONE* **2019**, *14*, e0212823. [CrossRef] [PubMed]
9. Zoccoli, G.; Amici, R. Sleep and autonomic nervous system. *Curr. Opin. Physiol.* **2020**, *15*, 128–133. [CrossRef]
10. Bertsch, K.; Hagemann, D.; Naumann, E.; Schächinger, H.; Schulz, A. Stability of heart rate variability indices reflecting parasympathetic activity. *Psychophysiology* **2012**, *49*, 672–682. [CrossRef] [PubMed]
11. Xiao, M.; Yan, H.; Song, J.; Yang, Y.; Yang, X. Sleep stages classification based on heart rate variability and random forest. *Biomed. Signal Process. Control* **2013**, *8*, 624–633. [CrossRef]
12. Mitsukura, Y.; Fukunaga, K.; Yasui, M.; Mimura, M. Sleep stage detection using only heart rate. *Health Inform. J.* **2020**, *26*, 376–387. [CrossRef] [PubMed]
13. Shaffer, F.; Ginsberg, J.P. An Overview of Heart Rate Variability Metrics and Norms. *Front. Public Health* **2017**, *5*, 258. [CrossRef]
14. Kovács, L.; Kézér, F.L.; Jurkovich, V.; Kulcsár-Huszenicza, M.; Tözsér, J.; Hillmann, E. Heart rate variability as an indicator of chronic stress caused by lameness in dairy cows. *PLoS ONE* **2015**, *10*, e0134792. [CrossRef] [PubMed]
15. Stewart, M.; Stafford, K.J.; Dowling, S.K.; Schaefer, A.L.; Webster, J.R. Eye temperature and heart rate variability of calves disbudded with or without local anaesthetic. *Physiol. Behav.* **2008**, *93*, 789–797. [CrossRef] [PubMed]
16. Laister, S.; Stockinger, B.; Regner, A.M.; Zenger, K.; Knierim, U.; Winckler, C. Social licking in dairy cattle-Effects on heart rate in performers and receivers. *Appl. Anim. Behav. Sci.* **2011**, *130*, 81–90. [CrossRef]
17. Frondelius, L.; Järvenranta, K.; Koponen, T.; Mononen, J. The effects of body posture and temperament on heart rate variability in dairy cows. *Physiol. Behav.* **2015**, *139*, 437–441. [CrossRef]
18. Hunter, L.B.; O'Connor, C.; Haskell, M.J.; Langford, F.M.; Webster, J.R.; Stafford, K.J. Lying Posture Does Not Accurately Indicate Sleep Stage in Dairy Cows. **2021**, submitted.
19. Bourdillon, N.; Schmitt, L.; Yazdani, S.; Vesin, J.M.; Millet, G.P. Minimal window duration for accurate HRV recording in athletes. *Front. Neurosci.* **2017**, *11*, 456. [CrossRef]
20. Hunter, L.B.; Baten, A.; Haskell, M.J.; Langford, F.M.; O'Connor, C.; Webster, J.R.; Stafford, K. Machine learning prediction of sleep stages in dairy cows from heart rate and muscle activity measures. *Sci. Rep.* **2021**, *11*, 10938. [CrossRef]

21. Ruckebusch, Y.; Bell, F.R.; Barbey, P.; Guillemot, P.; Serthelon, J.P. Étude polygraphique et comportementale des états de veille et de sommeil chez la vache (bos taurus). *Ann. Rech. Vet.* **1970**, *1*, 41–62.

22. Iber, C.; American Academy of Sleep Medicine. *The AASM Manual for the Scoring of Sleep and Associated Events: Rules Terminology and Technical Specifications*, 1st ed.; AASM: Westchester, NY, USA, 2007.

23. Gamer, M.; Lemon, J.; Fellows, I.; Singh, P. irr: Various Coefficients of Interrater Reliability and Agreement. R Package Version 0.84.1. 2019. Available online: https://CRAN.R-project.org/package=irr (accessed on 14 July 2021).

24. McHugh, M.L. Interrater reliability: The kappa statistic. *Biochem. Med.* **2012**, *22*, 276–282. [CrossRef]

25. R Core Team. *R: A Language and Environment for Statistical Computing*; R Foundation for Statistical Computing: Vienna, Austria, 2021. Available online: https://www.R-project.org/ (accessed on 14 July 2021).

26. Bates, D.; Maechler, M.; Bolker, B.; Walker, S. Fitting Linear Mixed-Effects Models Using lme4. *J. Stat. Softw.* **2015**, *67*. [CrossRef]

27. Kuznetsova, A.; Brockhoff, P.; Christensen, R. lmerTest Package: Tests in Linear Mixed Effects Models. *J. Stat. Softw.* **2017**, *82*, 1–26. [CrossRef]

28. Luo, D.; Ganesh, S.; Koolard, J. Predictmeans: Calculate Predicted Means for Linear Models. R Package Version 1.0.5. 2021. Available online: https://CRAN.R-project.org/package=predictmeans (accessed on 14 July 2021).

29. Trenk, L.; Kuhl, J.; Aurich, J.; Aurich, C.; Nagel, C. Heart rate and heart rate variability in pregnant dairy cows and their fetuses determined by fetomaternal electrocardiography. *Theriogenology* **2015**, *84*, 1405–1410. [CrossRef]

30. Hagen, K.; Langbein, J.; Schmied, C.; Lexer, D.; Waiblinger, S. Heart rate variability in dairy cows—Influences of breed and milking system. *Physiol. Behav.* **2005**, *85*, 195–204. [CrossRef]

31. Schwarzwald, C.C.; Kedo, M.; Birkmann, K.; Hamlin, R.L. Relationship of heart rate and electrocardiographic time intervals to body mass in horses and ponies. *J. Vet. Cardiol.* **2012**, *14*, 343–350. [CrossRef]

32. Webster, J.R.; Stewart, M.; Rogers, A.R.; Verkerk, G.A. Assessment of welfare from physiological and behavioural responses of New Zealand dairy cows exposed to cold and wet conditions. *Anim. Welf.* **2008**, *17*, 19–26.

33. Li, H.; Zhang, Y.; Li, R.; Wu, Y.; Zhang, D.; Xu, H.; Zhang, Y.; Qi, Z. Effect of seasonal thermal stress on oxidative status, immune response and stress hormones of lactating dairy cows. *Anim. Nutr.* **2021**, *7*, 216–223. [CrossRef]

34. Kitajima, K.; Oishi, K.; Miwa, M.; Anzai, H.; Setoguchi, A.; Yasunaka, Y.; Himeno, Y.; Kumagai, H.; Hirooka, H. Effects of Heat Stress on Heart Rate Variability in Free-Moving Sheep and Goats Assessed with Correction for Physical Activity. *Front. Vet. Sci.* **2021**, *8*, 658763. [CrossRef] [PubMed]

35. Cajochen, C.; Pischke, J.; Aeschbach, D.; Borbély, A.A. Heart rate dynamics during human sleep. *Physiol. Behav.* **1994**, *55*, 769–774. [CrossRef]

36. Varga, B.; Gergely, A.; Galambos, Á.; Kis, A. Heart rate and heart rate variability during sleep in family dogs (*Canis familiaris*). moderate effect of pre-sleep emotions. *Animals* **2018**, *8*, 107. [CrossRef] [PubMed]

37. Stein, P.K.; Pu, Y. Heart rate variability, sleep and sleep disorders. *Sleep Med. Rev.* **2012**, *16*, 47–66. [CrossRef]

38. Wendt, S.L.; Welinder, P.; Sorensen, H.B.; Peppard, P.E.; Jennum, P.; Perona, P.; Mignot, E.; Warby, S.C. Inter-expert and intra-expert reliability in sleep spindle scoring. *Clin. Neurophysiol.* **2015**, *126*, 1548–1556. [CrossRef]

39. Danker-Hopfe, H.; Anderer, P.; Zeitlhofer, J.; Boeck, M.; Dorn, H.; Gruber, G.; Heller, E.; Loretz, E.; Moser, D.; Parapatics, S.; et al. Interrater reliability for sleep scoring according to the Rechtschaffen & Kales and the new AASM standard. *J. Sleep Res.* **2009**, *18*, 74–84. [CrossRef] [PubMed]

40. Schulz, H. Rethinking Sleep Analysis Comment on the AASM Manual for the Scoring of Sleep and Associated Events. *J. Clin. Sleep Med.* **2008**, *4*, 99–103. [CrossRef] [PubMed]

41. Wu, L.; Shi, P.; Yu, H.; Liu, Y. An optimization study of the ultra-short period for HRV analysis at rest and post-exercise. *J. Electrocardiol.* **2020**, *63*, 57–63. [CrossRef]

42. Thong, T.; Li, K.; McNames, J.; Aboy, M.; Goldstein, B. Accuracy of ultra-short heart rate variability measures. In Proceedings of the 25th Annual International Conference of the IEEE Engineering in Medicine and Biology Society, Cancun, Mexico, 17–21 September 2003; pp. 2424–2427. [CrossRef]

Article

Effects of Housing System on Anxiety, Chronic Stress, Fear, and Immune Function in Bovan Brown Laying Hens

Andrew M. Campbell, Alexa M. Johnson, Michael E. Persia and Leonie Jacobs *

School of Animal Sciences, Virginia Tech, Blacksburg, VA 24061, USA; drewc1@vt.edu (A.M.C.); alexaj3@vt.edu (A.M.J.); mpersia@vt.edu (M.E.P.)
* Correspondence: jacobsl@vt.edu; Tel.: +1-540-231-4735

Simple Summary: The objectives of this study were to determine if housing Bovan brown laying hens in conventional cages or enriched floor pens impacted novel physiological and behavioral markers for animal welfare and whether we can use these markers to assess animal welfare. We found that birds that were housed in conventional cages showed increased tonic immobility durations (indication of fearfulness), decreased fecal Immunoglobulin A (indicator of immune function), and increased feather corticosterone concentrations (indicator of chronic stress) compared to hens that were housed in enriched pens. These results indicate that caged birds are more stressed, have reduced immune function, and are more fearful than birds that are housed in pens. In contrast to expectations, we found that caged hens showed a shorter latency to feed during attention bias testing, indicating reduced anxiety compared to birds from pens. Overall, we found that conventional cages generally impacted animal welfare negatively, with the exception of anxiety. In addition, the results suggest that the chosen novel markers for animal welfare show appropriate contrast between long-term housing systems for laying hens. Yet, additional work needs to be done before these measures can be used more broadly.

Citation: Campbell, A.M.; Johnson, A.M.; Persia, M.E.; Jacobs, L. Effects of Housing System on Anxiety, Chronic Stress, Fear, and Immune Function in Bovan Brown Laying Hens. *Animals* **2022**, *12*, 1803. https://doi.org/10.3390/ani12141803

Academic Editor: Vera Baumans

Received: 8 June 2022
Accepted: 10 July 2022
Published: 14 July 2022

Publisher's Note: MDPI stays neutral with regard to jurisdictional claims in published maps and institutional affiliations.

Abstract: The scientific community needs objective measures to appropriately assess animal welfare. The study objective was to assess the impact of housing system on novel physiological and behavioral measurements of animal welfare for laying hens, including secretory and plasma Immunoglobulin (IgA; immune function), feather corticosterone (chronic stress), and attention bias testing (ABT; anxiety), in addition to the well-validated tonic immobility test (TI; fearfulness). To test this, 184 Bovan brown hens were housed in 28 conventional cages (3 birds/cage) and 4 enriched pens (25 birds/pen). Feces, blood, and feathers were collected 4 times between week 22 and 43 to quantify secretory and plasma IgA and feather corticosterone concentrations. TI tests and ABT were performed once. Hens that were from cages tended to show longer TI, had increased feather corticosterone, and decreased secretory IgA at 22 weeks of age. The caged hens fed quicker, and more hens fed during the ABT compared to the penned hens. Hens that were in conventional cages showed somewhat poorer welfare outcomes than the hens in enriched pens, as indicated by increased chronic stress, decreased immune function at 22 weeks of age but no other ages, somewhat increased fear, but reduced anxiety. Overall, these novel markers show some appropriate contrast between housing treatments and may be useful in an animal welfare assessment context for laying hens. More research is needed to confirm these findings.

Keywords: attention bias; conventional cage; environmental complexity; enrichment; feather corticosterone; IgA; laying hen

---

1. Introduction

Animal welfare is a multifaceted concept that involves an animal's ability to interact and cope with its environment. Good animal welfare would be achieved when an animal is allowed to display natural behaviors (natural living), be healthy and function normally (basic health and functioning), and experience a generally positive emotional state (affective state) [1]. For laying hens, commercial housing systems such as conventional cages can negatively impact animal welfare by restricting natural behaviors, causing health and functioning concerns such as cage layer fatigue, and likely results in worse affective states (anxiety and fear) [2]. While aspects of natural living and basic health and functioning are relatively easily measured and well-studied, few studies investigate the effects of housing system on the affective states of laying hens. Additionally, as nearly all measures of affective state rely on interpretations of animal behavior, there is need for additional physiological measurements to allow for a more cumulative assessment of animal affective states.

With limited physiological measures of emotion and affective state available, most insight comes from behavioral assessments of negative emotions such as fear. Fear is a short-term emotional response to a current threat and elicits either a freeze, fight, or flight response [3–5]. A common assessment of animal emotion is performed using fear tests, such as a novel object test, human approach test, or tonic immobility (TI) test [6]. Measuring fear using TI is well-documented in poultry and utilizes their natural prey-predator behavioral response. TI is a type of freezing behavior where birds will feign death and is used when captured by a predator [7]. Previous investigations of TI in relation to housing systems had variable results. For example, barren caging conditions have resulted in longer TI durations, thus greater levels of fear, in Hyline Brown laying hens when compared to more enriched conditions [8,9]. However, other studies have found no effect of housing system on TI duration in laying hens [10,11]. These variable results highlight the need for further investigations of TI and other measures to elucidate the effects of housing system on emotion and affect.

Affective state is a long-term mood state which comes from the culmination of experiences and emotional responses. An animal's life experiences elicit short-term emotional responses [12]. These responses culminate to form a mood, which can range from positive to negative in valence, and shapes the animal's affective state. Affective states influence how animals make decisions and can bias their cognitive reasoning, which can then be used to infer the affective state based on behaviors that indicate information-processing [12–14]. One cognitive bias test that was previously applied in chickens is the attention bias test (ABT). The ABT is a validated and well-used method to measure affective states, more specifically anxiety, in agricultural animals [13]. In an ABT, the level of vigilance or attention an animal allocates to a perceived threat is quantified [13]. This allocation of vigilance is differential and affect-mediated, with more anxious affective states resulting in more vigilance towards a threatening stimulus [3,13]. The ABT was validated in laying hens, with birds that were given anxiogenic drugs displaying increased vigilance compared to control birds [15]. Excess anxiety decreases the ability of commercial poultry to cope with changes in their environment such as transport, handling, and loud noises. In laying hens, the ABT reflected anxiety that was related to range usage [15,16]. Determining anxiety via the ABT was also successful in broilers [3], sheep [17,18], and pigs [19]. However, ABT has not yet been applied to assess the impact of housing system (cage vs. cage-free) environments on laying hen anxiety.

Feather corticosterone (CORT) is a potential promising physiological biomarker for chronic stress [20]. Concentrations of feather CORT can provide a retrospective view on stressful experiences during feather growth [21]. The calamus of a feather is highly vascularized which allows for the deposition of circulating CORT into the feather as it grows, allowing for quantification of hypothalamic-pituitary-adrenal (HPA) axis activity over a prolonged period of time [20]. The extraction and quantification of feather CORT has been validated for use in layers, broilers, turkeys, and non-domesticated bird species and requires no invasive sampling procedures [20–23]. Feather CORT could provide insight in chronic stress that is caused by housing systems in laying hens.

Secretory Immunoglobulin-A (IgA) is the most abundant antibody on mucosal surfaces including the intestinal tract and has an important role in mediating the adaptive humoral immune defense [24–26]. In addition, IgA circulates in the blood, yet the role is less understood and likely serves to help reduce inflammation [26]. Concentrations of IgA also seem to reflect the valence (positivity or negativity) of environmental stimuli [18]. For example, IgA concentrations are downregulated in response to physical or psychological stress [27–30]. Broiler chickens and laying hens under prolonged periods of heat stress (chronic negative stressor) showed decreased concentrations of plasma IgA when compared to the control treatment [31,32]. Mice that were exposed to restraint stress over four days (negative stressor) showed decreased concentrations of intestinal IgA compared to the control [29]. Shelter cats with access to enrichment (positive stimuli) had higher levels of secretory (fecal) IgA than cats without access to enrichments [33]. In mice, prolonged voluntary exercise (a high arousal-positive-valence activity) increased salivary IgA concentrations after 3 weeks, indicating that IgA concentrations can increase in response to positive activities [34]. In addition, forced prolonged exercise (a high arousal-negative-valence activity) had the opposite effect in horses, rats, and humans [26,35–37]. The valence-dependent response of circulating and secretory IgA indicates the potential for use as a marker for the affective state.

The combined use of novel and well-validated measures for emotion and affective state could provide a better understanding of the impacts of different housing conditions on laying hen welfare. In addition, it allows further confirmation of novel measures as we can compare novel test outcomes with well-validated test responses, such as the TI test. Therefore, the objectives of this study are to (1) determine if the housing system impacts laying hen welfare outcomes that are related to the affective state and emotion and (2) determine if these novel measures can be used to assess laying hen welfare. We hypothesized that birds that were housed in enriched floor pens (pen) would show decreased fear and anxiety, increased IgA concentrations, and decreased feather CORT concentrations than birds that were housed in traditional conventional caging, indicating that these novel measures can be used to assess laying hen welfare and that enriched housing systems contribute to positive affective states in pen laying hens.

2. Materials and Methods

2.1. Birds and Housing Treatments

This experiment was approved by the Virginia Tech Institutional Animal Care and Use Committee (IACUC protocol 18-205). Day-old Bovan Brown chicks (n = 184) were sourced from a commercial hatchery (Blackstone, VA, USA) where they were beak trimmed after hatch. The birds received a Salmonella vaccination at 16 weeks of age. From day one to week six of age, 84 chicks were reared in conventional cages and 100 chicks in floor pens as part of an unrelated study that was investigating the impact of dietary phosphorous on egg production.

In week six, all the birds were wing banded for individual identification, relocated to another facility, and regrouped, yet housing treatments (either cage or floor pen) remained consistent. The chicks from floor pens were randomly distributed over four enriched floor pens (pen of 16.7 m^2), with 25 birds per pen. The chicks from cages were distributed over 14 conventional cages (cage) of 0.093 m^2, with six birds per cage until week 12 of age. At week 12, three birds from each cage were moved to 14 unoccupied cages to reduce the stocking density, resulting in 28 cages total with three birds per cage. The penned birds were not regrouped at 12 weeks of age. All the pens contained pine wood shavings (7 cm depth), one trough feeder (92.7 cm^2/bird) and bell drinker (Plasson, Ma'agan Michael, Israel), 10 galvanized steel nest boxes that were arranged in two tiers of five boxes (30.5 cm long × 30.5 cm wide × 35.6 cm high), one hay bale (replaced as needed), and a head of cabbage that was provided twice per week, which was suspended from the ceiling at bird level. The birds had access to perches (5.49 m or 0.22 m/bird perch space) consisting of pressure-treated 5 cm × 10 cm boards for the frame and three 2.5 cm diameter PVC pipes that were mounted at 45 cm, 60 cm, and 90 cm heights.

There were four rows of stacked conventional cages (38 cm wide × 38 cm long × 46 cm high) with a sloped wire floor that were located in the same room as the pens, and contained two nipple drinkers per cage, a feed trough that was suspended outside the cage with a gap in the wire for birds to access feed, and an egg collection trough on the opposite side. The pen space allowance was 0.668 m^2/bird, compared to 0.031 m^2/bird in the cage housing after week 12. The birds had *ad libitum* access to feed and water. From hatch to week six, the pullets were fed an experimental diet as part of the phosphorous experiment. Following week six, the birds were phase fed diets that were formulated to meet their nutritional needs that were appropriate for their age and developmental status. Lighting included 12 h light and 12 h dark, with windows allowing for natural light exposure during the daytime. Daylight exposure was equal between the treatments. Temperatures within the house were managed by assessing bird comfort based on behavioral responses (huddling when cold/panting when warm), however, a cold period in the winter reduced the in-house temperatures to a minimum of 5 °C when the birds were 33–37 weeks of age.

2.2. Behavioral Measurements

A TI test was performed to assess fearfulness on six arbitrarily selected hens per pen (total of 24 birds) and on one hen per cage from 24 randomly selected cages (total of 24 hens) at 23 weeks of age. The test was performed by a single researcher in the hallway of the room in which the hens were housed as described in [3]. The hens were placed on their backs in a V-shaped wooden cradle and restrained by the researcher by placing one hand on the sternum and cupping the head with the other hand. After 15 s of restraint, the researcher stepped away without making eye contact. Following the induction of TI, the duration (s) was recorded to determine fearfulness. If the induction of TI was unsuccessful, the researcher attempted to induce TI again for a maximum of three times. If TI was not induced in three attempts, the latency to rightening was scored as 0 s. The maximum duration of TI was 300 s.

An attention bias test (ABT) was performed by two observers to assess anxiety using a modified method as described in [3] on nine hens per pen and three hens per cage for 12 arbitrarily selected cages at 30 weeks of age. One bird from each cage that was tested during ABT was also tested for TI. The inter-observer agreement was tested for latency to start feeding of 12 hens and was good among the two observers (Cronbach's α of 0.841). The test arena consisted of plastic paneling and rubber flooring (76.2 cm × 76.2 cm), and contained a trough feeder with feed, mealworms, and oats. The arena was located in a separate room near the hens' room but far enough to block out the alarm call in the hens' room. The birds were tested in familiar groups of three hens. After placement of three hens in the arena, a conspecific ground predator alarm call was played for 8 s. Following the alarm call, the number of birds and latencies to begin feeding (s) were recorded. Video recordings (EOS Rebel T7 DSLR Camera, Canon, Tokyo, Japan) were used to determine the occurrence (yes/no) of vigilant behaviors (freeze, neck stretches, looking around, and erect posture) within the first 30 s of testing. Each of the four vigilance behaviors were scored as a 1 (yes) or 0 (no) and combined to obtain a vigilance score from 0 to 4 [3,16]. The alarm call was replayed for 8 s if one of four scenarios occurred (Table 1). Birds that did not start feeding after the first alarm call received a maximum latency of 300 s. After the second alarm call, the number of birds that were feeding and the latency to resume feeding (s) were recorded.

Table 1. Method that was used during attention bias (ABT) testing. The birds were tested in groups of three and the testing procedure differed depending on the number of birds who began feeding following the first alarm call.

Scenario	Procedure	Total Test Duration	Data Recorded
Testing begins	Play first alarm call	300 s	n/a
No. birds begin feeding	Allow test to run 300 s	300 s	All birds receive 300 s maximum latency to begin feeding
One bird begins feeding	Play first alarm call and allow test to run for 300 s	300 s	Latency to begin feeding for bird that began feeding. Other two birds receive maximum latency of 300 s
Two birds begin feeding	Play first alarm call and allow test to run for 300 s. Play second alarm call at 300 s and allow test to run until 420 s.	420 s	Latencies to begin feeding for two birds. Third bird receives maximum latency of 300 s. Latencies to resume feeding for two birds that began feeding if they feed before 420 s.
All three birds begin feeding before 270 s	Play first alarm call and allow test to run until the third bird begins feeding. Allow birds to feed for 5 s and play second alarm call. Allow test to run until 300 s	300 s	Latencies to begin feeding. Latencies to resume feeding for all three birds if they resume feeding prior to 300 s.
All three birds begin feeding between 270–300 s	Play first alarm call and allow test to run until the third bird eats. Allow birds to feed for 5 s and play second alarm call. Extend testing duration to 420 s.	420 s	Latencies to begin feeding. Latencies to resume feeding for all three birds if they resume feeding before 420 s.

n/a: not applicable.

2.3. Molecular Measurements

Feces and Blood

A total of 4 fecal samples and 12 plasma samples per treatment per time point were collected during weeks 22, 25, 29, and week 43 of age to determine the fecal and plasma IgA concentrations. Fresh fecal samples were collected from the cage or pen floors and pooled in microcentrifuge tubes. The birds were arbitrarily selected from floor pens ($n = 3$/pen) or cages ($n = 1$ bird/cage) for blood sampling during each time point. Across ages, nine birds were selected twice, and 30 birds were selected once.

For fecal samples, visual inspection and observation of defecation were used to ensure the freshness of the samples and to prevent degradation of fecal IgA by fecal proteases. Following collection, the fecal samples were placed on ice and then stored at $-80\,^{\circ}$C. Fecal IgA was quantified using the total protein extraction with a saline extraction method that was similar to that described in [38–40]. A total of 10 mL of a saline extraction buffer (0.01 M phosphate-buffered saline, 0.5% Tween (Sigma-Aldrich, St. Louis, MO, USA), and 0.05% sodium azide) was added to each 1 g fecal sample, followed by homogenization. Fecal suspensions were centrifuged at $1500 \times g$ for 20 min at $5\,^{\circ}$C and the supernatant was removed and placed in microcentrifuge tubes. Then, 20 μL of protease inhibitor cocktail (Sigma-Aldrich, St. Louis, MO, USA) was added to the supernatant and homogenized before storage at $-20\,^{\circ}$C until analysis.

Blood (1 mL) was drawn from the brachial vein and collected in glass tubes containing 0.05% EDTA for anticoagulation. The sample collection times (s) were recorded from start of handling until the removal of the needle, to ensure that all the samplings occurred in under two minutes to minimize the potential effects of sampling stress. Prolonged handling can impact acute-stress-related blood parameters [41], although it is unknown whether this is the case for IgA. The mean sample collection time ($\pm$SD) was 77 ± 30 s and IgA concentrations showed no correlation with collection times ($R^2 = 0.0007$; $p = 0.80$). The sample vials were then lightly mixed via inversion before storage on ice. Then, the blood samples were centrifuged at $10,000\times g$ for 10 min at room temperature, after which the plasma was removed and aliquoted into sterile microcentrifuge tubes and stored at $-20\,^{\circ}$C until analysis. The plasma and fecal samples were analyzed for IgA concentrations [µg/µL] via a commercial ELISA kit (Abcam, Cambridge, MA, USA) following the manufacturer instructions. The intra-assay CV% were below 2% for all the samples (min: 0.005; max: 1.1%).

2.4. Feathers

Tail feathers ($n = 6$/treatment per timepoint; 48 samples total) were collected during weeks 22, 25, 29, and 43 of age to determine the feather CORT concentrations. The tail feather samples were collected by cutting the calamus as close to the skin as possible without contacting or damaging the skin. At each sampling timepoint, different tail feathers were collected, thus the same tail feather was never collected more than once. The birds were arbitrarily selected from floor pens (3 birds/pen) and conventional cages (1 bird/cage) for feather sampling at each timepoint. Across ages, six birds were selected twice, and 36 birds were selected once. Following collection, the feathers were stored in Whirl Pac bags (Nasco, Fort Atkinson, WI, USA) and stored at $-20\,^{\circ}$C until assay. Visual inspection ensured that the least damaged feathers were selected for the assay. The feather CORT concentrations were determined following an extraction procedure that was described in [20]. First, the feathers were weighed (mg) to standardize the feather CORT concentrations by feather weight. The feathers were finely minced (including vane and rachis) using surgical scissors (<5 mm sections) into 20 mL scintillation vials. Following mincing, 1 mL of methanol was added and the vials were placed in a sonicating water bath at room temperature for 30 min. The samples were then placed in a shaking hot water bath (Jouan Inc., Precision Sci. Div. Chicago, IL, USA) at 56 $^{\circ}$C overnight. The samples were filtered to remove feather material and the filtrate was transferred to scintillation vials. The methanol was allowed to evaporate completely under a fume hood and CORT was reconstituted in 1 mL ELISA buffer. Reconstituted CORT concentrations were assayed using a commercial ELISA kit (Abcam, Cambridge, MA, USA) following the manufacturers protocol. The intra-assay CV% were below 13% for all the feather CORT samples (range: 0.15–12.50%).

2.5. Statistical Analysis

All statistical analyses were performed in JMP Pro 15 (SAS institute, Cary, NC, USA). Pen or cage was considered the experimental unit for all the response variables. Bird was considered the observational unit for all response variables besides fecal IgA where pen or cage was the observational unit due to pooling of the samples. The distribution of residuals of all the dependent variables were visually inspected using a normal quantile plot to determine normalcy. Normally distributed dependent variables included TI duration (s), ABT latency to begin feeding (s), latency to resume feeding (s), plasma and fecal IgA concentrations (ng/mL), and feather CORT concentrations (ng/mg). Normally distributed data were analyzed using general linear mixed models. Fixed effects were housing system (pen or cage), age (weeks 22, 25, 29, 43) and their interaction. For all the response variables besides fecal IgA, mixed models were used with pen or cage number and bird ID as random effects so that the model identifies the unit (cage or pen) to which the treatment was randomly assigned and independently applied. Non-significant interactions ($p > 0.1$) were removed from the model. Post-hoc analysis was done using Tukey HSD testing. Dependent variables without normally distributed residuals were tested using a nonparametric Wilcoxon Rank sum test. These variables included ABT birds (%) which began and resumed feeding, total vigilance scores (1–4), and birds (%) displaying vigilance behaviors. Associations were deemed significant at $p \leq 0.05$ and trends at $p \leq 0.1$. The data are presented as LS means $\pm$ SEM unless otherwise noted.

3. Results

3.1. Behavioral Measures

TI duration (Figure 1) tended to be shorter for the pen hens (82.88 ± 35.86 s) compared to the caged hens (135.70 ± 21.78 s; $F_{1,46}$ = 3.16; p = 0.080). During ABT, 92.1% of birds (35/38) showed ≥1 vigilance behaviors. After the first alarm call, more cage hens tended to begin feeding compared to the pen birds (χ^2 = 3.55; p = 0.058; Table 2). Latencies to begin feeding were shorter in the cage hens compared to the pen hens ($F_{1,70}$ = −2.33; p = 0.022; Table 2). Following the second alarm call, more cage birds resumed feeding compared to the pen birds (χ^2 = 5.28; p = 0.020; Table 2). The latency to resume feeding did not differ between the cage and pen hens ($F_{1,50}$ = 1.20; p = 0.279; Table 2). The total vigilance behavior scores (χ^2 = 0.01; p = 0.967) and the frequency of observed vigilance behaviors (all p > 0.200) did not differ between the treatments (Table 2).

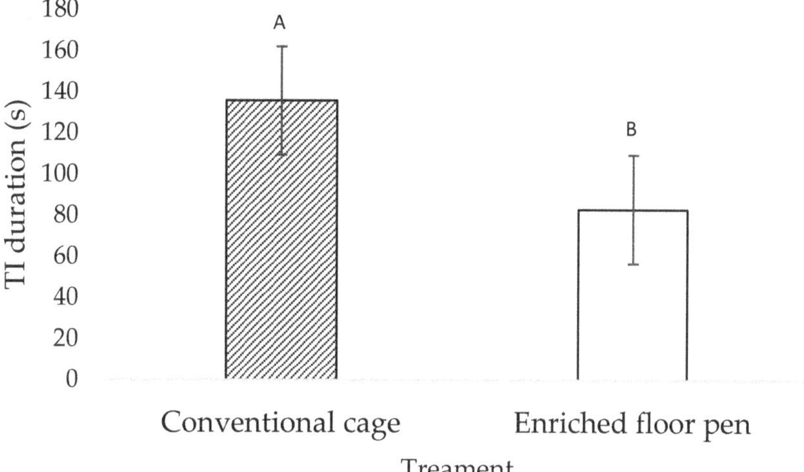

Figure 1. Least square mean estimates (±SEM) of tonic immobility (TI) duration at week 23 of age (n = 24 birds/treatment) for hens that were housed in conventional cages and hens that were housed in enriched floor pens. Bars lacking a common superscript tend to differ at p < 0.1.

Table 2. Responses in the attention bias test at 30 weeks of age for laying hens that were housed in conventional cages (n = 36) or enriched floor pens (n = 36).

Measure	Conventional Cage	Enriched Floor Pen
Latency to begin feeding (s)	99.17 ± 19.23 [a]	145.97 ± 24.52 [b]
Latency to resume feeding (s)	54.21 ± 13.63	54.13 ± 13.87
Birds begin feeding (%)	91.66 [A]	77.77 [B]
Birds resume feeding (%)	87.87 [a]	82.14 [b]
Vigilance behavior score (0–4 score) [1]	2.35 ± 0.26	2.33 ± 0.27
Freeze (% birds)	55	35
Erect (% birds)	55	44
Neck stretch (% birds)	65	55
Look (% birds)	80	72

[1] Birds were scored either 0 (not observed) or 1 (observed) for each of four vigilance behavior characteristics (erect posture, neck stretching, freezing, and looking around), resulting in a vigilance score between 0 (no vigilance behavior observed) and 4 (all vigilance behaviors observed). The data are displayed as raw means ± SEM. [a,b] Row values lacking a common superscript differ at p < 0.05. [A,B] Row values lacking a common superscript tend to differ at p < 0.10.

3.2. Molecular Measures

There was a treatment by age interaction effect on the fecal IgA concentrations ($F_{1,27} = 3.51$; $p = 0.035$; Figure 2). The fecal IgA concentrations were lower in the cage hens in week 22 ($F_{1,27} = -3.38$; $p = 0.049$) compared to the pen hens in that week (Figure 2). The fecal IgA concentrations did not differ between the treatments in weeks 25 ($F_{1,27} = -1.94$; $p = 0.540$), 29 ($F_{1,27} = 0.23$; $p = 0.999$), and 43 ($F_{1,27} = -0.37$; $p = 0.999$). The fecal IgA concentrations in the pen layers were higher in weeks 22 compared to week 29 ($F_{1,27} = 4.23$; $p = 0.008$) and week 43 ($F_{1,27} = 6.06$; $p = 0.002$). The fecal IgA concentrations were higher in the pen hens in week 25 when compared to week 43 ($F_{1,27} = 4.85$; $p = 0.002$). No difference in the fecal IgA concentrations were found in the cage hens over time ($p > 0.1$).

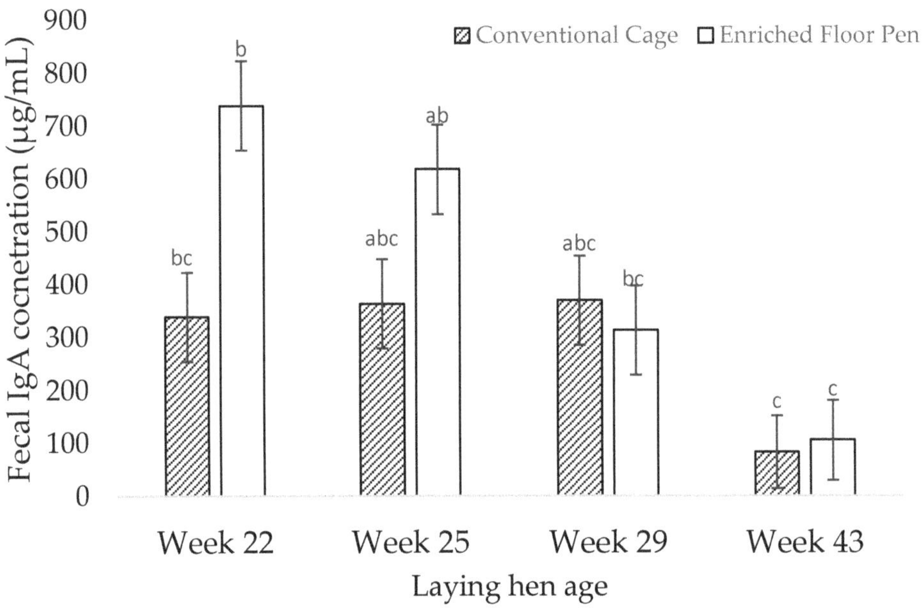

Figure 2. Least square mean estimates (±SEM) of fecal (secretory) immunoglobulin-A (IgA) concentrations in hens that were housed in conventional cages or enriched floor pens during weeks 22, 25, 29, and 43 of age. Bars lacking a common superscript differ at $p < 0.05$.

Plasma IgA concentrations were not impacted by treatment ($F_{1,88} = 0.66$; $p = 0.419$) or treatment by week interaction ($F_{1,48} = -1.63$; $p = 0.110$). The plasma IgA concentrations were lower in week 22 (87.25 ± 15.65 µg/mL) compared to week 29 (260.90 ± 15.65 µg/mL; $F_{1,88} = -7.69$; $p < 0.001$) and 43 (201.80 ± 14.45 µg/mL; $F_{1,88} = -5.36$; $p < 0.001$), and lower in week 25 (104.55 ± 16.45 µg/mL) compared to week 29 ($F_{1,88} = -7.00$; $p < 0.001$) and week 43 ($F_{1,88} = -4.44$; $p < 0.001$). The plasma IgA concentrations in week 29 were higher than in week 43 ($F_{1,88} = 2.82$; $p = 0.030$).

The feather CORT concentrations by feather weight (ng/mg) were higher in the cage hens compared to the pen hens ($F_{1,43} = 2.18$; $p = 0.004$; Figure 3). The feather CORT concentrations were higher in week 22 and 25 compared to week 29 ($F_{1,43} = 6.07$; $p = 0.015$; Figure 4), but lower in week 29 compared to week 43 ($F_{1,20} = -3.30$; $p = 0.012$; Figure 4). There was no treatment by week interaction effect on the feather CORT ($F_{1,37} = 1.43$; $p = 0.251$).

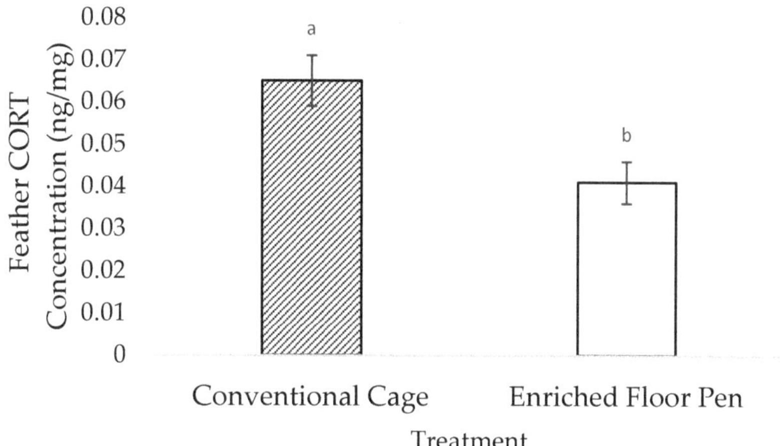

Figure 3. Least square means estimates (±SEM) of the total feather corticosterone (CORT) concentrations for laying hens that were housed in conventional cages or in enriched floor pens. n = 12 samples/timepoint (6 samples/treatment) from weeks 22, 25, 29, and 43 of age. Bars lacking a common superscript differ at $p < 0.05$.

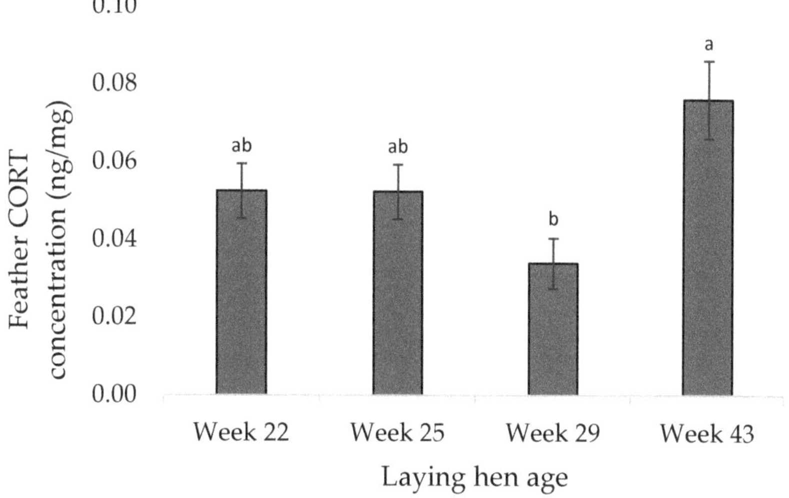

Figure 4. Least square means estimates (±SEM) of the total feather corticosterone (CORT) concentrations by age in weeks. Bars lacking a common superscript differ at $p < 0.05$.

4. Discussion

This study investigated the effects of housing system (enriched floor pens vs. conventional caging) on behavioral and molecular markers for laying hen welfare including TI, attention bias, fecal and plasma IgA concentrations, and feather CORT concentrations. Birds in conventional cages tended to be more fearful based on TI responses, were less anxious based on ABT responses, and experienced more chronic stress based on low IgA levels and high feather CORT levels. Overall, these results suggest that the cage layers experienced a decreased welfare status when compared to pen layers, except in terms of anxiety. This is the first study to show a relationship between housing conditions and IgA and feather CORT responses in laying hens.

4.1. Behavioral Measures

Cage housing tended to result in more fearful but less anxious birds compared to pen housing, which is in line with earlier findings [42,43]. Fearfulness is considered a negative emotion and frequent bouts of fear can be indicative of negative affective states in animals [3,44,45]. Similar to our results, cage housing increased the TI duration in 18-week-old (cages: 519 s vs. pens: 471 s) and 1-year-old (cages: 443 s; pens: 189 s) white leghorn hens when compared to floor pen housing [44]. However, these reported durations were longer than those in the current study in Bovan brown laying hens (cages: 136 s vs. pens: 88 s), potentially due to strain differences [46,47]. Although the difference in the current study was only a tendency, we suspect this is due to limited statistical power because of the relatively small sample size, as a large numerical difference in TI durations was observed. Previous studies have found no difference in fear levels related to housing system in commercial laying hens [11,48]. In ideal situations, fear-associated behavior is adaptive and aids in helping animals prevent injury during threatening encounters. However, fear-associated behavior in production housing systems is generally maladaptive, with animals having no ability to escape or appropriately respond to fearful stimuli [49]. Over time, this excessive fear response could lead to frustration, learned helplessness, and eventually negative affective states which can impact productivity. Increased fear can lead to decreased egg production [50,51], increased feather pecking [52–55], and increased injury rate [56]. In systems that have high fear levels, these impacts could erase, or reverse production gains that are generally attributed to intensive housing systems. The increased level of fear in cages compared to pens could have contributed to a negative affective state in hens in cages.

Cage housing conditions reduced anxiety in laying hens compared to pen housing conditions. Attention bias testing was validated for use in laying hens [15,16], and has been applied to test the effect of environmental conditions on anxiety in starlings [57] and broilers [3]. During the test, 92.1% of birds showed some form of vigilance behavior, indicating that the ABT was sufficiently threatening to achieve anxiety. Contrary to expectations, cage housing decreased the latency to begin feeding and resulted in a higher percentage of hens feeding after the first and second alarm call compared to pen housing. These results indicate that hens that were housed in cages did not bias their attention towards the threat, thus were less anxious than hens that were housed in pens, which is somewhat contradictory to earlier studies [3,15]. Broilers from high-complexity pens were quicker to begin feeding following an alarm call compared to broilers from low-complexity pens (high complexity: 160 s vs. low complexity 214 s; [3]). Outside-ranging laying hens showed shorter latencies to feed compared to hens that never went outside (outdoor: 86 s vs. indoor: 170 s; [16]). Although the treatment conditions were different in both studies compared to the current study, they both involve a level of environmental complexity. In the current study, as in Anderson et al. 2021 [3], the test was performed in groups of three birds, rather than testing birds individually (as in [15]). The cage hens were tested with their cage conspecifics, and pen hens with 3 out of 25 pen conspecifics. It is possible that pen hens experienced this temporary regrouping as more negative than the cage hens, depending on the social hierarchy within the pen. Further testing is needed to confirm the effect individual versus group testing of anxiety, and of these housing conditions on anxiety in laying hens.

Our results support earlier findings that fear and anxiety can be opposing [5,58] although some studies show they can be positively associated [59,60]. Fear is a generally fast adaptive state of vigilance to a present, negatively valanced stimulus which activates a defensive response such as fight, flight, or freeze [4]. Anxiety elicits vigilance and apprehension to non-existing or ambiguous threats [4]. Although the symptoms of fear and anxiety are similar [4], evidence suggests that they are distinct emotional experiences. Research in rodents shows that three stages of defense exist, which include the pre-encounter defense (apprehension to a place where a predator has been seen), a circa-strike defense (physical contact with a predator), and a post-encounter defense (predator is identified at a distance) [4,61–63]. Anxiety has been associated with the pre- and post-encounter stages, but not with the circa-strike phase [4]. As the TI test is performed, the researcher acts as a predator coming in physical contact with the hen, inducing a catatonic state. If similar stages of defense exist in chickens, the TI test most likely simulates the circa-strike defense phase. This suggests that the emotional experiences that are tested in the ABT and the TI test are associated with distinct phases of defense. Cage birds may be more fearful in the circa-strike phase and less anxious in the pre- and post-encounter stages compared to the pen birds. Thus, this could explain why the outcomes are opposing. Although these defense phases have not been confirmed in chickens, it is possible that a similar distinction exists. Therefore, it is possible that housing systems impact hen fear and anxiety differently. Overall, the hens' behavioral responses indicate that anxiety and fear are affected by housing system, with pens tending to reduce fear but not anxiety compared to cages.

4.2. Molecular Measures

The housing treatments impacted fecal IgA concentrations during week 22, resulting in decreased concentrations in the cage layers compared to the pen layers. Our results indicate that birds that are housed in complex, low density environments (enriched floor pens) are under less chronic stress (reflected by immune status) than birds that are housed in barren, highly confined environments (conventional cages) at 22 weeks of age. Additionally, our results show that fecal IgA could be a potential physiological indicator of animal welfare status. This is in line with past findings in swine, rodents, laying hens, and broilers [26,31,32,40,64]. Prolonged heat stress decreased plasma IgA concentrations in broilers and layers [31,32], which showed plasma IgA concentrations in similar ranges as in the current study (previous study: 0.162–0.290 mg/mL vs. current study: 0.087–0.260 mg/mL).

Secretory and plasma IgA concentrations are promising valence-dependent indicators of stress in certain mammals [28,29,40]. We did not find an impact of housing conditions on plasma IgA concentrations, showing at least in similar contexts, this measure could be less relevant for laying hen welfare assessments. However, to our knowledge this study is the first to investigate secretory and plasma IgA in an animal welfare context in laying hens. Access to litter, containing microorganisms and fecal material, could have impacted the higher secretory IgA concentrations in pens compared to the cage hens at 22 weeks of age. Alternative housing systems (aviary, pens) result in higher concentrations of bacteria and fungi in the air than conventional cage housing [65,66]. Cages are considered more hygienic than litter systems due to the separation of waste and animals [65,67]. However, all the birds were housed in the same space, so airborne bacteria and fungi could reach the cage hens too. Nevertheless, ingestion of litter and fecal material could have exposed the hens' intestinal tracts to pathogens and initiated an immune response. In turn, the pen hens could have had increased fecal IgA concentrations to mediate an immune response against those intestinal pathogens. Further research should determine the impact of litter access on intestinal immune challenges and include a chronic stressor that is unrelated to housing conditions. Then, the impact of litter access and chronic stress can be separated and the use of secretory IgA as a biomarker for animal welfare confirmed.

Even though housing conditions impacted the fecal IgA concentrations at 22 weeks of age, the plasma IgA concentrations did not differ. Secretory IgA may provide more insight in stress-induced immune system disruptions because IgA are most prominently active and present on mucosal surfaces including the intestinal tract [28]. IgA originating from the intestinal tract will be deposited in fecal matter as it passes through. Negative stressors, including heat stress [68–71] and stocking density [72–75] negatively impact the intestinal tract via damaged intestinal morphology and decreased microbiota diversity in commercial broilers and laying hens. These impacts on intestinal health could also impact the immune system, and, therefore, intestinal IgA via a reduction in IgA-secreting plasma cells (mature B-cells) or via direct interactions with intestinal glucocorticoids [26] indicating multiple possible mechanisms for the observed impacts of welfare status on IgA concentrations. Our results are the first to suggest an association between housing conditions and fecal IgA responses in laying hens, albeit only at 22 weeks of age. These results are a first step towards using secretory IgA as an indicator of welfare status in laying hens. Further research is needed to elucidate the impact of litter access and bird age on IgA concentrations.

Feather CORT assays have been used in several species including wild birds, broilers, and laying hens [20–23]. The feather CORT concentrations were higher in cage hens compared to the pen hens overall, however, feather CORT concentrations did not differ between the housing treatments at any individual timepoints. The latter is likely due to the low sample volume at individual timepoints. Our CORT concentrations indicate that layers from cages experienced more chronic stress than layers from pens. This difference possibly occurred due to repeated bouts of frustration or negative stress contributing to negative affective states. During these bouts of frustration or negative stress, the HPA axis releases glucocorticoids into the bloodstream, which are partially deposited into the feathers as they grow. The sum of these deposits is used to quantify stress over the whole feather growth period. Brown Nick laying hens that were housed at high stocking density, which is largely considered a chronic stressor in commercial poultry [75–79] had increased feather CORT at 10 weeks of age, when compared to birds that were housed at a lower stocking density [80]. In the current study, the stocking density differed considerably between the treatments, which could have contributed to the low feather CORT concentrations in pens compared to cage hens. While feather cover was not quantified as part of this experiment, birds that were housed in cages showed poorer feather coverage than birds that were housed in pen housing. It is possible that this poor feather coverage in cage birds would have made them more susceptible to cold stress during the cold period in the winter months (weeks 33–37) thereby increasing their feather CORT concentrations. Additional research should be done to determine if cold stress will impact feather CORT concentrations. Feather CORT concentrations also significantly decreased at week 29 of age compared to week 25 regardless of treatment. This was unexpected as feather CORT should increase continuously throughout life (as CORT keeps being deposited) unless the feather is regrown or replaced. It is possible that this observed decrease is methodologically related, however, all analyses were performed by the same researcher and the same protocol was followed. Feather CORT concentrations do rebound and reach a peak in week 43, which coincides with the cold spell that was observed from weeks 33 to 37 and likely reflects chronic stress that was caused during this period to birds in both housing systems. Overall, these results indicate that feather CORT shows promise as a viable physiological biomarker for chronic stress in laying hens. However, additional research is needed to confirm these results in commercial production systems and in other genetic strains.

5. Conclusions

The birds' responses in the current study suggest that conventional cages may induce negative affective states and emotions, reflected in somewhat increased fear and chronic stress responses, compared to enriched floor pens. However, anxiety was reduced in hens that were housed in conventional cages compared to the enriched pens. Simlar to previous research, these results highlight the importance of including a range of measures when assessing the impact of housing conditions on animal welfare. The novel measures of animal welfare that were tested in this study (Immunoglobulin A and feather corticosterone) indicate a downregulated immune response at 22 weeks of age (decreased fecal Immunoglobulin A) and increased chronic stress response (increased feather corticosterone) in conventionally caged hens compared to hens that were housed in enriched floor pens. However, fecal Immunoglobulin A concentrations require more research to confirm that the difference was not because of a mounted immune response due to exposure to litter. Overall, enriched pens resulted in improved laying hen immune responses at 22 weeks of age, decreased chronic stress, somewhat decreased fear, but increased anxiety indicating potential improvements in affective states (fear and chronic stress) in some aspects, but worsening in others (anxiety), and an improvement in the basic health and functioning (immune responses) when compared to conventionally-housed birds. Fecal (secretory) Immunoglobulin A and feather corticosterone quantification show some contrasting responses in line with expectations and behavioral outcomes, yet need further confirmation before application as a routine measure for emotion and affective state in laying hens.

Author Contributions: Conceptualization, L.J. and A.M.C.; methodology, L.J. and A.M.C.; formal analysis, L.J. and A.M.C.; investigation, L.J., A.M.C. and A.M.J.; resources, L.J. and M.E.P.; data curation, L.J. and A.M.C.; writing—original draft preparation, A.M.C.; writing—review and editing, A.M.C., L.J. and M.E.P.; visualization, A.M.C.; supervision, L.J., M.E.P. All authors have read and agreed to the published version of the manuscript.

Funding: This research received no external funding.

Institutional Review Board Statement: The animal study protocol was approved by the Institutional Care and Use Committee of Virginia Polytechnic and State University (protocol code 18-205: Approved 9/24/2021).

Informed Consent Statement: Not applicable.

Data Availability Statement: Data are available from the corresponding author upon reasonable request.

Acknowledgments: We would like to thank Shelly Underwood, Noah Goldfarb, Jessica Duvall, Katie Kirkpatrick, and Kyla Poe for their valuable help during the experiment and the VT farm staff for their assistance.

Conflicts of Interest: The authors declare no conflict of interest.

References

1. Fraser, D. Understanding animal welfare. *Acta Vet. Scand.* **2008**, *50*, S1–S7. [CrossRef] [PubMed]
2. Hartcher, K.M.; Jones, B. The welfare of layer hens in cage and cage-free housing systems. *World's Poult. Sci. J.* **2017**, *73*, 767–782. [CrossRef]
3. Anderson, M.G.; Campbell, A.M.; Crump, A.; Arnott, G.; Newberry, R.C.; Jacobs, L. Effect of Environmental Complexity and Stocking Density on Fear and Anxiety in Broiler Chickens. *Animals* **2021**, *11*, 2383. [CrossRef]
4. Davis, M.; Walker, D.L.; Miles, L.; Grillon, C. Phasic vs Sustained Fear in Rats and Humans: Role of the Extended Amygdala in Fear vs. Anxiety. *Neuropsychopharmacology* **2010**, *35*, 105–135. [CrossRef] [PubMed]
5. De Haas, E.N.; Kops, M.S.; Bolhuis, J.E.; Groothuis, T.G.; Ellen, E.D.; Rodenburg, T.B. The relation between fearfulness in young and stress-response in adult laying hens, on individual and group level. *Physiol. Behav.* **2012**, *107*, 433–439. [CrossRef] [PubMed]
6. Forkman, B.; Boissy, A.; Meunier-Salaün, M.-C.; Canali, E.; Jones, R. A critical review of fear tests used on cattle, pigs, sheep, poultry and horses. *Physiol. Behav.* **2007**, *92*, 340–374. [CrossRef]
7. Fogelholm, J.; Inkabi, S.; Höglund, A.; Abbey-Lee, R.; Johnsson, M.; Jensen, P.; Henriksen, R.; Wright, D. Genetical Genomics of Tonic Immobility in the Chicken. *Genes* **2019**, *10*, 341. [CrossRef] [PubMed]
8. Li, X.; Chen, D.; Li, J.; Bao, J. Effects of Furnished Cage Type on Behavior and Welfare of Laying Hens. *Asian-Australas. J. Anim. Sci.* **2015**, *29*, 887–894. [CrossRef]
9. Shi, H.; Tong, Q.; Zheng, W.; Tu, J.; Li, B. Effects of nest boxes in natural mating colony cages on fear, stress, and feather damage for layer breeders 1,2,3. *J. Anim. Sci.* **2019**, *97*, 4464–4474. [CrossRef]
10. Tauson, R.; Wahlström, A.; Abrahamsson, P. Effect of Two Floor Housing Systems and Cages on Health, Production, and Fear Response in Layers. *J. Appl. Poult. Res.* **1999**, *8*, 152–159. [CrossRef]
11. Yilmaz Dikmen, B.; Ipek, A.; Şahan, Ü.; Petek, M.; Sözcü, A. Egg production and welfare of laying hens kept in different housing systems (conventional, enriched cage, and free range). *Poult. Sci.* **2016**, *95*, 1564–1572. [CrossRef] [PubMed]
12. Mendl, M.; Burman, O.H.P.; Parker, R.M.A.; Paul, E.S. Cognitive bias as an indicator of animal emotion and welfare: Emerging evidence and underlying mechanisms. *Appl. Anim. Behav. Sci.* **2009**, *118*, 161–181. [CrossRef]
13. Crump, A.; Arnott, G.; Bethell, E.J. Affect-Driven Attention Biases as Animal Welfare Indicators: Review and Methods. *Animals* **2018**, *8*, 136. [CrossRef] [PubMed]
14. Košťál, Ľ.; Skalná, Z.; Pichová, K. Use of cognitive bias as a welfare tool in poultry. *J. Anim. Sci.* **2020**, *98*, S63. [CrossRef]
15. Campbell, D.L.; Taylor, P.; Hernandez, C.E.; Stewart, M.; Belson, S.; Lee, C. An attention bias test to assess anxiety states in laying hens. *PeerJ* **2019**, *10*, e3707. [CrossRef] [PubMed]
16. Campbell, D.L.; Dickson, E.J.; Lee, C. Application of open field, tonic immobility, and attention bias tests to hens with different ranging patterns. *PeerJ* **2019**, *7*, e8122. [CrossRef]

17. Monk, J.E.; Belson, S.; Lee, C. Pharmacologically-induced stress has minimal impact on judgement and attention biases in sheep. *Sci. Rep.* **2019**, *9*, 11446. [CrossRef] [PubMed]

18. Monk, J.E.; Lee, C.; Belson, S.; Colditz, I.G.; Campbell, D.L. The influence of pharmacologically-induced affective states on attention bias in sheep. *PeerJ* **2019**, *7*, e7033. [CrossRef]

19. Luo, L.; Reimert, I.; De Haas, E.N.; Kemp, B.; Bolhuis, J.E. Effects of early and later life environmental enrichment and personality on attention bias in pigs (Sus scrofa domesticus). *Anim. Cogn.* **2019**, *22*, 959–972. [CrossRef]

20. Bortolotti, G.R.; Marchant, T.A.; Blas, J.; German, T. Corticosterone in feathers is a long-term, integrated measure of avian stress physiology. *Funct. Ecol.* **2008**, *22*, 494–500. [CrossRef]

21. Häffelin, K.; Lindenwald, R.; Kaufmann, F.; Döhring, S.; Spindler, B.; Preisinger, R.; Rautenschlein, S.; Kemper, N.; Andersson, R. Corticosterone in feathers of laying hens: An assay validation for evidence-based assessment of animal welfare. *Poult. Sci.* **2020**, *99*, 4685–4694. [CrossRef]

22. Harris, C.M.; Madliger, C.L.; Love, O.P. An evaluation of feather corticosterone as a biomarker of fitness and an ecologically relevant stressor during breeding in the wild. *Oecologia* **2017**, *183*, 987–996. [CrossRef] [PubMed]

23. Voit, M.; Merle, R.; Baumgartner, K.; Von Fersen, L.; Reese, L.; Ladwig-Wiegard, M.; Will, H.; Tallo-Parra, O.; Carbajal, A.; Lopez-Bejar, M.; et al. Validation of an Alternative Feather Sampling Method to Measure Corticosterone. *Animals* **2020**, *10*, 2054. [CrossRef] [PubMed]

24. Corthésy, B. Multi-Faceted Functions of Secretory IgA at Mucosal Surfaces. *Front. Immunol.* **2013**, *4*, 185. [CrossRef] [PubMed]

25. Mantis, N.J.; Rol, N.; Corthésy, B. Secretory IgA's complex roles in immunity and mucosal homeostasis in the gut. *Mucosal Immunol.* **2011**, *4*, 603–611. [CrossRef]

26. Staley, M.; Conners, M.G.; Hall, K.; Miller, L.J. Linking stress and immunity: Immunoglobulin A as a non-invasive physiological biomarker in animal welfare studies. *Horm. Behav.* **2018**, *102*, 55–68. [CrossRef]

27. Ablimit, A.; Kühnel, H.; Strasser, A.; Upur, H. Abnormal Savda syndrome: Long-term consequences of emotional and physical stress on endocrine and immune activities in an animal model. *Chin. J. Integr. Med.* **2012**, *19*, 603–609. [CrossRef]

28. Guhad, F.; Hau, J. Salivary IgA as a marker of social stress in rats. *Neurosci. Lett.* **1996**, *216*, 137–140. [CrossRef]

29. Jarillo-Luna, A.; Rivera-Aguilar, V.; Garfias, H.R.; Lara-Padilla, E.; Kormanovsky, A.; Campos-Rodríguez, R. Effect of repeated restraint stress on the levels of intestinal IgA in mice. *Psychoneuroendocrinology* **2007**, *32*, 681–692. [CrossRef]

30. Rammal, H.; Bouayed, J.; Falla, J.; Boujedaini, N.; Soulimani, R. The impact of high anxiety level on cellular and humoral immunity in mice. *NeuroImmunoModulation* **2009**, *17*, 1–8. [CrossRef]

31. Quinteiro-Filho, W.; Calefi, A.; Cruz, D.; Aloia, T.; Zager, A.; Astolfi-Ferreira, C.; Ferreira, J.P.; Sharif, S.; Palermo-Neto, J. Heat stress decreases expression of the cytokines, avian β-defensins 4 and 6 and Toll-like receptor 2 in broiler chickens infected with Salmonella Enteritidis. *Vet. Immunol. Immunopathol.* **2017**, *186*, 19–28. [CrossRef] [PubMed]

32. Li, D.; Tong, Q.; Shi, Z.; Li, H.; Wang, Y.; Li, B.; Yan, G.; Chen, H.; Zheng, W. Effects of chronic heat stress and ammonia concentration on blood parameters of laying hens. *Poult. Sci.* **2020**, *99*, 3784–3792. [CrossRef] [PubMed]

33. Gourkow, N.; Phillips, C.J. Effect of cognitive enrichment on behavior, mucosal immunity and upper respiratory disease of shelter cats rated as frustrated on arrival. *Prev. Vet. Med.* **2016**, *131*, 103–110. [CrossRef] [PubMed]

34. Kurimoto, Y.; Saruta, J.; To, M.; Yamamoto, Y.; Kimura, K.; Tsukinoki, K. Voluntary exercise increases IgA concentration and polymeric Ig receptor expression in the rat submandibular gland. *Biosci. Biotechnol. Biochem.* **2016**, *80*, 2490–2496. [CrossRef]

35. Kimura, F.; Aizawa, K.; Tanabe, K.; Shimizu, K.; Kon, M.; Lee, H.; Akimoto, T.; Akama, T.; Kono, I. A rat model of saliva secretory immunoglobulin: A suppression caused by intense exercise. *Scand. J. Med. Sci. Sports* **2007**, *18*, 367–372. [CrossRef]

36. Souza, C.; Miotto, B.; Bonin, C.; Camargo, M. Lower serum IgA levels in horses kept under intensive sanitary management and physical training. *Animal* **2010**, *4*, 2080–2083. [CrossRef]

37. Robson, P.J.; Alston, T.D.; Myburgh, K.H. Prolonged suppression of the innate immune system in the horse following an 80 km endurance race. *Equine Vet. J.* **2010**, *35*, 133–137. [CrossRef]

38. Peters, I.R.; Calvert, E.L.; Hall, E.; Day, M.J. Measurement of Immunoglobulin Concentrations in the Feces of Healthy Dogs. *Clin. Vaccine Immunol.* **2004**, *11*, 841. [CrossRef]

39. Ricci, F.G.; Terkelli, L.R.; Venancio, E.J.; Justino, L.; Dos Santos, B.Q.; Baptista, A.A.S.; Oba, A.; Souza, B.D.D.O.; Bracarense, A.P.F.R.L.; Hirooka, E.Y.; et al. Tryptophan Attenuates the Effects of OTA on Intestinal Morphology and Local IgA/IgY Production in Broiler Chicks. *Toxins* **2020**, *13*, 5. [CrossRef]

40. Royo, F.; Lyberg, K.; Abelson, K.; Carlsson, H.; Hau, J. Effect of repeated confined single housing of young pigs on faecal excretion of cortisol and IgA. *Scand. J. Lab. Anim. Sci.* **2005**, *31*, 33–37. [CrossRef]

41. Romero, L.M.; Reed, J.M. Collecting baseline corticosterone samples in the field: Is under 3 min good enough? *Comp. Biochem. Physiol. Part A Mol. Integr. Physiol.* **2005**, *140*, 73–79. [CrossRef] [PubMed]

42. Anderson, K.; Adams, A. Effects of Floor Versus Cage Rearing and Feeder Space on Growth, Long Bone Development, and Duration of Tonic Immobility in Single Comb White Leghorn Pullets. *Poult. Sci.* **1994**, *73*, 958–964. [CrossRef] [PubMed]

43. Jones, R.B.; Faure, J.M. Tonic immobility ("righting time") in laying hens housed in cages and pens. *Appl. Anim. Ethol.* **1981**, *7*, 369–372. [CrossRef]

44. Monk, J.E.; Doyle, R.E.; Colditz, I.G.; Belson, S.; Cronin, G.M.; Lee, C. Towards a more practical attention bias test to assess affective state in sheep. *PLoS ONE* **2018**, *13*, e0190404. [CrossRef]

45. Steimer, T. The biology of fear- and anxiety-related behaviors. *Dialog Clin. Neurosci.* **2002**, *4*, 231–249. [CrossRef]

46. Albentosa, M.; Kjaer, J.; Nicol, C. Strain and age differences in behaviour, fear response and pecking tendency in laying hens. *Br. Poult. Sci.* **2003**, *44*, 333–344. [CrossRef]

47. Fraisse, F.; Cockrem, J.F. Corticosterone and fear behaviour in white and brown caged laying hens. *Br. Poult. Sci.* **2006**, *47*, 110–119. [CrossRef]

48. Campo, J.L.; Prieto, M.T.; Dávila, S.G. Effects of Housing System and Cold Stress on Heterophil-to-Lymphocyte Ratio, Fluctuating Asymmetry, and Tonic Immobility Duration of Chickens. *Poult. Sci.* **2008**, *87*, 621–626. [CrossRef]

49. Jones, R.B. Fear and adaptability in poultry: Insights, implications and imperatives. *World's Poult. Sci. J.* **1996**, *52*, 131–174. [CrossRef]

50. Sefton, A. The Interactions of Cage Size, Cage Level, Social Density, Fearfulness, and Production of Single Comb White Leghorns. *Poult. Sci.* **1976**, *55*, 1922–1926. [CrossRef]

51. Sefton, A.E.; Crober, D.C. Social and physical environmental influences on caged single comb white leghorn layers. *Can. J. Anim. Sci.* **1976**, *56*, 733–738. [CrossRef]

52. Craig, J.; Craig, T.; Dayton, A. Fearful behavior by caged hens of two genetic stocks. *Appl. Anim. Ethol.* **1983**, *10*, 263–273. [CrossRef]

53. Hughes, B.O.; Duncan, I.J.H. The influence of strain and environmental factors upon feather pecking and cannibalism in fowls. *Br. Poult. Sci.* **1972**, *13*, 525–547. [CrossRef] [PubMed]

54. Okpokho, N.A.; Craig, J.V.; Milliken, G.A. Density and group size effects on caged hens of two genetic stocks differing in escape and avoidance behavior. *Poult. Sci.* **1987**, *66*, 1905–1910. [CrossRef]

55. Ouart, M.D.; Adams, A.W. Effects of Cage Design and Bird Density on Layers. *Poult. Sci.* **1982**, *61*, 1606–1613. [CrossRef]

56. Hansen, R.S. Nervousness and Hysteria of Mature Female Chickens. *Poult. Sci.* **1976**, *55*, 531–543. [CrossRef]

57. Brilot, B.O.; Bateson, M. Water bathing alters threat perception in starlings. *Biol. Lett.* **2012**, *8*, 379–381. [CrossRef]

58. Nordquist, R.E.; Heerkens, J.L.; Rodenburg, T.B.; Boks, S.; Ellen, E.D.; van der Staay, F.J. Laying hens selected for low mortality: Behaviour in tests of fearfulness, anxiety and cognition. *Appl. Anim. Behav. Sci.* **2011**, *131*, 110–122. [CrossRef]

59. Engel, O.; Müller, H.W.; Klee, R.; Francke, B.; Mills, D.S. Effectiveness of imepitoin for the control of anxiety and fear associated with noise phobia in dogs. *J. Vet. Intern. Med.* **2019**, *33*, 2675–2684. [CrossRef]

60. Peixoto, M.R.L.V.; Karrow, N.A.; Newman, A.; Widowski, T.M. Effects of Maternal Stress on Measures of Anxiety and Fearfulness in Different Strains of Laying Hens. *Front. Vet. Sci.* **2020**, *7*, 128. [CrossRef]

61. Blanchard, R.J.; Yudko, E.B.; Rodgers, R.; Blanchard, D. Defense system psychopharmacology: An ethological approach to the pharmacology of fear and anxiety. *Behav. Brain Res.* **1993**, *58*, 155–165. [CrossRef]

62. Bolles, R.C.; Fanselow, M.S. A perceptual-defensive-recuperative model of fear and pain. *Behav. Brain Sci.* **1980**, *3*, 291–301. [CrossRef]

63. Fanselow, M.S. Associative vs topographical accounts of the immediate shock-freezing deficit in rats: Implications for the response selection rules governing species-specific defensive reactions. *Learn. Motiv.* **1986**, *17*, 16–39. [CrossRef]

64. Eriksson, E.; Royo, F.; Lyberg, K.; Carlsson, H.-E.; Hau, J. Effect of metabolic cage housing on immunoglobulin A and corticosterone excretion in faeces and urine of young male rats. *Exp. Physiol.* **2004**, *89*, 427–433. [CrossRef] [PubMed]

65. Rodenburg, T.B.; Tuyttens, F.A.M.; Sonck, B.; De Reu, K.; Herman, L.; Zoons, J. Welfare, Health, and Hygiene of Laying Hens Housed in Furnished Cages and in Alternative Housing Systems. *J. Appl. Anim. Welf. Sci.* **2005**, *8*, 211–226. [CrossRef] [PubMed]

66. Jones, D.R.; Cox, N.A.; Guard, J.; Fedorka-Cray, P.; Buhr, R.J.; Gast, R.K.; Abdo, Z.; Rigsby, L.L.; Plumblee, J.R.; Karcher, D.M.; et al. Microbiological impact of three commercial laying hen housing systems. *Poult. Sci.* **2015**, *94*, 544–551. [CrossRef]

67. Hofmann, T.; Schmucker, S.S.; Bessei, W.; Grashorn, M.; Stefanski, V. Impact of Housing Environment on the Immune System in Chickens: A Review. *Animals* **2020**, *10*, 1138. [CrossRef]

68. Calefi, A.S.; Honda, B.T.B.; Costola-De-Souza, C.; de Siqueira, A.; Namazu, L.B.; Quinteiro-Filho, W.M.; Fonseca, J.G.D.S.; Aloia, T.P.A.; Piantino-Ferreira, A.J.; Palermo-Neto, J. Effects of long-term heat stress in an experimental model of avian necrotic enteritis. *Poult. Sci.* **2014**, *93*, 1344–1353. [CrossRef]

69. Calefi, A.; Quinteiro-Filho, W.; Ferreira, A.; Palermo-Neto, J. Neuroimmunomodulation and heat stress in poultry. *World's Poult. Sci. J.* **2017**, *73*, 493–504. [CrossRef]

70. Slawinska, A.; Mendes, S.; Dunislawska, A.; Siwek, M.; Zampiga, M.; Sirri, F.; Meluzzi, A.; Tavaniello, S.; Maiorano, G. Avian model to mitigate gut-derived immune response and oxidative stress during heat. *Biosystems* **2019**, *178*, 10–15. [CrossRef]

71. Rostagno, M.H. Effects of heat stress on the gut health of poultry. *J. Anim. Sci.* **2020**, *98*, skaa090. [CrossRef] [PubMed]

72. Sohail, M.U.; Ijaz, A.; Yousaf, M.S.; Ashraf, K.; Zaneb, H.; Aleem, M.; Rehman, H. Alleviation of cyclic heat stress in broilers by dietary supplementation of mannan-oligosaccharide and Lactobacillus-based probiotic: Dynamics of cortisol, thyroid hormones, cholesterol, C-reactive protein, and humoral immunity. *Poult. Sci.* **2010**, *89*, 1934–1938. [CrossRef]

73. Guardia, S.; Konsak, B.; Combes, S.; Levenez, F.; Cauquil, L.; Guillot, J.-F.; Moreau-Vauzelle, C.; Lessire, M.; Juin, H.; Gabriel, I. Effects of stocking density on the growth performance and digestive microbiota of broiler chickens. *Poult. Sci.* **2011**, *90*, 1878–1889. [CrossRef]

74. Wu, Y.; Li, J.; Qin, X.; Sun, S.; Xiao, Z.; Dong, X.; Shahid, M.S.; Yin, D.; Yuan, J. Proteome and microbiota analysis reveals alterations of liver-gut axis under different stocking density of Peking ducks. *PLoS ONE* **2018**, *13*, e0198985. [CrossRef] [PubMed]

75. Wang, J.; Zhang, C.; Zhang, T.; Yan, L.; Qiu, L.; Yin, H.; Ding, X.; Bai, S.; Zeng, Q.; Mao, X.; et al. Dietary 25-hydroxyvitamin D improves intestinal health and microbiota of laying hens under high stocking density. *Poult. Sci.* **2021**, *100*, 101132. [CrossRef]

76. Saki, A.A.; Zamani, P.; Rahmati, M.; Mahmoudi, H. The effect of cage density on laying hen performance, egg quality, and excreta minerals. *J. Appl. Poult. Res.* **2012**, *21*, 467–475. [CrossRef]
77. Kang, H.K.; Park, S.B.; Kim, S.H.; Kim, C.H. Effects of stock density on the laying performance, blood parameter, corticosterone, litter quality, gas emission and bone mineral density of laying hens in floor pens. *Poult. Sci.* **2016**, *95*, 2764–2770. [CrossRef]
78. Li, W.; Wei, F.; Xu, B.; Sun, Q.; Deng, W.; Ma, H.; Bai, J.; Li, S. Effect of stocking density and alpha-lipoic acid on the growth performance, physiological and oxidative stress and immune response of broilers. *Asian-Australas. J. Anim. Sci.* **2019**, *32*, 1914–1922. [CrossRef]
79. Weimer, S.L.; Robison, C.I.; Tempelman, R.J.; Jones, D.R.; Karcher, D.M. Laying hen production and welfare in enriched colony cages at different stocking densities. *Poult. Sci.* **2019**, *98*, 3578–3586. [CrossRef]
80. Von Eugen, K.; Nordquist, R.E.; Zeinstra, E.; van der Staay, F.J. Stocking Density Affects Stress and Anxious Behavior in the Laying Hen Chick During Rearing. *Animals* **2019**, *9*, 53. [CrossRef]

Article

The "Real Welfare" Scheme: Changes in UK Finishing Pig Welfare since the Introduction of Formal Welfare Outcome Assessment

Fanny Pandolfi [1,*], Claire Barber [2] and Sandra Edwards [3]

1   Independent Researcher, 75000 Paris, France
2   Animal Health & Welfare, Agricultural and Horticultural Development Board, Kenilworth CV8 2TL, UK; claire.barber@ahdb.org.uk
3   School of Natural and Environmental Sciences, Newcastle University, Newcastle upon Tyne NE1 7RU, UK; sandra.edwards@newcastle.ac.uk
*   Correspondence: fanny.pandolfi@gmail.com

Simple Summary: Farm animal welfare is an important issue for both farmers and the general public. In order to monitor animal welfare on a large population of finishing pig farms in the UK, the "Real Welfare" project developed a methodology for on-farm assessment based on regular measurement of some key animal-based measures, so-called welfare outcomes. This paper presents estimates of the percentage of these different welfare outcomes over the years since the inception of the scheme. Between 2016 and 2019, the mean percentage of pigs in the mainstream herd showing four mandatory welfare outcomes (pigs requiring hospitalization, lame pigs, pigs with severe tail lesions, and pigs with severe body marks) was very low and a decreasing trend was observed for all outcomes except severe tail lesions. These data give good representation of the overall situation in the UK finishing pig population.

Abstract: Farm animal welfare is an increasingly important issue, leading to the need for an efficient methodology to deliver accurate benchmarking. The "Real Welfare" project developed a methodology based on regular recording of a limited number of animal-based measures, so-called welfare outcomes, which allows faster and easier on-farm assessment of finishing pig welfare. The objective of this paper is to estimate, with sufficient robustness and confidence, the prevalence of different mandatory and optional welfare outcomes in the mainstream herd of the finishing farms in the UK based on the "Real Welfare" scheme data and to assess the changes in prevalence over time, inspection visits and seasons. The mean overall prevalence of the four mandatory welfare outcomes (pigs requiring hospitalization, lame pigs, pigs with severe body marks, and pigs with severe tail lesions) was very low ($\leq 0.2\%$) and a significant decreasing trend was observed for the first three of these mandatory welfare outcomes since the inception of the scheme. This result might reflect either a reduction in factors giving rise to welfare problems in the mainstream herd or increasing awareness about management of compromised pigs. Additional data are required to clarify these possibilities, but both represent improved pig welfare.

Keywords: enrichment; hospitalization requirement; lameness; lesions; pigs; tail biting

Citation: Pandolfi, F.; Barber, C.; Edwards, S. The "Real Welfare" Scheme: Changes in UK Finishing Pig Welfare since the Introduction of Formal Welfare Outcome Assessment. *Animals* **2022**, *12*, 607. https://doi.org/10.3390/ani12050607

Academic Editor: Vera Baumans

Received: 21 January 2022
Accepted: 21 February 2022
Published: 28 February 2022

Publisher's Note: MDPI stays neutral with regard to jurisdictional claims in published maps and institutional affiliations.

1. Introduction

As a result of a growing body of scientific research and awareness-raising campaigns, farm animal welfare has become an increasingly important issue for consumers, leading to progressive developments in product marketing and in national and European legislation [1,2]. These changes have highlighted the need for an efficient methodology to accurately benchmark animal welfare. There has been a growing trend in the adoption for this purpose of animal-based measures, sometimes called welfare outcomes (WO), which are now recognized as a better alternative to assess animal welfare compared to the assessment of features in the environment [3,4]. The EU Welfare Quality® project pioneered the development of a methodology for on-farm assessment of animal welfare with animal-based measures [5]. However, the large set of measures and the time required for completion of their comprehensive protocol made it impractical to utilize for a routine assessment which could be applicable for repeated regular use on a large sample of farms. The creation of simpler alternatives became possible thanks to the development of a limited number of "iceberg indicators", which can be defined as single measures which are reflective of several different aspects of animal welfare. These act as a signal of multiple health or welfare issues and, therefore, allow faster and easier on-farm assessment [6].

Previous studies have indicated that the prevalence of such WO is often low [3,7,8]. Consequently, accurate estimation of the prevalence of different WO requires a very large sample which is also representative of the population of interest [2]. Such large samples are particularly difficult to obtain in scientific research based on primary data analysis, due to the difficulty to enroll a large number of farms, the restricted time available and evident cost limitations. Secondary analysis of the data arising from mandatory inspections or schemes which encompass a large part the national herd (i.e., analysis of large datasets not initially collected for the purpose of a study) might therefore represent a good alternative to accurately estimate the current status and trends in health or welfare indicators with low prevalence.

The "Real Welfare" scheme was developed in order to assess pig welfare on finishing farms using animal-based measures [9]. These "iceberg indicators" were chosen after stakeholder consultation to capture the most important welfare issues for the pigs and the industry, using protocols developed and piloted in a previous research project [9]. The "Real Welfare" scheme data represent a unique database which is representative of the mainstream herd of finishing pig farms in the UK and now includes data collected on repeat visits over many years. The results presented in this paper represent the third in a series of publications based on a secondary analysis of the "Real Welfare" scheme data. The first paper focused on the description of the methodology for collection of "Real Welfare" data and the assessment of the initial trends in the prevalence of the welfare outcomes during the period of introduction of the scheme [9]. The second paper described farm characteristics and the associations between farm features and management and the prevalence of the different welfare outcomes over this initial period [10]. The objective of this third paper is to extend the analysis to the period after full stabilization of the scheme, estimating the long-term trends in the prevalence of different WO and the extent to which they can be considered as representative of the mainstream herd of finishing pigs in the UK. It includes assessment of the trend over time and the changes over successive farm inspection visits and seasons of the different WO for the whole six-year period (2013 to 2019) covered by the scheme. To further extend the information presented in the previous papers, we combined different methodologies to increase the robustness of our analysis and conducted a power analysis to estimate the confidence in the trend of the prevalence of the different WO and to assess to possibility to extend the results to the mainstream herd of all finishing farms in the UK.

2. Materials and Methods

Based on the data described below, we made basic descriptive analyses of the number of farms, pens and pigs assessed in this study and described the prevalence of different WO which have been measured during the assessment. Moreover, we used specific statistical analyses to measure the association between different WO and the changes of these WO over time. Additional precision analysis was carried out to strengthen our analysis and its interpretation.

2.1. Data

The analyses in this report were carried out using all data collected from on-farm assessments conducted in the context of the "Real Welfare" scheme between the 4 April 2013 and the 31 December 2019. This includes the data collected between 2013 and 2016 used for the two previous publications [9,10] and additional data collected from 2017 to 2019. Some farms were followed over the whole period covered by the scheme, whilst others were present only during a part of this period. The farms assessed by the "Real Welfare" scheme produce approximately 95% of all commercially raised pigs in the UK [11]. A random sampling of pens and pigs was conducted during each visit by a trained vet, targeting to be representative of the pig herd (Figure 1, Supplementary Material S1). An earlier publication, based only on the farm visited between April 2013 and May 2016, described in detail the training of veterinarians performing the assessment, the sampling methodology, the standardized on-farm assessment protocol, and data collection, all of which are owned and managed by the Agriculture and Horticulture Development Board [9]. Earlier publications also described the farming systems (indoor pens exclusively, both indoor and outdoor pens, outdoor exclusively), group size (small (<30pigs), medium (30–200 pigs), large (>200)) and the percentage of tail-docked pigs for the farms of the "Real Welfare" scheme [9,10]. Two databases were used to conduct our analysis. These comprised a database at the farm level, with information from 2616 different farms, and a database at the pen level, with information from 253,713 different pen inspections. In total, 182 pens were excluded from the analysis because they lacked a unit code (farm identification). Data were collected only from production pens, with hospital pens excluded, as was the case for the Welfare Quality® scheme from which the protocol of the "Real Welfare" scheme was derived [5]. This was because the focus was placed on the identification and appropriate treatment of welfare problems in mainstream pens (excluding hospital pens). From the date of the assessment, the calendar year and the season were extracted. Four seasons representing spring (March, April, May), summer (June, July, August), autumn (September, October, November) and winter (December, January, February) were identified from the date of assessment.

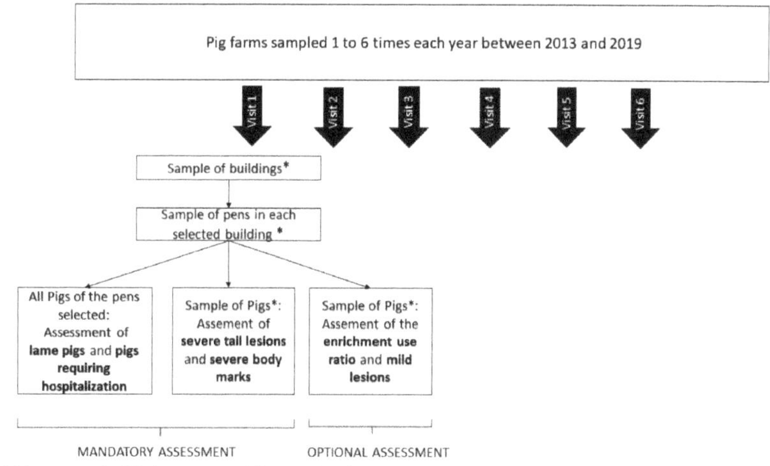

Figure 1. Flowchart of the yearly "Real Welfare" scheme assessment between 2013 and 2019.

In total, nine different WO were recorded for some period of time at the pen and farm levels: pigs requiring hospitalization (pigs who are sick, injured or lame and are unable to compete for resources, being bullied/tail bitten or would benefit from access to bedding that is more comfortable than that available in the mainstream pen), lame pigs, pigs with severe tail lesions, pigs with severe body marks, enrichment use ratio (number of pigs interacting with enrichment/number of pigs interacting with enrichment or other pen features or pen mates), pigs with mild tail lesions, pigs with dirty tail, pigs with mild body marks, and pigs with dirty body. The enrichment could be natural enrichment (straw, wood) or objects [9,10]. The recording of tail lesions and body marks changed during the implementation of the "Real Welfare" scheme. After an initial 8 months period, a review of the scheme's operation determined that recording the enrichment use ratio, minor tail lesions (dirty tails and mild tail lesions), and minor body marks (dirty bodies and mild body marks) should become optional to facilitate the implementation of the scheme and focus on the most important WO to assess pig welfare. Minor lesions and enrichment use were consequently recorded for only a subsample of farms between January 2014 and December 2019. In total, 32,024,840 pigs were present on the farms visited between 2013 and 2019. In order to assess the percentage of pigs requiring hospitalization and lameness, 13,678,660 pigs were assessed individually, while 7,112,777 pigs were assessed individually for tail lesions, and 7,110,608 for body marks. In total, 65,023 different pens grouped in 10,549 different buildings were assessed. However, as some pens were visited several times and were considered different between the different visits, the actual number of pen visits between 2013 and 2019 was 253,713.

For each type of lesion, the total number of pigs assessed consisted of the number of pigs with a specific lesion and the number of pigs in the same pen without this lesion. This allowed calculation of the percentage of assessed pigs exhibiting a given WO in each pen and each farm. The enrichment use ratio records the proportion of actively investigating pigs in the pen which are directing this behavior to the enrichment provided rather than to other pigs or pen components and was expressed on a 0–100 scale.

2.2. Number of Pigs, Pens and Farms Visited

An initial description was made of the sample available for the analysis: number of pigs, farms, buildings and pens, number of assessment visits, and average herd size for the full period of the "Real Welfare" scheme. We also described the average percentage of pigs assessed per visit (based on the number of pig places) and the average number of pens and buildings assessed per visit. For 10 farms, the number of pigs sampled exceeded the number of pigs reported as present in the farm, then the percentage of pigs assessed was replaced by a missing value.

Given that farms and pens were visited several times, the total number of pens assessed over the entire period may have included the same pen numbers and farm IDs multiple times, but they were treated as different pens for the descriptive analysis because the pig population would be different at each visit. Moreover, the farms were generally visited several times, but the same pens were not always sampled.

2.3. Associations between the Different Welfare Outcomes and Prevalence

The sample size of "Real Welfare" scheme data, the population coverage and the random sampling applied at the pen and pig levels allowed calculating estimates for the mean percentage of the different WO of the mainstream herd of finishing pig farms in the UK with 95% confidence interval. The mean percentage was estimated based on either the proportion of pigs with lesions among all pigs in the pen (for lame pigs and pigs requiring hospitalization at the pen level) or the proportion of pigs with lesions among a sample of pigs present in the pen or the farm (Supplementary Material S1). The mean percentage, standard deviation, median, minimum and maximum values, 10% and 90% percentiles at the pen level, and the mean, minimum and maximum values for the annual rolling average per pen are presented for each WO. The mean percentage, standard deviation, median, minimum and maximum values, 10% and 90% percentiles were also calculated at the farm level based on the annual rolling average of each farm. The number of pens assessed per year for mandatory and optional WO were separately calculated.

In order to understand the associations between the different WO at the pen level, the correlations between them were calculated using Spearman's rank correlation. The correlation was considered significant if $r > 0.3$ and $p < 0.05$ [12]. For mandatory WO, the correlations were calculated for the whole sample. The correlation between all WO were calculated only for pens for which optional WO were recorded.

2.4. Assessment of the Changes over Time, Visits and Seasons of the Different Welfare Outcomes

Two complementary approaches were used to assess changes in the prevalence of the different WO over time: time series analysis, and regression. Time series analyses were conducted for the WO in order to identify an underlying trend in the data. We converted the data at the pen level in monthly time series for all WO. The time series was decomposed into three components, which were the underlying trend component, a seasonal component (pattern that repeats with fixed period of time) and an irregular component (residuals after allocation into trend and seasonal components). Visual interpretation of the plot was used as well as the augmented Dickey–Fuller (ADF) test (unit root test) and autocorrelation function (ACF) to assess, respectively, the stationarity and the autocorrelation of the time series (Supplementary Methods S1).

Additional analyses were conducted to identify associations between specific years, calendar seasons or visits and the prevalence of WO, taking into consideration the data structure associated with the specific farm and assessor identity. The changes over calendar years in the mean values of all the measures of welfare were assessed with generalized linear mixed models (glmm) (one for each WO) in an analysis performed at the pen level. The dependent variable was the proportion of pigs exhibiting each WO, or the enrichment use ratio. The variable 'year' was considered as a fixed effect. The pen nested in the farm unit was considered as a random effect as different pens could belong to the same farm. The interaction between the farm unit and the vet practice that performed the assessment was also used as a random effect in order to consider any potential inter-observer variations. In order to make all the pair-wise comparisons between years, we used Tukey's Honest Significant Difference (Tukey's HSD) method and the Compact Letter Display (CLD) function was applied to the results from Tukey's HSD method to identify sets of years which were detectably different (Supplementary Methods S1). The same analyses were performed with 'calendar season' as the fixed effect instead of year for the mandatory WO and the enrichment use ratio.

Since farms entered and left the "Real Welfare" scheme at different points in time, the changes over successive inspection visits for each individual farm were also assessed. The mean percentage of pigs exhibiting each WO was calculated for each farm visit (from the first to the last date of the visits). Based on this value, the changes over the visits were assessed using Kendall's tau b correlation between visits for each individual farm and the mean percentage for all farms for each visit was plotted on a graph in order to visualize the general trend of the changes that occurred with the progressive number of assessment visits.

2.5. Precision of the Estimates and Trend over Time

Firstly, to calculate the precision of the estimates and trends over time, the sample size was adjusted for each year in order to consider the design effect.

- The intraclass correlation (ICC) for each welfare outcome was calculated with the R package ICC.
- The design effect (Deff) was calculated based on the ICC and the average number of pens per farm (m) as follows:

$$Deff = 1 + ICC(m - 1)$$

- The estimated sample size (n), considering the design effect (*Deff*) and the actual sample size (n'), was calculated as follows:

$$n = n'/Deff$$

Secondly, the margin of error (e) for the various WO was calculated for each individual year based on the following equation:

$$n = \frac{Z^2\left((Z\alpha/2 + Z\beta)2\sigma^2\right)}{e^2}$$

where n is the sample size, σ^2 is the variance of the WO, and Z is the value from a standard normal distribution, corresponding to the desired confidence level as the type I error Alpha (Z = 1.96 for 95 percent confidence interval (CI)), and Beta the power of the analysis set at 80%, which is considered acceptable. Finally, we assessed if the margins of error overlapped between years in order to estimate the confidence in the trend identified over the years for the different WO.

2.6. Software

Data processing and data analyses were carried out using Microsoft Excel Office Professional Plus 2010 (Microsoft, Redmond, Washington, WA, USA) and RStudio (4.0.3 for Window 64 bit, Boston, MA, USA). The lme4, mvtnorm and multcomp R Packages were used for the glmm analysis, Tukey's HSD and cld. The forecast, tseries and ggplot2 R packages were used for the time series analyses.

3. Results

3.1. Number of Pigs, Pens and Farms Visited

The farm visits were repeated with a minimum of 1 visit per year and a maximum of 6 visits per year, dependent on scheme entry date and the presence of finishing pigs on the farm, so these farms were visited between one and maximum 29 times since the "Real Welfare" scheme was first implemented. In total, 7.8% of the farms had only one visit which occurred between 4 April 2013 and 30 December 2019. Moreover, 5.0% of the farms had only one visit which occurred at least one year ago (before 1 January 2019), suggesting that these farms might no longer be part of the "Real Welfare" scheme. It is likely that many of these were finishing pig sites temporarily contracted by integrators. The rest of the farms with one visit might have joined the scheme recently and completed only one visit by the end of 2019. In total, 20,240 farm visits have been conducted in 2616 different farms.

The average number of finishing pig places was 1583 pigs per farm and the average number of finishing pigs present in the farm at the time of assessment was 1213 pigs per farm. Ninety percent of the farms had 400 to 3060 pig places in the farm. Based on the number of pig places, the average percentage of pigs assessed per visit was 23.0% for severe lesions and 44.6% for hospitalization requirement or lameness. In total, 0.1% of the farms had missing values for the percentage of pigs assessed. The number of pens selected largely depended on the farming system and the herd size. The average number of pigs per pens was 56. On average, 9.6 pens per farm were assessed per visit, with a minimum of one and a maximum of 98 pens. On average, 2.2 buildings were visited per farm and per visit with a minimum of 1 and a maximum of 13. The number of farms, pens and pigs assessed for each lesion (including the mild lesions and enrichment use ratio) in each year of the "Real Welfare" scheme is documented in Supplementary Table S1. Approximately 2,000,000 pigs were assessed per year for mandatory WO.

3.2. Association between the Different Welfare Outcomes and Overall Prevalence

The mean percentage of each WO was calculated at the pig level (Table 1), the pen level (Table 2) and the farm level (based on annual rolling average) (Table 3). While the number of pens assessed per year increased in 2014 and remained stable in the subsequent years for the mandatory WO, the number of pens assessed per year constantly decreased for the optional WO (minor lesions and enrichment use ratio) (Figure 2). The only significant correlation among the mandatory WO was between the percentage of pigs requiring hospitalization and the percentage of pigs with lameness ($p < 0.05$, $r = 0.34$). When only the farms with minor lesions recorded were considered, the correlation between the percentage of pigs requiring hospitalization and the percentage of pigs with lameness remained unchanged, and the percentage of dirty tail was moderately correlated to the percentage of dirty body ($p < 0.05$, $r = 0.49$). No significant correlations were identified between minor and severe lesions ($r < 0.30$).

Table 1. Description of the welfare outcomes at the pig level for all pigs assessed between 2013 and 2019: mean percentage and 95% confidence interval [CI].

Welfare Outcomes	Mean [CI] [2]
Pigs requiring hospitalization	0.04 [0.04–0.04]
Lameness	0.12 [0.12–0.12]
Severe tail lesions	0.15 [0.15–0.16]
Severe body marks	0.18 [0.17–0.18]
Mild tail lesions [1]	1.07 [1.05–1.09]
Mild body marks [1]	8.55 [8.49–8.60]
Dirty tail [1]	5.82 [5.77–5.86]
Dirty body [1]	3.96 [3.92–3.99]
Enrichment use ratio	54.4 [54.1–54.7]

[1] Includes only the pens where minor lesions were assessed; [2] the value is the percentage of pigs exhibiting each outcome among the pigs sampled, or the percentage of pigs directing their exploratory behavior to the enrichment item provided among pigs sampled directing their exploratory behavior toward enrichment item provided or other pen features or pen mates.

Table 2. Description of the welfare outcomes at the pen level for all pens visited between 2013 and 2019: Mean percentage, standard deviation (SD), minimum (MIN) and maximum (MAX) values, median, 10% percentile (P10), and 90% percentile (P90).

Welfare Outcomes	Mean [CI] [3]	SD	Min	P10	Median	P90	Max	Min [2]	Median [2]	Max [2]
Pigs requiring hospitalization	0.06 [0.06–0.06]	0.72	0	0	0	0	100	0	0	50
Lameness	0.15 [0.15–0.16]	1.11	0	0	0	0	100	0	0	66.7
Severe tail lesions	0.17 [0.16–0.18]	1.67	0	0	0	0	100	0	0	100
Mild tail lesions [1]	1.25 [1.12–1.29]	4.80	0	0	0	4	100	0	0	100
Dirty tail [1]	5.40 [5.35–5.56]	16.2	0	0	0	16.7	100	0	0	100
Severe body marks	0.20 [0.19–0.21]	1.62	0	0	0	0	100	0	0	100
Mild body marks [1]	9.55 [9.41–9.69]	14.7	0	0	4	27.8	100	0	4	100
Dirty body [1]	3.53 [3.39–3.66]	13.6	0	0	0	5	100	0	0	100
Enrichment use ratio	48.9 [48.2–49.5]	35.6	0	0	50	100	100	0	50	100

[1] Includes only the pens where mild lesions were assessed; [2] values based on annual rolling averages; [3] the value is the percentage of pigs in the pen exhibiting each outcome among the pigs sampled, or the percentage of pigs directing their exploratory behavior to the enrichment item provided among pigs sampled directing their exploratory behavior toward enrichment item provided or other pen features or pen mates.

Table 3. Description of the welfare outcomes at the farm level (% of pigs or ratio) for all farms visited between 2013 and 2019: mean percentage and 95% confidence interval [CI], standard deviation (SD), minimum (MIN) and maximum (MAX) values, median, 10% percentile (P10), and 90% percentile (P90).

Welfare Outcomes [2]	Mean [CI] [3]	SD	Min	P10	Median	P90	Max
Pigs requiring hospitalization	0.05 [0.05–0.05]	0.18	0	0	0	0	5.11
Lameness	0.15 [0.15–0.16]	0.47	0	0	0	0	18.7
Severe tail lesions	0.15 [0.14–0.16]	0.60	0	0	0	0	23.9
Mild tail lesions [1]	1.12 [1.03–1.22]	2.25	0	0	0.18	4	29.2
Dirty tail [1]	6.98 [6.32–7.64]	16.2	0	0	0	16.7	100
Severe body marks	0.18 [0.17–0.20]	0.68	0	0	0	0	14.6
Mild body marks [1]	9.68 [9.20–10.2]	11.6	0	0	5.65	27.8	82.7
Dirty body [1]	5.06 [4.47–5.64]	14.0	0	0	0	5	97.7
Enrichment use ratio	51.7 [51.2–52.2]	25.9	0	0	52.5	100	100

[1] Includes only the pens where mild lesions were assessed; [2] values based on annual rolling averages; [3] the value is the percentage of pigs in the pen exhibiting each outcome among the pigs sampled, or the percentage of pigs directing their exploratory behavior to the enrichment item provided among pigs sampled directing their exploratory behavior toward enrichment item provided or other pen features or pen mates.

Figure 2. Number of pens assessed per year for the mandatory welfare outcome (WO) (pigs requiring hospitalization, lame pigs, severe tail lesions, severe body marks) and optional welfare outcomes (WO) (minor: mild tail lesions and dirty tail, mild body marks and dirty body; enrichment: enrichment use ratio) for the data collected between 4 April 2013 and 31 December 2019.

3.3. Assessment of the Changes over the Years, Visits and Seasons of the Different Welfare Outcomes

3.3.1. Change over the Years

Following the time series analyses, "seasonal" patterns were identified for all WO (Figure 3). For all WO, the *p*-value of the ADF test was greater than 0.05, indicating non-stationary time series and inconsistent change over time. For the time series of pigs requiring hospitalization, lame pigs and pigs with severe body marks, we could identify a slow and irregular decrease in the ACF as the lags increased; this is due to the trend and the seasonality observed in these time series (Supplementary Figure S1). We observed a decreasing trend in the percentages of pigs requiring hospitalization, lame pigs, and severe body marks, but not for the percentage with severe tail lesions and the enrichment use ratio (Figure 3).

Firstly, using generalized linear mixed models, all subsequent years were compared to 2013 (the first year of the "Real Welfare" scheme). Secondly, a pair-wise comparison between years, based on Tukey's HSD method and cld function, was used to identify sets of years that were not detectably different from each other. Finally, the results were used to conclude changes in prevalence over time of these WO.

The proportion of pigs requiring hospitalization was significantly lower in 2019 and 2018 compared to 2015, 2014 and 2013 ($p < 0.001$) (Table 4, Supplementary Tables S2 and S4). The proportion of pigs requiring hospitalization was significantly lower in 2019 compared to 2017 and 2016 ($p < 0.005$). This suggests a smaller mean prevalence in most recent years compared to the first years of the period covered by the scheme.

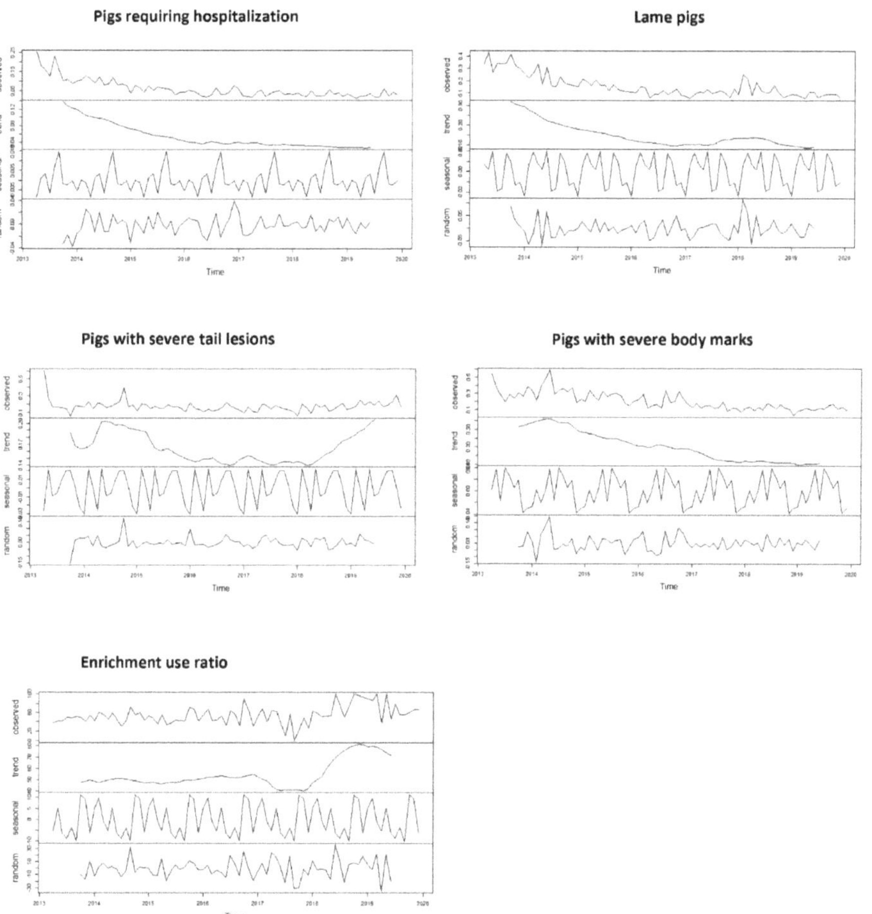

Figure 3. Decomposition of the time series of the percentage of pigs requiring hospitalization, lame pigs, pigs with severe tail lesions, severe body marks and enrichment use ratio from April 2013 to December 2019 (observed = gross pattern, trend = underlying trend, seasonal = seasonal component, and random = residual variations).

Table 4. Mean percentage and 95% confidence interval [CI] for each year (without considering random effect) and Compact Letter Display (cld) for mandatory welfare outcomes and enrichment use ratio. Values with a different cld are significantly different ($p < 0.05$).

	Pigs Requiring Hospitalization		Lame Pigs		Severe Tail Lesions		Severe Body Marks		Enrichment Use Ratio	
Year	Mean [CI]	cld	Mean [CI]	cld	Mean [CI]	cld	Mean [CI]	cld	Mean [CI]	cld
2013	0.14 [0.13–0.15]	e	0.34 [0.32–0.37]	e	0.16 [0.14–0.18]	A	0.29 [0.26–0.32]	e	47.1 [46.6–47.6]	a
2014	0.10 [0.09–0.11]	d	0.22 [0.20–0.23]	d	0.20 [0.18–0.22]	Bcd	0.34 [0.31–0.36]	e	49.8 [49.5–50.2]	a
2015	0.06 [0.05–0.06]	c	0.16 [0.15–0.17]	c	0.16 [0.14–0.17]	bc	0.25 [0.23–0.27]	d	43.9 [43.6–44.3]	b
2016	0.04 [0.03–0.05]	b	0.11 [0.10–0.12]	b	0.15 [0.13–0.17]	ac	0.20 [0.18–0.22]	c	50.3 [49.9–50.6]	b
2017	0.03 [0.03–0.04]	b	0.10 [0.09–0.11]	ab	0.14 [0.13–0.16]	ab	0.13 [0.12–0.14]	a	39.4 [39.1–39.7]	ab
2018	0.03 [0.03–0.04]	ab	0.14 [0.12–0.15]	bc	0.15 [0.13–0.17]	c	0.12 [0.11–0.13]	ab	64.1 [63.8–64.4]	c
2019	0.03 [0.02–0.03]	a	0.09 [0.08–0.10]	a	0.20 [0.19–0.22]	d	0.11 [0.10–0.12]	b	64.1 [63.8–64.4]	c

The proportion of lame pigs was significantly lower in 2019 compared to 2016 ($p < 0.05$), 2018, 2015, 2014 and 2013 ($p < 0.001$). The proportion of lame pigs was significantly lower in 2018 compared to 2015, 2014 and 2013 ($p < 0.001$) (Table 4, Supplementary Tables S2 and S4). This suggests a smaller mean prevalence in most recent years compared to the first years of the period covered by the scheme.

The proportion of pigs with severe tail lesions was significantly higher in 2019 compared to 2013, 2015, 2016, 2017, and 2018 but was not significantly higher than 2014. The proportion of pigs with severe tail lesions was significantly higher in 2018 compared to 2017 and 2013 but not the other years (Table 4, Supplementary Tables S2 and S4). This suggests no significant differences in the mean prevalence between the first years and last years of the period covered by the scheme.

The proportion of pigs with severe body marks was significantly higher in 2013, 2014, 2015, and 2016 compared to 2018 and 2019 and significantly higher in 2017 compared to 2019 ($p < 0.001$) (Table 4, Supplementary Tables S2 and S4). This suggests a smaller mean prevalence in most recent years compared to the first years of the period covered by the scheme.

The enrichment use ratio was significantly lower in 2013, 2014, 2015, and 2016 compared to 2018 and 2019 and significantly lower in 2013, 2014 compared to 2015, 2016, 2017, 2018, 2019 ($p < 0.001$) (Table 4, Supplementary Tables S2 and S4). This may suggest an increasing trend which appears mainly over the two last years. However, the diminishing sample size once this outcome became optional must be borne in mind.

For the WO with optional recording, the results are summarized in Table 5 and Supplementary Tables S3 and S5 but would need to be confirmed by further analyses with mandatory recording. The results suggest an initial decline of mild tail lesions between 2013 and 2014, which cannot be interpreted as the recording of mild lesion became optional in 2014. However, there is a further decline between 2014 and 2016 which tends to flatten out in the subsequent years. The results also suggest a progressive increase in the proportion of pigs with dirty tails between 2014 and 2019, a significant decline each year of the proportion of pigs with mild body marks, and an increase between 2017 and 2019 of the proportion of pigs with dirty bodies.

Table 5. Mean percentage and 95% confidence interval [CI] for each year (without considering random effect) and Compact Letter Display (cld) for optional welfare outcomes. Values with a different cld are significantly different ($p < 0.05$).

	Mild Tail Lesions		Dirty Tails		Mild Body Marks		Dirty Bodies	
Year	Mean [CI]	cld	Mean [CI]	cld	Mean [CI]	cld	Mean [CI]	cld
2013	1.76 [1.68–1.83]	e	5.77 [5.55–6.00]	abc	13.1 [12.9–13.3]	g	3.06 [2.88–3.23]	abc
2014	0.98 [0.89–1.06]	d	3.27 [2.99–3.55]	a	5.36 [5.14–5.59]	f	2.17 [1.93–2.40]	ab
2015	0.68 [0.59–0.77]	c	3.99 [3.61–4.37]	a	5.35 [5.06–5.63]	e	2.76 [2.42–3.10]	a
2016	0.61 [0.48–0.74]	ab	5.85 [5.27–6.44]	b	5.74 [5.40–6.09]	d	3.61 [3.17–4.06]	b
2017	0.87 [0.66–1.09]	bcd	8.31 [7.48–9.14]	b	5.19 [4.83–5.55]	c	4.59 [4.03–5.16]	c
2018	0.42 [0.32–0.53]	a	4.68 [3.94–5.42]	c	3.74 [3.41–4.06]	b	4.68 [3.93–5.43]	d
2019	0.56 [0.36–0.76]	a	4.77 [3.88–5.65]	c	1.74 [1.46–2.03]	a	4.82 [3.93–5.71]	d

3.3.2. Change over Calendar Seasons

The results of the generalized linear models suggest differences between calendar seasons in the pigs requiring hospitalization, lame pigs, pigs with severe tail lesions, pigs with severe body marks and enrichment use ratio (Table 6, Supplementary Table S6). The different outcomes did not show a similar seasonal pattern. The proportion of pigs requiring hospitalization was significantly higher in autumn and winter compared to spring ($p < 0.001$, $p = 0.006$). The proportion of lame pigs was significantly higher in winter compared to summer. The proportion of pigs with severe tail lesions was significantly higher in spring and autumn compared to summer and winter ($p < 0.001$). The proportion of pigs with severe body marks was significantly lower in autumn compared to spring ($p < 0.001$). The enrichment use ratio was significantly higher in winter compared to summer, autumn and spring ($p < 0.001$). The enrichment use ratio was significantly higher in autumn compared to spring ($p < 0.001$) (Supplementary Table S7).

Table 6. Mean percentage and 95% confidence interval [CI] for each calendar season (without considering random effect) and Compact Letter Display (cld) for mandatory welfare outcomes and enrichment use ratio. Values with a different cld are significantly different ($p < 0.05$).

Season	Pigs Requiring Hospitalization Mean [CI]	cld	Lame Pigs Mean [CI]	cld	Severe Tail Lesions Mean [CI]	cld	Severe Body Marks Mean [CI]	cld	Enrichment Use Ratio Mean [CI]	cld
Spring	0.11 [0.09–0.13]	a	0.38 [0.34–0.42]	ab	0.20 [0.16–0.25]	b	0.23 [0.18–0.27]	a	41.6 [40.9–42.4]	b
Summer	0.13 [0.11–0.15]	ab	0.30 [0.28–0.33]	a	0.15 [0.13–0.17]	a	0.19 [0.16–0.22]	ab	47.5 [46.9–48.1]	bc
Autumn	0.12 [0.10–0.13]	b	0.36 [0.33–0.39]	ab	0.16 [0.12–0.19]	b	0.21 [0.19–0.24]	b	49.3 [48.7–49.9]	c
Winter	0.09 [0.07–0.11]	b	0.30 [0.25–0.34]	b	0.16 [0.12–0.20]	a	0.18 [0.13–0.22]	ab	54.8 [54.1–55.6]	a

3.3.3. Changes over the Sequential Assessment Visits

The general trend within farm over successive visits is visualized in Figure 4. The percentages for the last visits cannot be interpreted, as very few farms were included in the sample. While a decreasing trend can be observed over the visits for the mean percentage of pigs requiring hospitalization, lame pigs and pigs with severe body marks, the other WO remained more stable.

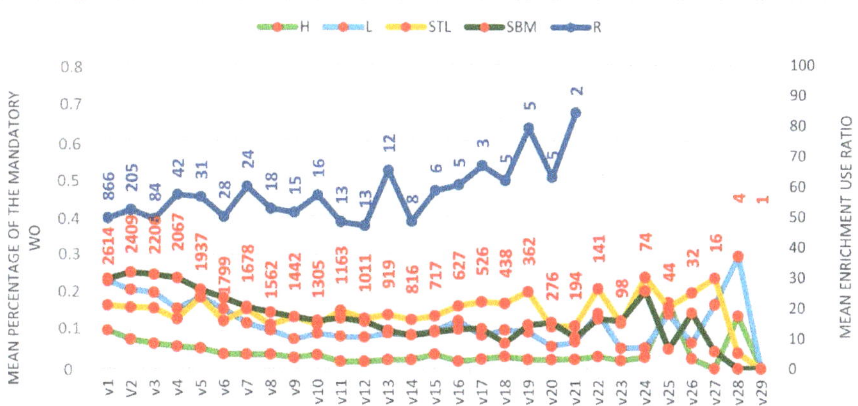

Figure 4. Mean percentage of the mandatory welfare outcomes (WO) (Pigs requiring hospitalization (H), lame pigs (L), pigs with severe tail lesions (STL), pigs with severe body marks (SBM) and enrichment use ratio (R) from the 1st to the 29th visit (21st visit for enrichment use ratio). The number of farms included in the sample to calculate the mean for each visit is reported on the graph in red for the mandatory welfare outcomes and in blue for the enrichment use ratio.

There was no Kendall tau-b correlation between the different visits for the percentage of pigs requiring hospitalization, confirming changes over the visits for this WO with an overall downward trend. Moderate correlations were identified for the percentage of lame pigs, mainly between consecutive visits. The correlations were weaker for visits with wider interval, suggesting that changes occurred but could be slower compared to the percentage of pigs requiring hospitalization. Indeed, a downward trend over visits was also identified. The Kendall's tau-b correlation showed no correlation between the different visits for the percentage of pigs with severe tail lesions confirming changes over the visits. However, looking at the graph summarizing the mean WO prevalence per visit, no trend could be identified. Indeed, the mean percentage of pigs with severe tail lesions tend to go up and down around a mean value, suggesting that some farms experienced an increase in this WO while others experienced the opposite. Moderate correlations for the percentage of pigs showing severe body marks, mainly between consecutive visits, indicated small changes between these visits. Moreover, the lack of correlation between other visits with wider intervals suggests that changes occurred but could be slower compared to the percentage of pigs requiring hospitalization. Indeed, a downward trend over visits was also identified. The correlation between several visits, even with a longer interval between visits, suggests only moderate changes in the enrichment use ratio over the visits. No specific trend could be observed in the graph summarizing the mean enrichment use ratio per visit (Figure 4) (Supplementary Table S8).

3.4. Precision of the Estimates

Considering the sampling size, we can see that the margin of error is not always overlapping for pigs requiring hospitalization, lameness, severe body marks, mild body marks and enrichment use ratio (Figure 5) (Supplementary Table S9). This indicates that the sampling size allows good confidence in the trend observed over the years, keeping in mind that the recording of enrichment use ratio and mild body marks was not mandatory.

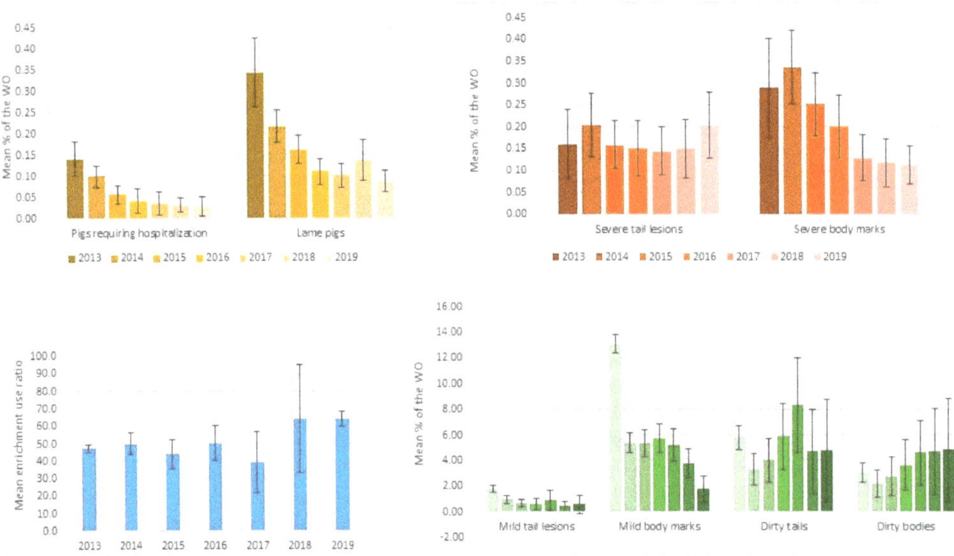

Figure 5. Annual mean and margin of error considering the sampling size (accounting for the design effect) and 80% power for all welfare outcomes (WO).

4. Discussion

4.1. Study Limitations

Despite a very low prevalence of some measures, the large amount of data available from the "Real Welfare" scheme allows more accurate analysis compared to studies based on primary data, where data are often collected on a restricted number of farms. However, absence of data related to hospital pens makes it difficult to separate a real decrease in the prevalence of the different WO over time from a better management of compromised pigs. Since the WO were only recorded in the mainstream herd, the decreasing trend in several WO could be explained by either a real reduction in these WO or better management of transfer to hospital pens or euthanasia of the most severely injured pigs, but both would contribute to overall improvement in pig herd welfare. The possible change in management of compromised pigs could be due to the impact of the vet assessor on farmer perception of sick pigs, whereby regular visits of the vet operating welfare assessment might have changed the perception of the farmer regarding sick pigs and hospital pen management [13]. Another bias, due to repeating assessments over time and the absence of "update training" of the vets might have been an impact on inter- and intra-observer reliability. However, several studies have shown good inter-observer reliability of similar WO data recorded by trained assessors [14,15]. We also included the interaction of the vet with the farm in the models that assessed the changes over time and season in order to reduce any information bias in our analysis. Regarding the additional time series analysis, which confirms the decreasing trend of some WO, the autoregressive model used in this analysis might appear too simplistic. While this model considers its previous terms as the predictors, there is likely more complexity than a simple linear model of past observation and a more complex model should be used to create forecasts.

Using a large sample size and conducting an assessment of pigs randomly selected at the pen level appear to be the best options to evaluate WO [16]. Indeed, high variability can exist between pens of the same farm, which makes it difficult to accurately assess farm prevalence with reduced sample size [17]. Moreover, assessing the prevalence of the different WO at the pen level appeared to be more informative in another study [18]. The sample of pig farms assessed across each year by the "Real Welfare" scheme produces approximately 95% of all commercially raised pigs in the UK and, for each visit, the assessment was conducted on a representative sample of pigs and pens selected by trained vets [11]. This allowed us to conduct the analysis on a large sample representative of the mainstream herd of finishing pigs in the UK. Following the analysis of the margin of error for each individual year, considering 80% power, we could conclude that the sampling allows accurate estimation of the prevalence of these different WO and allows accurate assessment of the changes in prevalence over time. However, the recording of enrichment use ratio and minor lesions was not mandatory after 2013 and this may have led to selection bias and lower precision in the estimates. The results should therefore be interpreted with caution because the number of pens which were recorded for these lesions drastically dropped in recent years. For the same reason, the trend over visits (using visualization of the graph in Figure 4 and Kendall's tau-b correlation) was not interpreted beyond eight visits because of the declining sample size. However, we decided to share the results on enrichment use ratio and minor lesions because the sample remains exceptionally large compared to most other studies. Moreover, highlighting the decreasing number of pens assessed over time (when the measures became optional) in itself constitutes interesting information for the pig sector and future projects. Pen selection, pen size and the variability between pens will impact the precision of the estimate (margin of error) [19,20]. In the "Real Welfare" scheme, only a subset of the pens and subset of the pigs according to farm size were sampled in order to be representative of pigs present in the farm, but also to reduce time and commitment required on the operational aspects [9]. To get exactly the same level of precision in each farm, a specific sampling calculation which considers farm design could be applied in each individual farm but, at a large scale, this calculation would represent a constraint that could impair the simplicity of implementation of the assessment.

4.2. Prevalence of the Welfare Outcomes

The mean prevalence of the mandatory WO was very low at the pig, pen and farm levels. Similar results regarding the prevalence of the different WO were obtained in previous studies [21–23]. The standard deviation was higher than the mean for all WO, indicating that the percentage of WO in the different pens and farms spreads out over a wide range of values. The percentage of pigs with a welfare issue could be very high in certain individual pens, but the mean percentage of pigs with these welfare issues at the farm level was much lower. This confirms that these welfare issues remain rare and sporadic [18,22]. A previous paper, based on the data recorded between 2013 and 2016, identified several risk factors associated to higher prevalence of the different WO [10]. However, more detailed analysis could be done on farms and pens with high prevalence of WO to better understand the combination of factors that lead to welfare issues and identify other risk factors. Furthermore, such high prevalence cases should be the subject of more detailed investigation and remedial action by the farmer and veterinarian whenever they are observed.

Based on the European Food Safety Authority (EFSA) [24] report, the prevalence of tail lesions on farm may vary widely (of the order of 1–5%). In our study, the percentage of pigs with severe tail lesions and severe body marks was 0.04% and 0.12%, respectively. Some studies which evaluated the prevalence of tail damage reported a prevalence of severe tail biting at the pig level of 1.3% in Finland [25] and 2.2% in France (0.9% for the most severe lesions) [7]. The mean percentage at the batch level from farrow-to-finish herds (representing 12% of the Irish herd) was 1.2% for severe tail lesions and 1.2% severe skin lesions [8]. Comparisons are difficult to make because the precise definition of a lesion differs between studies. A similar sampling methodology and classification should be adopted to enable valid comparison between studies.

A higher percentage of pigs requiring hospitalization was associated with a higher percentage of lame pigs. This may be due to similar risk factors within a pen for both of these welfare issues, or to the fact that individual pigs may fall into both categories (lameness could be a common reason for moving pigs in hospital pens). Some correlation between these two WO was also identified in a previous study [26]. Different results were found by Munsterhjelm et al. [27], who also excluded hospital pens from their analysis and found a connection between wounds and lameness. However, both samples were much smaller in these studies. The absence of other correlations between WO suggests that all WO are specific and complementary in welfare assessment and, therefore, could reveal different aspects of animal welfare. Indeed, it has been shown that pigs in hospital pens are associated with both respiratory and locomotion problems [16], lame pigs are associated with both flooring and the detection of *Mycoplasma hyosynoviae* [28], and body and tail lesions are associated with various stressors and negative social behavior [29,30]; all acting as signals of impaired animal well-being.

4.3. Changes over Time and Season of the Welfare Outcomes

Following our analysis, we could conclude that pigs requiring hospitalization, lame pigs, severe body marks and mild body marks decreased over the years and enrichment use ratio increased over time. No trend could be identified for severe tail lesions, as similar prevalence was observed in 2014 and 2019. Dirty tails and dirty body tended to increase in recent years covered by the "Real Welfare" scheme, though the sample size became very small and the trend observed should be interpreted with care. Considering the very low percentage prevalence of all WO in the most recent years, we could easily expect a stabilization of the prevalence, as a complete absence of welfare issues seems to be unrealistic. While the "Real Welfare" scheme might be considered to have provided good support to decrease lameness, severe body marks or improve hospital pen management, severe tail lesions may be greatly influenced by other less controllable factors, such as the housing infrastructure of the farm, diet formulation or climatic variation, making changes over time less visible. The identification by the farmer of the different farm-specific risk factors is essential. Tools such as "WebHAT" or "SchwIP" have been created to raise awareness amongst farmers about the risk factors and may offer a better help to reduce tail biting [21,31]. It is known that benchmarking of health and welfare measures can lead to greater awareness and motivation to improve [32] but our study suggests that this is not uniformly successful for all WO, especially for severe tail lesions. While other welfare assessment initiatives exist and measure similar WO, such a large-scale assessment as reported here is not documented in the literature for other countries. The implementation of a similar scheme in other countries would allow comparisons of the prevalence and the trend of the different WO between countries.

While a decline over repeated assessment visits to individual farms could be clearly seen for the percentage of pigs requiring hospitalization, the percentage of lameness and severe body marks, no trend over visits was identified for the percentage of pigs with severe tail lesions and enrichment use ratio. Only the percentage of pigs requiring hospitalization showed a decreasing trend with no correlation between years, suggesting quicker changes compared to the percentage of lame pigs and the percentage of severe body marks, which showed correlations between consecutive visits. No correlations between visits were identified for severe tail lesions, suggesting both increases and decreases according to the farm, without identifying a general trend. In contrast, the enrichment use ratio showed significant correlations between visits, suggesting no particular changes over visits.

'Seasonal' changes identified in the time series analysis are difficult to interpret. Indeed, seasonal changes probably act as a proxy for other factors related to environmental or management differences or may be connected to other seasonal health issues. For example, some health issues have been reported to be more common during winter [33,34]. Severe tail lesions were more prevalent in autumn and spring, corresponding to times with greater temperature fluctuations. Higher enrichment use ratio was observed in autumn and winter, possibly associated with greater bedding supply at these times.

5. Conclusions

The estimate of the mean percentage of the different WO in the mainstream herd of the UK finishing pig population was $\leq 0.2\%$ when assessed at the pig, pen and farm levels, but a high prevalence was detected sporadically. In the period since the implementation of the "Real Welfare" scheme, our analysis indicates an improvement in welfare in the mainstream herd of finishing pig farms in the UK, reflected in the reduction in the prevalence of the main WO (except for severe tail lesions) over both calendar years and sequential assessment visits at a given farm. The decreasing trend over both years and successive assessment visits tended to flatten out, suggesting that further decline from the present low values would be difficult to achieve. The "Real Welfare" scheme might have contributed to a decrease in the prevalence of certain WO in the mainstream herd by stimulating a reduction in causal factors for these welfare problems and/or an improvement in the management of compromised pigs. The possible influence of the scheme should be confirmed by further studies. Other initiatives, such as increasing farmer awareness, identification and understanding of risk factors for tail biting and other outcomes, and support for changes in farm management should also be considered.

Supplementary Materials: The following supporting information can be downloaded at: https://www.mdpi.com/article/10.3390/ani12050607/s1. Material S1; Methods S1; Figure S1: Autocorrelation function (ACF) (with lag= fixed amount of passing time) for the percentage of pigs requiring hospitalization, lame pigs from, pigs with severe tail lesions, severe body marks and enrichment use ratio April 2013 to December 2019; Table S1: The number of farms, pens, pen visits and pigs assessed for each lesion (including the mild lesions and enrichment use ratio) in each year of the scheme (For the calculations for each year, pens without a visit date have been excluded), Table S2: Changes over years compared to 2013 for the different welfare outcomes: Odds ratio, confidence intervals and P-value for all pens included in the study, Table S3: Changes over years compared to 2013 for the minor welfare outcomes: Odds ratio, confidence intervals and P-value for all pens with minor lesions recorded (42,108 pens), Table S4: *p*-values for pair-wise comparison between years with Tukey's HSD for the proportion of pigs requiring hospitalization, lame pigs, pigs with severe tail lesions, pigs with severe body marks and enrichment use ratio, Table S5: *p*-values for pair-wise comparison between years with Tukey's HSD for the proportion of pigs with mild tail lesions, dirty tails, mild body marks, dirty bodies, Table S6: Changes over seasons compared to winter for the mandatory welfare outcomes and enrichment use ratio: Odds ratio, confidence intervals and P-value for all pens included in the study, Table S7: values for pair-wise comparison between seasons with Tukey's HSD for the proportion of pigs requiring hospitalization, lame pigs, pigs with severe tail lesions, pigs with severe body marks and enrichment use ratio, Table S8: Kendall's tau-b correlation coefficient between the average percentages of welfare outcomes (H: pigs requiring hospitalization, L: lame pigs, STL: pigs with severe tail lesions, SBM; pigs with severe body marks, R: enrichment use ratio) for individual farms in each visit (1st to 8th visits). ($p < 0.05$, Values > 0.3 represent a significant correlation), Table S9: Annual mean and calculated margin of error considering the sampling size of each year (accounting for the design effect) and 80% power for all welfare outcomes.

Author Contributions: Conceptualization, C.B., F.P. and S.E.; methodology, F.P. and S.E.; software, F.P.; validation, C.B., F.P. and S.E.; formal analysis, F.P.; investigation, F.P. and S.E.; resources, C.B., F.P. and S.E.; data curation C.B., F.P. and S.E.; writing—original draft preparation, F.P. and S.E.; writing—review and editing, F.P. and S.E.; visualization, C.B., F.P. and S.E.; supervision, C.B., F.P. and S.E.; project administration, C.B.; funding acquisition, C.B. All authors have read and agreed to the published version of the manuscript.

Funding: The research was funded by the Agriculture and Horticulture Development Board (AHDB).

Institutional Review Board Statement: Ethical review and approval were waived for this study, which is an anonymized secondary data analysis that does not directly involve humans or animals.

Informed Consent Statement: Membership of the Red Tractor Farm Assurance scheme provides consent for the collection and anonymized use of welfare outcome data.

Data Availability Statement: The datasets generated and analyzed during the current study are not publicly available as the ownership of the non-anonymized data is that of each individual farmer.

Acknowledgments: The authors thank all farmers and vets for their contribution to the "Real Welfare" Scheme and for the data collection. The authors also thank the "Real Welfare" project Steering Group.

Conflicts of Interest: Although one author is a member of AHDB staff, this organization did not influence the methodology of analyses and did not interfere in the interpretation of the results. We therefore declare no competing interest.

References

1.	Horgan, R.; Gavinelli, A. The expanding role of animal welfare within EU legislation and beyond. *Livest. Sci.* **2006**, *103*, 303–307. [CrossRef]
2.	Johansson-Stenman, O. Animal Welfare and Social Decisions: Is It Time to Take Bentham Seriously? *Ecol. Econom.* **2018**, *145*, 90–103. [CrossRef]
3.	European Food Safety Authority (EFSA). Statement on the use of animal based measures to assess the welfare of animals. *EFSA J.* **2012**, *10*, 2767.
4.	Whay, H.R.; Main, D.C.J.; Green, L.E.; Webster, A.J.F. Animal-based measures for the assessment of welfare state of dairy cattle, pigs and laying hens: Consensus of expert opinion. *Anim. Welf.* **2005**, *12*, 205–217.
5.	Blokhuis, H.J.; Veissier, I.; Miele, M.; Jones, B. The Welfare Quality® project and beyond: Safeguarding farm animal well-being. *Acta Agric. Scand. A Anim. Sci.* **2010**, *60*, 129–140. [CrossRef]
6.	Heath, C.A.E.; Browne, W.J.; Mullan, S.; Main, D.C.J. Navigating the iceberg: Reducing the number of parameters within the Welfare Quality® assessment protocol for dairy cows. *Animals* **2014**, *8*, 978–1986. [CrossRef]
7.	Courboulay, V.; Drouet, A. Evaluation de la prévalence de caudophagie par la notation des carcasses en abattoir. *JRP* **2018**, *50*, 333–334.
8.	Van Staaveren, N.; Doyle, B.; Manzanilla, E.G.; Calderón Díaz, J.A.; Hanlon, A.; Boyle, L.A. Validation of carcass lesions as indicators for on-farm health and welfare of pigs. *J. Anim. Sci.* **2017**, *95*, 1528–1536. [CrossRef]
9.	Pandolfi, F.; Stoddart, K.; Wainwright, N.; Kyriazakis, I.; Edwards, S.A. The Real Welfare Scheme: Benchmarking welfare outcomes for commercially farmed pigs. *Animals* **2017**, *11*, 1816–1824. [CrossRef]
10.	Pandolfi, F.; Stoddart, K.; Wainright, N.; Kyriazakis, I.; Edwards, S. The "Real Welfare" Scheme: Identification of risk and protective factors for welfare outcomes in commercial pig farms in the UK. *Prev. Vet. Med.* **2017**, *146*, 34–43. [CrossRef]
11.	AHDB. Real Welfare Baseline Report 2013–2016. Available online: https://ahdb.org.uk/knowledge-library/real-welfare-baseline-report-2013-2016 (accessed on 27 January 2021).
12.	Mukaka, M.M. Statistics corner: A guide to appropriate use of correlation coefficient in medical research. *Malawi Med. J.* **2012**, *24*, 69–71. [PubMed]
13.	Thomsen, P.T.; Klottrup, A.; Steinmetz, H.; Herskin, M.S. Attitudes of Danish pig farmers towards requirements for hospital pens. *Res. Vet. Sci.* **2016**, *106*, 45–47. [CrossRef] [PubMed]
14.	Main, D.C.J.; Clegg, J.; Spatz, A.; Green, L.E. Repeatability of a lameness scoring system for finishing pigs. *Vet. Rec.* **2000**, *147*, 574–576. [CrossRef] [PubMed]

15. Mullan, S.; Edwards, S.A.; Butterworth, A.; Whay, H.R.; Main, D.C. Inter-observer reliability testing of pig welfare outcome measures proposed for inclusion within farm assurance schemes. *Vet. J.* **2011**, *190*, e100–e109. [CrossRef] [PubMed]
16. Pfeifer, M.; Schmitt, A.O.; Hessel, E.F. Animal Welfare Assessment of Fattening Pigs: A Case Study on Sample Validity. *Animals* **2020**, *10*, 389. [CrossRef]
17. Mullan, S.; Browne, W.J.; Edwards, S.A.; Butterworth, A.; Whay, H.R.; Main, D.C.J. The effect of sampling strategy on the estimated prevalence of welfare outcome measures on finishing pig farms. *Appl. Anim. Behav. Sci.* **2009**, *119*, 39–48. [CrossRef]
18. Hampton, J.O.; MacKenzie, D.I.; Forsyth, D.M. How many to sample? Statistical guidelines for monitoring animal welfare outcomes. *PLoS ONE* **2019**, *14*, e0211417. [CrossRef]
19. Hoffman, A.C.; Moore, D.A.; Wenz, J.R.; Vanegas, J. Comparison of modeled sampling strategies for estimation of dairy herd lameness prevalence and cow-level variables associated with lameness. *J. Dairy Sci.* **2013**, *96*, 5746–5755. [CrossRef]
20. Lauer, S.A.; Kleinman, K.P.; Reich, N.G. The effect of cluster size variability on statistical power in cluster-randomized trials. *PLoS ONE* **2015**, *10*, e0119074. [CrossRef]
21. Grümpel, A.; Krieter, J.; Veit, C.; Dippel, S. Factors influencing the risk for tail lesions in weaner pigs (*Sus scrofa*). *Livest. Sci.* **2018**, *216*, 219–226. [CrossRef]
22. Temple, D.; Courboulay, V.; Velarde, A.; Dalmau, A.; Manteca, X. The welfare of growing pigs in five different production systems in France and Spain: Assessment of health. *Anim. Welf.* **2012**, *21*, 257–271. [CrossRef]
23. Van Staaveren, N.; Vale, A.P.; Manzanilla, E.G.; Teixeira, D.L.; Leonard, F.C.; Hanlon, A.; Boyle, L.A. Relationship between tail lesions and lung health in slaughter pigs. *Prev. Vet. Med.* **2016**, *127*, 21–26. [CrossRef] [PubMed]
24. European Food Safety Authority (EFSA). Scientific report on the risks associated with tail biting in pigs and possible means to reduce the need for tail docking considering the difference housing and husbandry systems. *EFSA J.* **2007**, *611*, 1–98.
25. Valros, A.; Ahlström, S.; Rintala, H.; Häkkinen, T.; Saloniemi, H. The prevalence of tail damage in slaughter pigs in Finland and associations to carcass condemnations. *Acta Agric. Scand. A Anim Sci.* **2004**, *54*, 213–219. [CrossRef]
26. Mullan, S.; Edwards, S.A.; Butterworth, A.; Whay, H.R.; Main, D.C.J. Interdependence of welfare outcome measures and potential confounding factors on finishing pig farms. *Appl. Anim. Behav. Sci.* **2009**, *121*, 25–31. [CrossRef]
27. Munsterhjelm, C.; Heinonen, M.; Valros, A. Application of the Welfare Quality (R) animal welfare assessment system in Finnish pig production, part II: Associations between animal-based and environmental measures of welfare. *Anim. Welf.* **2015**, *24*, 161–172. [CrossRef]
28. Pillman, D.; Nair, N.M.; Schwartz, J.; Pieters, M. Detection of Mycoplasma hyorhinis and Mycoplasma hyosynoviae in oral fluids and correlation with pig lameness scores. *Vet. Microbiol.* **2019**, *239*, 108448. [CrossRef]
29. Bulens, A.; Biesemans, C.; Van Beirendonck, S.; Van Thielen, J.; Driessen, B. The effect of a straw dispenser on behavior and lesions in weanling pigs. *J. Vet. Behav.* **2016**, *12*, 49–53. [CrossRef]
30. Lingling, F.; Bo, Z.; Huizhi, L.; Schinckel, A.P.; Tingting, L.; Qingpo, C.; Yuan, L.; Feilong, X. Teeth clipping, tail docking and toy enrichment affect physiological indicators, behaviour and lesions of weaned pigs after re-location and mixing. *Livest. Sci.* **2018**, *212*, 137–142.
31. AHDB. AHDB Tail Biting WebHAT. Available online: https://webhat.ahdb.org.uk/ (accessed on 27 January 2021).
32. Tremetsberger, L.; Leeb, C.; Winckler, C. Animal health and welfare planning improves udder health and cleanliness but not leg health in Austrian dairy herds. *J. Dairy Sci.* **2015**, *98*, 6801–6811. [CrossRef]
33. Scollo, A.; Gottardo, F.; Contiero, B.; Mazzoni, C.; Leneveu, P.; Edwards, S.A. Benchmarking of pluck lesions at slaughter as a health monitoring tool for pigs slaughtered at 170kg (heavy pigs). *Prev. Vet. Med.* **2017**, *144*, 20–28. [CrossRef] [PubMed]
34. Swanenburg, M.; Gonzales, J.L.; Bouwknegt, M.; Boender, G.J.; Oorburg, D.; Heres, L.; Wisselink, H.J. Large-scale serological screening of slaughter pigs for Toxoplasma gondii infections in The Netherlands during five years (2012–2016): Trends in seroprevalence over years, seasons, regions and farming systems. *Vet. Parasitol.* **2019**, *2*, 100017. [CrossRef] [PubMed]

Article

The Effect of Environmental Enrichment on Laboratory Rare Minnows (*Gobiocypris rarus*): Growth, Physiology, and Behavior

Chunsen Xu [1,2], Miaomiao Hou [1,2], Liangxia Su [1], Ning Qiu [1], Fandong Yu [1,2], Xinhua Zou [1,2], Chunling Wang [1], Jianwei Wang [1,3,*] and Yongfeng He [1,3,*]

[1] Institute of Hydrobiology, Chinese Academy of Sciences, Wuhan 430072, China; 18227588944@163.com (C.X.); houmiao006@163.com (M.H.); suliangxia027@126.com (L.S.); qiuning181@163.com (N.Q.); yufd666@163.com (F.Y.); 18370085282@163.com (X.Z.); chwang@ihb.ac.cn (C.W.)

[2] University of Chinese Academy of Sciences, Beijing 100049, China

[3] National Aquatic Biological Resource Center, Institute of Hydrobiology, Chinese Academy of Sciences, Wuhan 430070, China

* Correspondence: wangjw@ihb.ac.cn (J.W.); yfhe@ihb.ac.cn (Y.H.); Tel.: +86-027-6878-0033 (J.W. & Y.H.)

Simple Summary: Environmental enrichment is an important part of animal welfare. In this study, rare minnow in different rearing conditions underwent comprehensive evaluation regarding growth, anxiety-like behavior, and physiology parameters. Results showed that there were no differences in SGR, anxiety-like behavior, DA, DOPAC, and 5-HIAA levels between control and enriched groups. However, the enriched group had higher cortisol and 5-HT levels. Therefore, researchers should focus on the effect of environmental enrichment regarding the welfare of rare minnow and how it effects the validity of data from laboratory studies.

Abstract: Environmental enrichment is a method to increase environmental heterogeneity, which may reduce stress and improve animal welfare. Previous studies have shown that environmental enrichment can increase the growth rate, decrease aggressive and anxiety-like behaviors, improve learning ability and agility, and reduce cortisol levels in animals. These effects usually differ between species. Unfortunately, habitat enrichment on laboratory fish is poorly studied and seldom adopted in care guidance. Rare minnows (*Gobiocypris rarus*) have been cultured as a native laboratory fish in China in barren banks without environmental enrichment since 1990; they have been widely used in studies on ecotoxicology, environmental science, and other topics. The purpose of this study was to investigate the effect of environment enrichment on the growth, physiological status, and anxiety-like behavior of laboratory rare minnows. We observed and analyzed SGR, cortisol levels, DA, DOPAC, 5-HT and 5-HIAA, and anxiety-like behavior indexes after one month of treatment in barren (control) and enrichment tanks. We found that there were no significant differences in SGR, anxiety-like behavior, DA, DOPAC, or 5-HIAA levels between the two treatments. However, higher cortisol and 5-HT levels were observed in the enrichment tanks. This study suggests that rare minnows might be influenced by their living environment, and future related studies should consider their environmental enrichment.

Keywords: rare minnow; environmental enrichment; growth; physiology; anxiety-like behavior; welfare

Citation: Xu, C.; Hou, M.; Su, L.; Qiu, N.; Yu, F.; Zou, X.; Wang, C.; Wang, J.; He, Y. The Effect of Environmental Enrichment on Laboratory Rare Minnows (*Gobiocypris rarus*): Growth, Physiology, and Behavior. *Animals* **2022**, *12*, 514. https://doi.org/10.3390/ani12040514

Academic Editor: Vera Baumans

Received: 21 January 2022
Accepted: 18 February 2022
Published: 18 February 2022

Publisher's Note: MDPI stays neutral with regard to jurisdictional claims in published maps and institutional affiliations.

1. Introduction

The physical and mental status of animals is often affected by the environment. In fact, a lack of shelter in artificial environments may elicit a variety of stress responses in fish, including physiological, psychological, and behavioral effects [1]. In addition, environment homogeneity may decrease fishes' learning opportunities and learning ability [1]. Enrichment is a method to increase environmental heterogeneity and provide shelter by introducing plants, cobbles, and other physical structures into the rearing environment [1,2].

As a result of simulating the natural habitat, enrichment usually has a positive impact on fish because it provides the fish diverse physical or hydraulic environments and opportunities to choose its habitat. Many studies have shown that environmental enrichment can enhance growth rate [3–7], decrease aggressive and anxiety-like behaviors [8–10], enhance learning ability [11–13] and agility [14], and reduce cortisol (an indicator of stress) levels [8]. However, some studies have identified cases in which enrichment elicited negative or neutral effects [15–21]. The differences in these results are not only related to differences in species or life history stages, but also with the type and level of enrichment [2,22].

Owing to their small size, high fecundity, short generation, rapid sexual maturity, and simple facility, laboratory fish are widely used in the life sciences [23]. The welfare of laboratory fish is becoming a new and urgent topic and, as it is an important part of animal welfare, environmental enrichment of laboratory fish should be given more attention. Unfortunately, environmental enrichment is less adopted in care guidance instructions for laboratory fish [21]. For instance, studies have shown that environmental enrichment has certain benefits, such as reducing zebrafish anxiety-like behavior and increasing proliferating cell nuclear antigen positive nuclei and spatial learning ability [10,15,24], but most laboratories still use barren tanks to raise zebrafish. Enriching laboratory fish not only improves the fish's welfare [22] but may also impact the validity of the data collected [25,26] and feeding management [27].

The rare minnow (*Gobiocypris rarus*) is a small cyprinid fish that is widely used as a native laboratory fish for chemical testing and research on disease, toxicology, behavior, and genetics [28–34]. For its use as a laboratory animal, serial standard drafts have been established on topics including controls on pathogens, heredity, environment, and nutrition. However, environmental enrichment is not considered essential for animal welfare, although previous study showed rare minnow had a significant preference for an enriched environment [35]. The aim of this study was to investigate the effect of environmental enrichment on the growth, anxiety-like behavior, and physiological status of laboratory rare minnows.

2. Material and Methods

2.1. Fish Culturing and Environmental Enrichment

Rare minnows (*Gobiocypris rarus*) were provided by the National Aquatic Biological Resource Center (NABRC). The whole experiment was done in NABRC's laboratory. In order to ensure animal welfare and meet the sampling requirements, 72 adult fish aged 6 months with an average total length of 33.5 ± 2.44 mm and body weight of 0.39 ± 0.05 g were randomly, equally placed into six plastic tanks (length: 40.0 cm, width: 25.0 cm, height: 20.0 cm), exhibiting 12 rare minnows per tank. All individuals were in the same recirculating aquatic housing system equipped with multistage filtration including activated filter stone, filter sponge, and UV sterilization. The tanks were arranged into two treatments: environmental enrichment and control. In the three enriched replicate tanks, the whole bottom was covered with faint yellow gravel, ~1 cm deep and ~0.4 cm in diameter, and plastic plants (three artificial plants that mimic *Potamogeton perfuliatus*, signal leaf length: 2.2 cm, plant length: 16 cm, color: green). Two artificial plants mimicking *Hydrilla verticillata* (signal leaf length: 1.5 cm, plant length: 16 cm, color: green) were added to occupy approximately 40% of the surface area (Figure 1). The other three barren replicate tanks without any enrichment were set as the control group. Test fish were reared in the tanks for one month until they were sampled or used for the behavior experiments.

Figure 1. Photographs of the enrichment group.

During the rearing period, we maintained the water depth at 16 cm and the water flow rate at 700 mL/min. The light/dark cycle was 12:12 h. Water temperature was maintained at 25.6 ± 0.2 °C. The pH was 7.49–7.80. Fish were fed commercial dry pellets (crude protein ≥ 35%, crude fat ≥ 3.0%, crude fiber ≤ 8.0%, crude ash ≤ 15%, moisture ≤ 10%, calcium ≥ 12%, phosphorus ≥ 0.6%, lysine ≥ 1.5%) twice daily at 10:00 a.m. and 4:00 p.m. No injuries or deaths were found during the experiment, and the remaining individuals were transferred back to the NABRC after the experiment ended.

2.2. Sampling and Measuring of Physiological Parameters

At the end of the month, four fish from the same tanks (three replicates enriched and control tanks) were euthanized in ice water [36]. Immediately after the total body length and body weight were measured, each fish was dissected on ice. The body weight and body length of the remaining fish were measured after the anxiety-like behavior test. All the brain tissue was used to determine the levels of neurotransmitters, including Dopamine (DA), Serotonin (5-HT), 3,4-dihydroxyphenylacetic acid (DOPAC), 5-hydroxyindoleacetic acid (5-HIAA), and brain protein. The rest of the body was used to determine the cortisol level. All procedures and the experimental protocols were approved by the Institutional Animal Care and Use Committee of the Institute of Hydrobiology, Chinese Academic of Sciences (IHB/LL/2020025).

Four fish brains from the same tank were mixed into one sample and two bodies from the same tank were mixed into one sample. Samples of brain or body were homogenized in cold PBS (9× weight, pH 7.4) and centrifuged at 3000 rpm for 20 min in a refrigerated centrifuge (4 °C). The supernatants were collected and used for analysis. The cortisol, DA, DOPAC, 5-HT, and 5-HIAA levels were measured using a fish-specific commercial ELISA Assay Kit (Jiangsu Meimian Industrial Co., Ltd., Yancheng, China) according to the manufacturer's guidelines. Brain protein was measured using a commercial BCA Assay Kit (Jiangsu Meimian Industrial Co., Ltd., Yancheng, China) according to the manufacturer's guidelines.

2.3. Anxiety-like Behavioral Studies

2.3.1. Light-Dark Test

The light-dark test was used for the anxiety-like behavior test in fish [10,37]. The light-dark experiment was designed based on a previous study [10], based on scototaxis. Rare minnows prefer a dark environment and are suitable for this light-dark test. The reduction in the number of entries to the light part and the time spent in the light part were considered indicators of anxiety-like behavior [37]. A rectangular glass tank was divided into two equal size parts, which were affixed with a white or black sticker separately (Figure 2). The 12 enriched and 12 control fish (four from each replicate enriched and control tank) were randomly selected and observed one by one, with one enriched and one control fish observed simultaneously. Each fish was only observed once to prevent any impacts from multiple measurements. Test fish were gently put into the dark part of the light-dark tank. The fish were allowed to adapt to the tank for 2 min, then data were collected with a camera (TTQ, China, Resolution: 1920 × 1080) for 8 min. The number of entries to the light part and total time spent in the light part were counted manually by playback video. In order to minimize human interference, we did not set up any partitions during the adaptation period. Fresh system water was added after each trial.

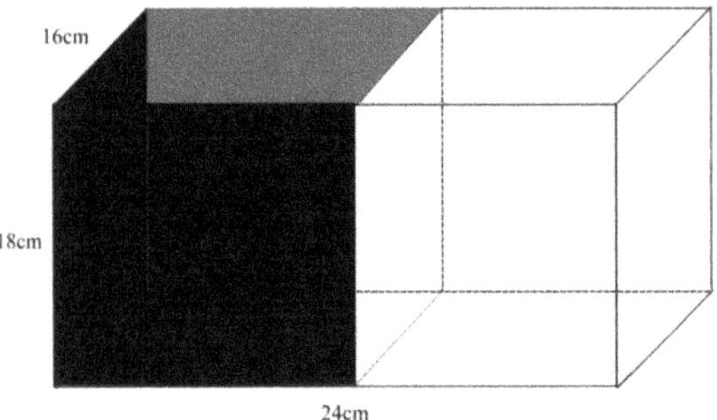

Figure 2. Diagram of the light–dark test tank (24 cm × 16 cm × 18 cm high, water level is 10 cm).

2.3.2. Novel Tank Test

The novel tank test was used for anxiety-like behavior test in fish, and mainly utilized their natural instinct to dive to the bottom of the tank and explore its environment [38,39]. The apparatus consisted of a trapezoidal glass tank (Figure 3). The tank was divided into two parts of the same height: a bottom layer and an upper layer. Except for the front camera (TTQ, China, Resolution:1920 × 1080), all parts of the test tank were covered with white frosted stickers to prevent any external interference. All behavior parameters were counted manually by playback video.

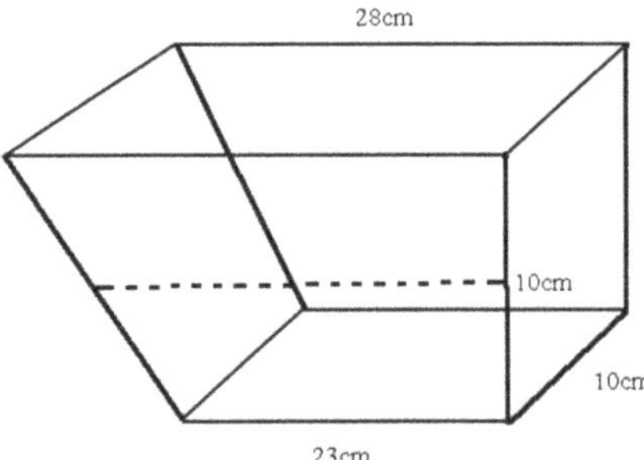

Figure 3. Front view of the novel tank test. Water level is 10 cm.

The 12 enriched and 12 control naïve fish (four from each replicate enriched and control tanks) were randomly selected, and one fish was gently put into the tank for each test at the same time. The fish were allowed to adapt to the tank for 2 min, then data were collected with a camera for 8 min. Each fish was only observed once to prevent the impact of multiple measurements. Different fish were used for the different behavioral tests. The tank was filled with system water, which was replaced at the end of each trial. The behavior parameters included the number of entries to the upper layer and total time spent in the upper layer. The reduction in the number of entries to the upper layer and the time spent in the upper layer are considered indicators of anxiety [39].

2.4. Data Analysis

The specific growth rate (SGR) was used to evaluate growth. SGR% = 100 × (ln (BW$_f$) − ln (BW$_i$))/T, where BW$_i$ represents initial body weight in mg, BW$_f$ represents final body weight in mg; T(d) represents experimental period in days. DA, DOPAC, 5-HT, and 5-HIAA levels were normalized to total brain protein weight (stated as ng/g of brain protein) and cortisol levels were normalized to body weight (stated as ng/g body weight).

The Shapiro–Wilk Test was used to determine whether the data obey normal distribution. If the data met the assumption of normal distribution, then an independent samples *t* test was used to analyze the differences between the two groups. If the data did not meet the assumption of normal distribution, then a Mann-Whitney U test was used to analyze the differentiation between the two groups. There were no data excluded. All statistical analyses were performed using SPSS 25.0. Significance level was set at $p < 0.05$, and extremely significant level set at $p < 0.001$.

3. Results

3.1. Growth

After one month of treatment in different environments, there was no significant difference between the two treatments (t = −0.1, $p = 0.924$) (Figure 4 and Table 1).

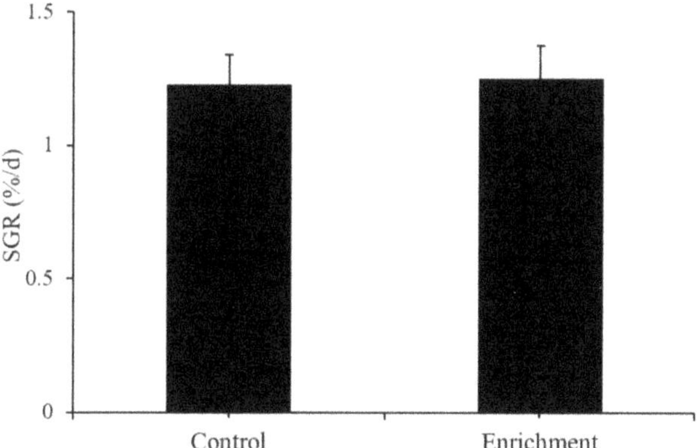

Figure 4. Special growth rate of body weight (SGR, mean ± S.D.) in different treatments (*n* = 3).

Table 1. Growth and physiology parameters of different rearing conditions.

	SGR (%/d)	Cortisol (ng/g)	DA (ng/g Protein)	DOPAC (ng/g Protein)	5-HT (ng/g Protein)	5-HIAA (ng/g Protein)
Control	1.13 ± 0.11	19.46 ± 0.11	7.9 ± 0.57	10.53 ± 0.62	15.03 ± 0.93	11.01 ± 0.62
Enrichment	1.25 ± 0.12	22.07 ± 0.5	7.03 ± 0.3	10.78 ± 0.58	18.06 ± 0.96	10.32 ± 0.64

3.2. Physiological Parameters

Fish in the enriched treatment had significantly higher cortisol levels than fish in the control treatment (t = −12.93; $p < 0.001$) (Figure 5). There were no significant differences in DA (t = 1.923, $p = 0.127$), DOPAC (t = −0.432, $p = 0.694$), or 5-HIAA (t = 1.1, $p = 0.335$) levels between the two treatments, except that fish in the enriched treatment had higher 5-HT levels (t = −3.22; $p = 0.032$) (Figure 6 and Table 1).

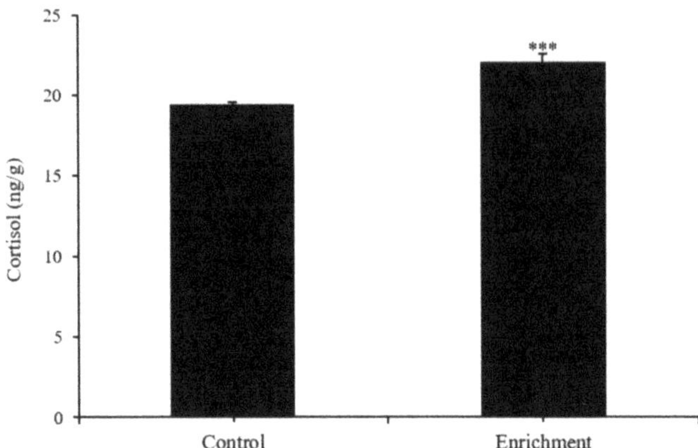

Figure 5. Cortisol level (mean ± S.D.) of the different environmental treatments (*n* = 3, three pooled samples of two fish each from each treatment). "***" represents extremely significant difference (*p* < 0.001).

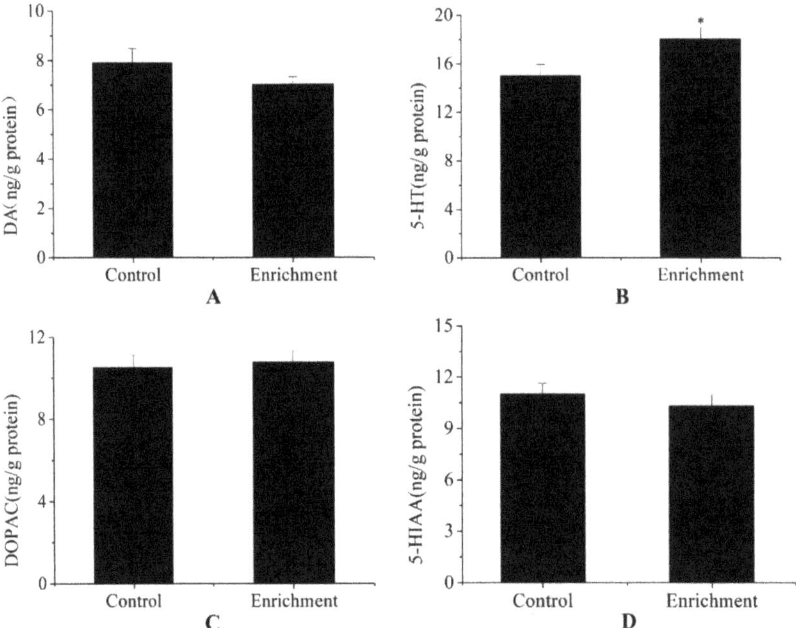

Figure 6. DA, DOPAC, 5-HT, and 5-HIAA levels (mean ± S.D.) of the different environmental treatments (*n* = 3, three pooled samples of four fish each from each treatment). (**A**) DA, (**B**) 5-HT, (**C**) DOPAC, and (**D**) 5-HIAA concentrations of the two groups. "*" represents a significant difference (*p* < 0.05).

3.3. Anxiety-like Behavior

No significant differences were found between the two treatments in terms of the time distribution between the light and dark areas (Mann–Whitney U test, U = 67, N = 12, *p* = 0.799) or number of entries to the light areas. (Mann–Whitney U test, U = 68.5, N = 12, *p* = 0.843) (Figure 7 and Table 2).

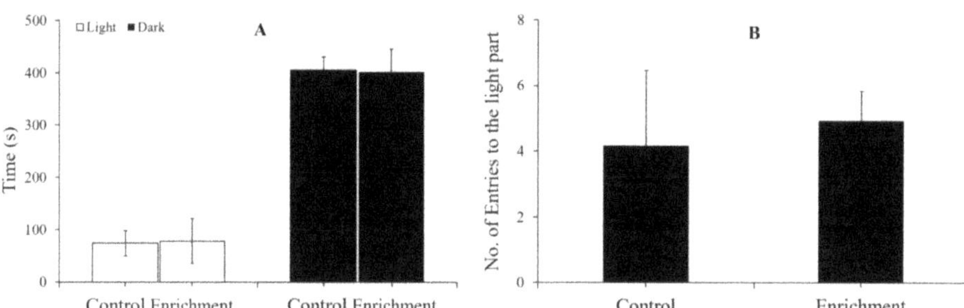

Figure 7. Time distribution and number of entries to the light part (mean ± S.E.) in the light-dark test (*n* = 12). (**A**) Time spent in the light and dark parts. (**B**) Numbers of entries to the light part.

Table 2. Anxiety-like parameters of different rearing condition.

	Light-Dark Test			Novel Tank Test		
	Light Time (s)	**Dark Time (s)**	**No. of Entries**	**Upper Time (s)**	**Bottom Time (s)**	**No. of Entries**
Control	74.17 ± 24.82	405.83 ± 24.82	4.17 ± 2.29	65.17 ± 40.43	414.83 ± 40.43	16.5 ± 10.23
Enrichment	77.92 ± 43.21	402.08 ± 43.21	4.92 ± 0.92	41.08 ± 14.41	438.92 ± 14.41	9.8 ± 2.45

4. Discussion

In the novel tank test, there were no significant differences between the two treatments in the time distribution in the upper and lower layers of the tank (Mann–Whitney U test, U = 90, N = 12, *p* = 0.932) or the number of entries to the upper layer. (Mann–Whitney U test, U = 72, N = 12, *p* = 1) (Figure 8 and Table 2).

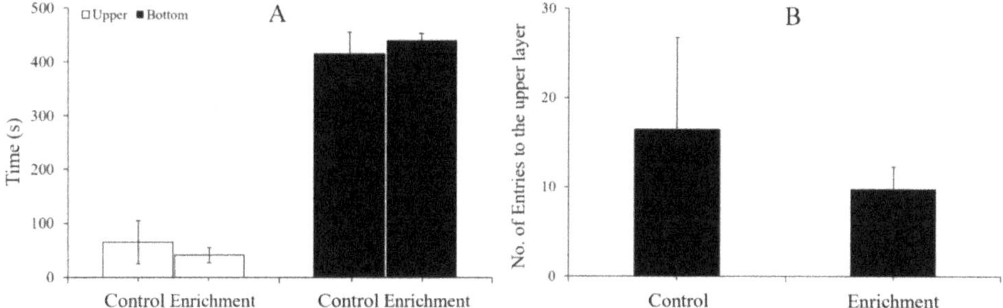

Figure 8. Time distribution and number of entries to the upper layer (mean ± S.E.) in the novel tank test (*n* = 12). (**A**) Time spent in the upper and bottom layer. (**B**) Numbers of entries to the upper layer.

4.1. Effect of Enrichment on Growth

Growth is one of the most important physiological processes in the life history of fish. Any factor that affects animal feeding, digestion, activity, metabolism, and other physiological processes may affect animal growth. Growth is a simple index that reflects the comprehensive performance of many eco-physiological processes. When considering animal welfare, growth and survival are basic "natural demands". If an animal's growth is significantly affected by adverse conditions, then its welfare could deteriorate significantly [40]. Previous studies have shown that the factors affecting the growth of rare minnows include nutrients [41,42], water temperature [43] and other physical and chemical characteristics of their water environment, feeding, and other culturing strategies [43–45]. Therefore, these impact factors are restricted or controlled in the standard drafts of rare minnows to ensure their basic welfare. Although some studies revealed that environmental enrichment may help fish grow [3–7], our results showed no significant difference in SGR between the environmental enrichment and control fish, indicating that the barren tank (control) did not result in significant welfare deficiency. A small fish, the rare minnow seems to be well adapted to aquarium breeding. Rare minnow growth was not affected by complex environments; this may be one of the reasons why the species is used so widely as a laboratory animal.

4.2. Effect of Enrichment on Physiological Parameters

Poor conditions are often accompanied by stress, so stress level is a common metric for judging animal welfare. Stress, which usually refers to a change in one's condition in response to a changing environment, is a physiological process of adapting to and maintaining homeostasis in the internal environment [46]. In the face of long-term stress, fish feeding, growth, immunity, and other factors suffer, and thus so does the fish's welfare [40]. When fish encounter stress, their hypothalamic-pituitary-renal tissue (HPI) axis and neurotransmitter activity are generally considered to be the primary neuroendocrine regulators, and subsequent secondary reactions affect immune function, enzyme activity, blood parameters, etc., and finally lead to changes in behavior, reproductive ability, growth, and survival throughout the organism [47–50]. Brain serotonergic system activity, dopaminergic system activity, and cortisol levels are considered to be accurate and commonly used indicators of stress and welfare in fish [22,51–54]. Previous studies have shown that environmental enrichment may increase, decrease, or have no effect on fish cortisol levels [15,22,54–57]. In the present study, cortisol level in the environmental enrichment treatment did not decrease significantly. On the contrary, it was 22.07 (ng/g), 13% higher than in the control. This result is somewhat unexpected. It may be that the disturbances that resulted from removing artificial aquatic plants from the aquarium before euthanasia resulted in the higher cortisol level in the enrichment fish. On the other hand, hormonal response to stressors in general and chronic stressors in particular can be highly context dependent, and there is currently a lack of consensus as to the direction and magnitude of this response [58]. The reasons for the higher cortisol level in the enriched group need further investigation.

The dopaminergic system is usually correlated with reward and motor functions [59–62]. In the present study, there were no significant differences in DA and DOPAC levels between the control and environmental enrichment, indicating that environmental enrichment has no effect on the above-mentioned functions of rare minnows. This result is similar to that of the black rockfish (*Sebastes schlegelii*) [22].

The serotonergic system is usually involved in the regulation of various physiological functions in animals—such as aggressive behavior, anxiety, and depression—and plays especially important roles in emotion regulation [63–66]. For instance, rodent studies have demonstrated that serotonin depletion results in increased aggression, and serotonin augmentation results in decreased aggression [67]. Environmental enrichment rescued the depressive phenotype of female HD mice, corresponding with increased gene expression of specific 5-HT receptors in the hippocampus and cortex [65]. In the present study, significantly higher 5-HT levels were found in the environmental enrichment group, indicating that the enriched environment benefits the rare minnow.

4.3. Effect of Enrichment on Anxiety-like Behavior

When fish face environmental stress, their intuitive response is reflected in behavior. If the fish are unable to escape the threat, then their behavior may change. Enriched environments create a complex habitat that allows animals to choose suitable hiding places, thus protecting them from fear and stress. Studies have shown that environmental enrichment can reduce anxiety-like behaviors and improve wellbeing [10,68,69]. However, in the present study, there was no significant difference in the time distribution or the numbers of entries between the two treatments in either the light-dark or novel tank tests. In fact, rare minnows, which live in small water bodies gregariously in the wild, are easily domesticated for aquariums. Rare minnows reared in barren tanks rarely display panic behavior. For instance, when breeders step close to the aquarium, fish tend to approach—a sign that they are waiting to be fed—instead of hiding or escaping. Therefore, environmental enrichment might have a limited effect on anxiety-like behavior.

5. Conclusions

The present study showed that environmental enrichment has no significant effect on the growth, anxiety-like behavior, or dopamine level of rare minnow, but it does significantly affect 5-HT level and cortisol levels. In the present study, we used stone bedding and aquatic plants to simulate a natural environment. The bottom of the aquarium box was fully covered with stone bedding, and the coverage of aquatic plants reached 40%; these conditions were too inconvenient to use in daily cultures. In this study, whether the environment is enriched or not probably had a limited effect on fish growth and anxiety-like behaviors. However, higher 5-HT level might be benefiting the rare minnow. Therefore, the barren tank may not meet some of the welfare demands of the laboratory rare minnow. It is important to note that the type and level of environmental enrichment are diverse, and there might be different results. More research is needed to meet the welfare demands of the laboratory rare minnow; at the same time, it is also convenient for daily management.

Author Contributions: J.W. and C.X. conceived and designed the experiments; C.X. and M.H. performed the experiments; C.X., L.S., N.Q. and F.Y. analyzed the data; Y.H., C.W. and X.Z. contributed to the writing of the manuscript. All authors have read and agreed to the published version of the manuscript.

Funding: This research was funded by National Natural Science Foundation of China (No. 31802023).

Institutional Review Board Statement: The animal study protocol was approved by Institutional Animal Care and Use Committee of the Institute of Hydrobiology, Chinese Academic of Sciences (IHB/LL/2020025).

Data Availability Statement: Data and associated calculation tools are available from the corresponding author upon reasonable requestion (wangjw@ihb.ac.cn).

Acknowledgments: This research was funded by National Natural Science Foundation of China (No. 31802023).

Conflicts of Interest: The authors declare no conflict of interest.

References

1. Näslund, J.; Johnsson, J.I. Environmental enrichment for fish in captive environments: Effects of physical structures and substrates. *Fish Fish.* **2016**, *17*, 1–30. [CrossRef]
2. Jones, N.; Webster, M.; Salvanes, A. Physical enrichment research for captive fish: Time to focus on the DETAILS. *J. Fish Biol.* **2021**, *99*, 704–725. [CrossRef] [PubMed]
3. Kientz, J.L.; Barnes, M.E. Structural complexity improves the rearing performance of rainbow trout in circular tanks. *N. Am. J. Aquacult.* **2016**, *78*, 203–207. [CrossRef]
4. Kientz, J.L.; Crank, K.M.; Barnes, M.E. Enrichment of circular tanks with vertically suspended strings of colored balls improves Rainbow Trout rearing performance. *N. Am. J. Aquacult.* **2018**, *80*, 162–167. [CrossRef]
5. Crank, K.M.; Kientz, J.L.; Barnes, M.E. An evaluation of vertically suspended environmental enrichment structures during Rainbow Trout rearing. *N. Am. J. Aquacult.* **2019**, *81*, 94–100. [CrossRef]
6. White, S.C.; Krebs, E.; Huysman, N.; Voorhees, J.M.; Barnes, M.E. Use of suspended plastic conduit arrays during Brown Trout and Rainbow Trout rearing in circular tanks. *N. Am. J. Aquacult.* **2019**, *81*, 101–106. [CrossRef]
7. Voorhees, J.M.; Huysman, N.; Krebs, E.; Barnes, M.E. Influence of water velocity and vertically-suspended structures on rainbow trout rearing performance. *Open J. Anim. Sci.* **2020**, *10*, 152–161. [CrossRef]
8. Zhang, Z.H.; Xu, X.W.; Wang, Y.H.; Zhang, X.M. Effects of environmental enrichment on growth performance, aggressive behavior and stress-induced changes in cortisol release and neurogenesis of black rockfish Sebastes schlegelii. *Aquaculture* **2020**, *528*, 735483. [CrossRef]
9. Batzina, A.; Karakatsouli, N. The presence of substrate as a means of environmental enrichment in intensively reared gilthead seabream Sparus aurata: Growth and behavioral effects. *Aquaculture* **2012**, *370*, 54–60. [CrossRef]
10. Maximino, C.; de Brito, T.M.; Dias, C.A.G.D.; Gouveia, A.; Morato, S. Scototaxis as anxiety-like behavior in fish. *Nat. Protoc.* **2010**, *5*, 209–216. [CrossRef]

11. Salvanes, A.G.V.; Moberg, O.; Ebbesson, L.O.E.; Nilsen, T.O.; Jensen, K.H.; Braithwaite, V.A. Environmental enrichment promotes neural plasticity and cognitive ability in fish. *Proc. R. Soc. B* **2013**, *280*, 20131331. [CrossRef] [PubMed]

12. Roy, T.; Bhat, A. Learning and memory in juvenile zebrafish: What makes the difference—Population or rearing environment? *Ethology* **2016**, *122*, 308–318. [CrossRef]

13. Strand, D.A.; Utne-Palm, A.C.; Jakobsen, P.J.; Braithwaite, V.A.; Jensen, K.H.; Salvanes, A.G.V. Enrichment promotes learning in fish. *Mar. Ecol. Prog. Ser.* **2010**, *412*, 273–282. [CrossRef]

14. Bergendahl, I.A.; Miller, S.; Depasquale, C.; Giralico, L.; Braithwaite, V.A. Becoming a better swimmer: Structural complexity enhances agility in a captive-reared fish. *J. Fish Biol.* **2017**, *90*, 1112–1117. [CrossRef] [PubMed]

15. von Krogh, K.; Sorensen, C.; Nilsson, G.E.; Overli, O. Forebrain cell proliferation, behavior, and physiology of zebrafish, Danio rerio, kept in enriched or barren environments. *Physiol. Behav.* **2010**, *101*, 32–39. [CrossRef] [PubMed]

16. White, S.C.; Barnes, M.E.; Krebs, E.; Huysman, N.; Voorhees, J. Addition of vertical enrichment structures does not improve growth of three salmonid species during hatchery rearing. *J. Mar. Biol. Aquacult.* **2018**, *4*, 48–52.

17. Brydges, N.M.; Braithwaite, V.A. Does environmental enrichment affect the behavior of fish commonly used in laboratory work? *Appl. Anim. Behav. Sci.* **2009**, *118*, 137–143. [CrossRef]

18. Näslund, J.; Rosengren, M.; Johnsson, J.I. Fish density, but not environmental enrichment, affects the size of cerebellum in the brain of juvenile hatchery-reared Atlantic salmon. *Environ. Biol. Fishes* **2019**, *102*, 705–712. [CrossRef]

19. Toli, E.A.; Noreikiene, K.; DeFaveri, J.; Merila, J. Environmental enrichment, sexual dimorphism, and brain size in sticklebacks. *Ecol. Evol.* **2017**, *7*, 1691–1698. [CrossRef]

20. Woodward, M.A.; Winder, L.A.; Watt, P.J. Enrichment increases aggression in zebrafish. *Fishes* **2019**, *4*, 22. [CrossRef]

21. Wilkes, L.; Owen, S.F.; Readman, G.D.; Sloman, K.A.; Wilson, R.W. Does structural enrichment for toxicology studies improve zebrafish welfare? *Appl. Anim. Behav. Sci.* **2012**, *139*, 143–150. [CrossRef]

22. Zhang, Z.H.; Bai, Q.Q.; Xu, X.W.; Guo, H.Y.; Zhang, X.M. Effects of environmental enrichment on the welfare of juvenile black rockfish Sebastes schlegelii: Growth, behavior and physiology. *Aquaculture* **2020**, *518*, 734782. [CrossRef]

23. Lawrence, C. The husbandry of zebrafish (Danio rerio): A review. *Aquaculture* **2007**, *269*, 1–20. [CrossRef]

24. Spence, R.; Magurran, A.E.; Smith, C. Spatial cognition in zebrafish: The role of strain and rearing environment. *Anim. Cogn.* **2011**, *4*, 607–612. [CrossRef] [PubMed]

25. Williams, T.; Readman, G.; Owen, S.F. Key issues concerning environmental enrichment for laboratory-held fish species. *Lab. Anim.* **2009**, *43*, 107–120. [CrossRef] [PubMed]

26. Killen, S.S.; Marras, S.; Metcalfe, N.B.; McKenzie, D.J.; Domenici, P. Environmental stressors alter relationships between physiology and behaviour. *Trends Ecol. Evol.* **2013**, *28*, 651–658. [CrossRef] [PubMed]

27. Pounder, K.C.; Mitchell, J.L.; Thomson, J.S.; Pottinger, T.G.; Buckley, J.; Sneddon, L.U. Does environmental enrichment promote recovery from stress in rainbow trout? *Appl. Anim. Behav. Sci.* **2016**, *176*, 136–142. [CrossRef]

28. Su, L.X.; Xu, C.S.; Cai, L.; Qiu, N.; Hou, M.M.; Wang, J.W. Susceptibility and immune responses after challenge with Flavobacterium columnare and Pseudomonas fluorescens in conventional and specific pathogen-free rare minnow (*Gobiocypris rarus*). *Fish Shellfish Immunol.* **2020**, *98*, 875–886. [CrossRef]

29. Wang, J.J.; Li, P.; Zhang, Y.G.; Peng, Z.G. The complete mitochondrial genome of Chinese rare minnow, *Gobiocypris rarus* (Teleostei: Cypriniformes). *Mitochondrial DNA* **2011**, *22*, 178–180. [CrossRef]

30. Hua, J.H.; Han, J.; Guo, Y.Y.; Zhou, B.S. Endocrine disruption in Chinese rare minnow (*Gobiocypris rarus*) after long-term exposure to low environmental concentrations of progestin megestrol Check for acetate. *Ecotoxicol. Environ. Saf.* **2018**, *163*, 289–297. [CrossRef]

31. Lin, Y.S.; Wang, B.; Wang, N.H.; Ouyang, G.; Cao, H. Transcriptome analysis of rare minnow (*Gobiocypris rarus*) infected by the grass carp reovirus. *Fish Shellfish Immunol.* **2019**, *89*, 337–344. [CrossRef]

32. Hong, X.S.; Zha, J.M. Fish behavior: A promising model for aquatic toxicology research. *Sci. Total Environ.* **2019**, *686*, 311–321. [CrossRef]

33. Zhang, J.L.; Zhang, C.N.; Sun, P.; Shao, X. Tributyltin affects shoaling and anxiety behavior in female rare minnow (*Gobiocypris rarus*). *Aquat. Toxicol.* **2016**, *178*, 80–87. [CrossRef] [PubMed]

34. Qiu, N.; Xu, C.S.; Wang, X.Z.; Hou, M.M.; Xia, Z.J.; Wang, J.W. Chemicals Weaken Shoal Preference in the Rare Minnow Gobiocypris rarus. *Environ. Toxicol. Chem.* **2020**, *39*, 2018–2027. [CrossRef]

35. Qiu, N. Effect of Chemicals Exposure on Behaviors of Rare Minnow. Ph.D. Thesis, University of Chinese Academy of Sciences, Beijing, China, 2020.

36. Song, C.; Liu, B.P.; Zhang, Y.P.; Peng, Z.L.; Wang, J.J.; Collier, A.D.; Echevarria, D.J.; Savelieva, K.V.; Lawrence, R.F. Modeling consequences of prolonged strong unpredictable stress in zebrafish: Complex effects on behavior and physiology. *Prog. Neuro-Psychopharmacol. Biol. Psychiatry* **2017**, *81*, 384–394. [CrossRef] [PubMed]

37. Collymore, C.; Tolwani, R.J.; Rasmussen, S. The Behavioral Effects of Single Housing and Environmental Enrichment on Adult Zebrafish (*Danio rerio*). *J. Am. Assoc. Lab. Anim. Sci.* **2015**, *54*, 280–285. [PubMed]

38. Egan, R.J.; Bergner, C.L.; Hart, P.C.; Cachat, J.M.; Canavello, P.R.; Elegante, M.F.; Elkhayat, S.I.; Bartels, B.K.; Tien, A.K.; Tien, D.H.; et al. Understanding behavioral and physiological phenotypes of stress and anxiety in zebrafish. *Behav. Brain Res.* **2009**, *205*, 38–44. [CrossRef] [PubMed]

39. Tamilselvan, P.; Sloman, K.A. Developmental social experience of parents affects behaviour of offspring in zebrafish. *Anim. Behav.* **2017**, *133*, 153–160. [CrossRef]

40. Huntingford, F.A.; Adams, C.; Braithwaite, V.A.; Kadri, S.; Pottinger, T.G.; Sandoe, P.; Turnbull, J.F. Current issues in fish welfare. *J. Fish Biol.* **2006**, *68*, 332–372. [CrossRef]

41. Wu, B.L.; Luo, S.; Xie, S.Q.; Wang, J.W. Growth of rare minnows (*Gobiocypris rarus*) fed different amounts of dietary protein and lipids. *Lab Anim.* **2016**, *45*, 105–111. [CrossRef] [PubMed]

42. Wu, B.L.; Xiong, X.Q.; Xie, S.Q.; Wang, J.W. Dietary lipid and gross energy affect protein utilization in the rare minnow *Gobiocypris rarus*. *Chin. J. Oceanol. Limnol.* **2016**, *34*, 740–748. [CrossRef]

43. Wu, B.L.; Luo, S.; Wang, J.W. Effects of temperature and feeding frequency on ingestion and growth for rare minnow. *Physiol. Behav.* **2015**, *140*, 197–202. [CrossRef] [PubMed]

44. Su, L.X.; Luo, S.; Qiu, N.; Xu, C.S.; Hou, M.M.; Xiong, X.Q.; Wang, J.W. Comparative allometric growth of rare minnow (*Gobiocypris rarus*) in two culture environments. *Hydrobiologia* **2020**, *847*, 2083–2095. [CrossRef]

45. Luo, S.; Wu, B.L.; Xiong, X.Q.; Wang, J.W. Effects of Total Hardness and Calcium: Magnesium Ratio of Water during Early Stages of Rare Minnows (*Gobiocypris rarus*). *Comp. Med.* **2016**, *66*, 181–187. [PubMed]

46. Selye, H. Stress and the general adaptation syndrome. *Br. Med. J.* **1950**, *1*, 1383–1392. [CrossRef] [PubMed]

47. Barton, B.A.; Iwama, G.K. Physiological changes in fish from stress in aquaculture with emphasis on the response and effects of corticosteroid. *Annual. Rev. Fish Dis.* **1991**, *1*, 3–26. [CrossRef]

48. Oliveira, R.; Galhardo, L. Psychological stress and welfare in fish. *Annu. Rev. Biomed. Sci.* **2009**, *11*, 1–20. [CrossRef]

49. Mommsen, T.P.; Vijayan, M.M.; Moon, T.W. Cortisol in teleosts: Dynamics, mechanisms of action, and metabolic regulation. *Rev. Fish Biol. Fisher.* **1999**, *9*, 211–268. [CrossRef]

50. Sørensen, C.; Johansen, I.B.; Øverli, Ø. Neural plasticity and stress coping in teleost fishes. *Gen. Comp. Endocrinol.* **2013**, *181*, 25–34. [CrossRef] [PubMed]

51. Batzina, A.; Dalla, C.; Papadopoulou-Daifoti, Z.; Karakatsouli, N. Effects of environmental enrichment on growth, aggressive behaviour and brain monoamines of gilthead seabream Sparus aurata reared under different social conditions. *Comp. Biochem. Physiol. Part A Mol. Integr. Physiol.* **2014**, *169*, 25–32. [CrossRef] [PubMed]

52. Batzina, A.; Dalla, C.; Tsopelakos, A.; Papadopoulou-Daifoti, Z.; Karakatsouli, N. Environmental enrichment induces changes in brain monoamine levels in gilthead seabream Sparus aurata. *Physiol. Behav.* **2014**, *130*, 85–90. [CrossRef] [PubMed]

53. Batzina, A.; Kalogiannis, D.; Dalla, C.; Papadopoulou-Daifoti, Z.; Chadio, S.; Karakatsouli, N. Blue substrate modifies the time course of stress response in gilthead seabream Sparus aurata. *Aquaculture* **2014**, *420*, 247–253. [CrossRef]

54. Rosengren, M.; Kvingedal, E.; Naslund, J.; Johnsson, J.I.; Sundell, K. Born to be wild: Effects of rearing density and environmental enrichment on stress, welfare, and smolt migration in hatchery-reared Atlantic salmon. *Can. J. Fish. Aquat. Sci.* **2016**, *74*, 396–405. [CrossRef]

55. Cogliati, K.M.; Herron, C.L.; Noakes, D.L.G.; Schreck, C.B. Reduced stress response in juvenile Chinook Salmon reared with structure. *Aquaculture* **2019**, *504*, 96–101. [CrossRef]

56. Naslund, J.; Rosengren, M.; Del Villar, D.; Gansel, L.; Norrgard, J.R.; Persson, L.; Winkowski, J.J.; Kvingedal, E. Hatchery tank enrichment affects cortisol levels and shelter-seeking in Atlantic salmon (*Salmo salar*). *Can. J. Fish. Aquat. Sci.* **2013**, *70*, 585–590. [CrossRef]

57. Keck, V.A.; Edgerton, D.S.; Hajizadeh, S.; Swift, L.L.; Dupont, W.D.; Lawrence, C.; Boyd, K.L. Effects of habitat complexity on pair-housed zebrafish. *J. Am. Assoc. Lab. Anim. Sci.* **2015**, *54*, 378–383.

58. Dickens, M.J.; Romero, L.M.A. Consensus endocrine profile for chronically stressed wild animals does not exist. *Gen. Comp. Endocrinol.* **2013**, *191*, 177–189. [CrossRef] [PubMed]

59. Koob, G.F. Hedonic valence, dopamine and motivation. *Mol. Psychiatry* **1996**, *1*, 186–189. [PubMed]

60. Phillips, A.G.; Vacca, G.; Ahn, S.A. Top-down perspective on dopamine, motivation and memory. *Pharmacol. Biochem. Behav.* **2008**, *90*, 236–249. [CrossRef] [PubMed]

61. Beninger, R.J. The role of dopamine in locomotor activity and learning. *Brain Res. Rev.* **1983**, *6*, 173–196. [CrossRef]

62. Salamone, J.D. Complex motor and sensorimotor functions of striatal and accumbens dopamine: Involvement in instrumental behavior processes. *Psychopharmacology* **1992**, *107*, 160–174. [CrossRef] [PubMed]

63. Saudou, F.; Amara, D.A.; Dierich, A.; Lemeur, M.; Ramboz, S.; Segu, L.; Buhot, M.C.; Hen, R. Enhanced aggressive-behavior in mice lacking 5-HT1B receptor. *Science* **1994**, *265*, 1875–1878. [CrossRef] [PubMed]

64. Mann, J.J. Role of the serotonergic system in the pathogenesis of major depression and suicidal behavior. *Neuropsychopharmacology* **1999**, *21*, 99–105. [CrossRef]

65. Pang, T.Y.C.; Du, X.; Zajac, M.S.; Howard, M.L.; Hannan, A.J. Altered serotonin receptor expression is associated with depression-related behavior in the R6/1 transgenic mouse model of Huntington's disease. *Hum. Mol. Genet.* **2009**, *18*, 753–766. [CrossRef] [PubMed]

66. Paterson, N.E.; Markou, A. Animal models and treatments for addiction and depression co-morbidity. *Neurotoxic. Res.* **2007**, *11*, 1–32. [CrossRef]
67. Mann, J.J. Violence and aggression. In *Psychopharmacology: The Fourth Generation of Progress*; Bloom, F.E., Kupfer, D.J., Eds.; Raven Press: New York, NY, USA, 1995; pp. 1919–1928.
68. Benaroya-Milshtein, N.; Hollander, N.; Apter, A.; Kukulansky, T.; Raz, N.; Wilf, A.; Yaniv, I.; Pick, C.G. Environmental enrichment in mice decreases anxiety, attenuates stress responses and enhances natural killer cell activity. *Eur. J. Neurosci.* **2004**, *20*, 1341–1347. [CrossRef]
69. Gortz, N.; Lewejohann, L.; Tomm, M.; Ambree, O.; Keyvani, K.; Paulus, W.; Sachser, N. Effects of environmental enrichment on exploration, anxiety, and memory in female TgCRND8 Alzheimer mice. *Behav. Brain Res.* **2008**, *191*, 43–48. [CrossRef]

Article

Computer Vision for Detection of Body Posture and Behavior of Red Foxes

Anne K. Schütz [1,*], E. Tobias Krause [2], Mareike Fischer [3], Thomas Müller [4], Conrad M. Freuling [5], Franz J. Conraths [1], Timo Homeier-Bachmann [1,†] and Hartmut H. K. Lentz [1,†]

[1] Friedrich-Loeffler-Institut (FLI), Federal Research Institute for Animal Health, Institute of Epidemiology, Südufer 10, 17493 Greifswald-Insel Riems, Germany; Franz.Conraths@fli.de (F.J.C.); Timo.Homeier@fli.de (T.H.-B.); Hartmut.Lentz@fli.de (H.H.K.L.)

[2] Friedrich-Loeffler-Institut, Federal Research Institute for Animal Health, Institute of Animal Welfare and Animal Husbandry, Dörnbergstr. 25/27, 29223 Celle, Germany; Tobias.Krause@fli.de

[3] Institute of Mathematics and Computer Science, University of Greifswald, Walther-Rathenau-Straße 47, 17487 Greifswald, Germany; mareike.fischer@uni-greifswald.de

[4] Friedrich-Loeffler-Institut, Federal Research Institute for Animal Health, Institute of Molecular Virology and Cell Biology, Südufer 10, 17493 Greifswald-Insel Riems, Germany; Thomas.Mueller@fli.de

[5] Friedrich-Loeffler-Institut, Federal Research Institute for Animal Health, Südufer 10, 17493 Greifswald-Insel Riems, Germany; Conrad.Freuling@fli.de

* Correspondence: Anne.Schuetz@fli.de

† Coequal senior authors.

Simple Summary: Monitoring animal behavior provides an indicator of their health and welfare. For this purpose, video surveillance is an important method to get an unbiased insight into behavior, as animals often show different behavior in the presence of humans. However, manual analysis of video data is costly and time-consuming. For this reason, we present a method for automated analysis using computer vision—a method for teaching the computer to see like a human. In this study, we use computer vision to detect red foxes and their body posture (lying, sitting, or standing). With this data we are able to monitor the animals, determine their activity, and identify their behavior.

Abstract: The behavior of animals is related to their health and welfare status. The latter plays a particular role in animal experiments, where continuous monitoring is essential for animal welfare. In this study, we focus on red foxes in an experimental setting and study their behavior. Although animal behavior is a complex concept, it can be described as a combination of body posture and activity. To measure body posture and activity, video monitoring can be used as a non-invasive and cost-efficient tool. While it is possible to analyze the video data resulting from the experiment manually, this method is time consuming and costly. We therefore use computer vision to detect and track the animals over several days. The detector is based on a neural network architecture. It is trained to detect red foxes and their body postures, i.e., 'lying', 'sitting', and 'standing'. The trained algorithm has a mean average precision of 99.91%. The combination of activity and posture results in nearly continuous monitoring of animal behavior. Furthermore, the detector is suitable for real-time evaluation. In conclusion, evaluating the behavior of foxes in an experimental setting using computer vision is a powerful tool for cost-efficient real-time monitoring.

Keywords: YOLOv4; computer vision; animal monitoring; animal behavior; animal activity; animal welfare; body posture

Citation: Schütz, A.K.; Krause, E.T.; Fischer, M.; Müller, T.; Freuling, C.M.; Conraths, F.J.; Homeier-Bachmann, T.; Lentz, H.H.K. Computer Vision for Detection of Body Posture and Behavior of Red Foxes. *Animals* **2022**, *12*, 233. https://doi.org/10.3390/ani12030233

Academic Editors: Luigi Faucitano and Maria José Hötzel

Received: 5 December 2021
Accepted: 14 January 2022
Published: 19 January 2022

Publisher's Note: MDPI stays neutral with regard to jurisdictional claims in published maps and institutional affiliations.

1. Introduction

Animal welfare is becoming increasingly important in animal experimentation and husbandry, and is often defined by the Five Freedoms concept [1], the Five Domains concept, or the complementary use of both [2,3]. Thus, due to its multidimensional character, it is influenced by many factors [4]. For its detection based on animal movements, different approaches have been applied in recent years. Sénèque et al. [5] found that altered welfare in horses was associated with their body postures. Besides active motion, sleeping behavior can be used as one indicator of animal welfare [6]. Furthermore, the monitoring of animal activities has been used to draw conclusions on animal welfare [7]. In particular, changes of behavioral activity can provide information about the welfare or disease status of an animal [8–10]. Observation, measurement, and evaluation of animal behavior provide important indicators for the determination of animal welfare [11]. Furthermore, animal behavior is often associated with certain postures and locomotion [12]. Fureix et al. [13] used horse postures (analyzed by geometric morphometrics) to characterize behavioral categories. Animal behavior is an important manifestation, which can be linked to the health and welfare status of animals. Therefore, conclusions about animal health and welfare can be drawn by tracking animals and detecting their posture and activity.

For monitoring, unbiased video observation of animal behavior is particularly suitable, since the presence of humans may change or influence the behavior of animals [9,14–16]. Moreover, manual observation has a number of disadvantages. It is time consuming, costly, and unsuitable for larger animal populations [17]. More importantly, the accuracy of the manual observation depends on the observer's experience and judgment [17], which may lead to observer bias [18].

An automated system may help to overcome at least some of the above-mentioned limitations and may be used to detect behavioral changes, e.g., unusual behavior caused by disease. Moreover, such a system could automatically alert laboratory personnel in case of unusual behavior and thus support animal welfare, health, and animal management [12]. Furthermore, automatic monitoring methods can be useful for continuous monitoring and detection of events [19] such as an animal caretaker in the room.

The application of sensors such as accelerometers or RFID (radio-frequency identification) chips is one way of measuring activity or locomotion. Data obtained using accelerometers, usually attached to the animals as leg sensors, have been used to classify cattle activities (e.g., walking, lying, standing) [20]. Kaler et al. [21] used data from accelerometer and gyroscope sensors to detect lameness in sheep. Furthermore, Diosdado et al. [22] used the accelerometer data for the classification of behaviors in dairy cows, i.e., feeding, standing, and lying. For the application of the RFID technology, chips can be either implanted or attached to collars, ear tags, or anklets [23]. There have been attempts for implementing automatic monitoring of animals using RFID technology [24,25]. However, the use of RFID is always associated with certain invasiveness of the animals, since the sensors (RFID chips) must be attached to the animals or implanted. In particular, the implantation may cause stress for the animals [26] and can thus affect their subsequent behavior. The application of video surveillance has the advantage that no sensors or tags need to be placed in or on animals, which avoids stress.

Computer vision based techniques are objective, contact-less, and low-cost methods. They offer an important basic tool in the study and monitoring [27] of animal behavior. Nasirahmadi et al. [28] tested three detector methods (region-based fully convolutional network, single shot multibox detector, and faster regions with convolutional neural network) for the detection of lying and not lying pigs. Yang et al. [29] applied an automatic recognition framework based on a fully convolutional network to detect the daily behavior of sows by analyzing motion (to detect movement, medium active, or inactive behavior) and image analyses (to detect drinking, feeding, and nursing). Another effective real-time object detection algorithm is YOLO (You Only Look Once) [30]. Wang et al. [31] used YOLOv3 (You Only Look Once version 3) to detect six different categories of behavior (e.g., drink, feed, stand) in group-housed hens in a self-breeding system and used the frequency of mating as a welfare indicator for the group. YOLOv4 was used by Jiang et al. [32] to detect goats and to recognize the behavior of group-housed goats (eating and drinking by position of the goats and active/inactive based on the movements). In addition, YOLOv4 has been applied to detect and monitor the motion and activity levels of red foxes [33], but without identification of specific behaviors.

The automated deduction of animal behavior patterns by a combined evaluation of the detection of body postures and the determination of activity levels is missing so far. For this purpose, we demonstrate the application of deep learning for the detection, tracking, activity, and behavior determination of red foxes (*Vulpes vulpes*) during an experimental study, which was being conducted to measure the long-time immunogenicity and efficacy of an oral rabies vaccine in these animals [34]. To this end, we used video surveillance data of the foxes generated as part of this experimental study. In particular, a convolutional neural network (CNN) (YOLOv4) is trained for red fox body posture detection, e.g., 'standing', 'lying', and 'sitting'. The results of this detection can be used to infer different activity levels [33], which can then be used in combined evaluation with the detected posture to determine different behavior. The presented technique can be applied to detect posture patterns of animals, including behavior determination.

2. Materials and Methods

2.1. Experimental Setup

The animals considered here are red foxes (*Vulpes vulpes*) of the fur color variant 'silver fox' [35]. The experimental study with 23 foxes was conducted over 450 days at the Friedrich-Loeffler-Institut (FLI), Greifswald, Insel Riems, Germany [34]. During that time the foxes were separately kept in cages, sized 3.18 m × 1.4 m × 1.75 m (length × width × height), equipped with a platform, 0.92 m × 1.4 m (length × width) at height of 0.8 m above the bottom of the cage. In addition, each cage was equipped with a hut. In the study, a novel oral vaccination regime against rabies was tested. This required animal blood to be sampled in regular intervals. To this end, the foxes were anesthetized [34]. All invasive procedures, i.e., blood sampling, infection, transponder application, euthanasia, were conducted under anesthesia by applying 0.5–1 mL of Zoletil (Virbac, France; 1 mL contains 50 mg Tiletamin, and 50 mg Zolazepam). It takes 7 to 10 min before you can start with the manipulations. Zoletil does not require an antidote, because it is a short time anesthesia. The increased activity after anesthesia may be reflective of the recovery phase. It is known that in canids, the half-life of tiletamine is 1.2 h and that of zolazepam is 1 h. During the recovery phase of canids from anesthesia with Zoletil®, tiletamine therefore has an even longer effect than zolazepam and excitation states and increased movement can occur [36]. Housing and maintenance of the animals complied with national and European legislation and guidelines for the veterinary care of laboratory animals [37]. Food and water were provided according to the species-specific requirements and were also individually adjusted. The foxes received enrichment (such as ball, kong) at irregular intervals to improve animal welfare, which were not considered in our study. The animals were visually inspected daily and the cages cleaned on a regular basis. The availability of external monitoring, including recording by video cameras, was one of the requirements for approval. Therefore, every fox was monitored via two cameras (ABUS IR HD TVIP61500, ABUS, Wetter, Germany). Each cage was equipped with two cameras hung on the opposite narrow sides of the cage. Together, 33 TB of video data was recorded discontinuously (due to memory requirements), on 73 different days. We have used image data that was extracted from video data of all foxes for training the algorithm. For the exemplary application shown below, we restrict ourselves to analyzing video data of 1 single red fox on 6 different days, spanning a period of 11 days. Table 1 shows the evaluated video data and the times of events such as anesthesia or cage cleaning.

On day 3, anesthesia is administered at 09:56, the red fox shows normal behavior until this time, then lies for about 1 h followed by a wake-up phase (approximately 11:00 to 11:30) in which the red fox repeatedly crawls a bit, then repeatedly starts to stand up, walks a few steps and sat down or lies down again. Then there follows a phase in which the red fox very excitedly walks around a lot in the cage; this lasts until about 14:15. From then on the sitting and lying times between the walking become longer and longer until the red fox lies down at 16:01 and sleeps until 17:08.

Table 1. Subset of days evaluated and of external events (anesthesia, cleaning). Video data does not exist for days not listed in the period and cannot be evaluated.

Day	Time	Event
1	11:30 to 11:55	Animal caretaker is in the room and cleans the cage
3	09:56	Anesthesia
	13:59 to 14:09	Animal caretaker is in the room and cleans the cage
4	09:18 to 09:31	Animal caretaker is in the room and cleans the cage
5	08:22 to 08:34	Animal caretaker is in the room and cleans the cage
7	11:30 to 11:55	Animal caretaker is in the room and cleans the cage
11	11:18 to 11:34	Animal caretaker is in the room and cleans the cage

2.2. Ethical Approval

The animal experiment was authorized by the local authority in Mecklenburg-Western Pomerania (Landesamt für Landwirtschaft, Lebensmittelsicherheit und Fischerei Mecklenburg-Vorpommern, # FLI-7221.3-1-087/16) and conducted in accordance with national and European legislation and guidelines for the veterinary care of laboratory animals [37].

2.3. Image and Video Data

The resolution of the video data was 1280 pixels (horizontal) × 720 pixels (vertical) and had a frame rate of 15 frames per second (fps). An image set was created by extracting single frames from the video data. For the image set, videos of all 23 red foxes with different body postures were used. No adjacent frames were extracted and frames with different illumination conditions (night and day) were used. The image set consisted of 8913 images. The red fox on each image of the image set was manually labeled using the software LabelImg [38], and attributed to one of the three classes 'sitting', 'standing', and 'lying'. Lying was defined as lying prone, lying on the side, lying curled up, or lying on the back. Standing was defined as a quadruped position. Sitting was regarded as an intermediate posture, i.e., the two hind legs not righted and the two forelegs righted on the floor.

The image set was split into a training (80%—7129 frames) and a test set (20%—1784 frames) (Table 2), maintaining the relation of the three labeled classes. The training set was used to train the YOLOv4 object detection algorithm and the test set to evaluate the trained model for the red fox posture detection.

Table 2. Splitting of the image set (8913 frames) into a training and a test set in the ratio 80% to 20%. Shown also is the number of frames per subset for each behavioral postures 'lying', 'sitting', and 'standing'.

	Total	Lying	Sitting	Standing
Training set	7129 frames	775 frames	3688 frames	2370 frames
Test set	1784 frames	194 frames	922 frames	593 frames

2.4. Environment Configuration

The processor used in the study was an Intel Xeon E5-2667 v4 with 3.20 GHz, 377 GB Ram, and NVIDIA K80 with 2 GPUs and 24 GB video RAM. The operating system was CentOS 7. The algorithm was developed by using a Jupyter notebook [39] and Python 3.6.8 [40].

2.5. Automatic Evaluation: Red Fox Detection and Posture Classification

The detection of red fox postures was implemented by using the deep learning algorithm YOLOv4. The algorithm YOLO is a one-stage object detection algorithm for real-time object detection based on CNN [30,41]. In this study, we used version 4 of YOLO, which consists of a 'head', a 'neck', and a 'backbone' [42]. The head is used to implement the object detection [42], and is the YOLOv3 algorithm [41]. The neck is used to collect feature maps from different stages and is based on a path aggregation network (PAN) and SPP spatial pyramid pooling (SPP) [42]. The backbone is used for training and feature extraction and it is a CSPDarknet53, which is an open source neural network framework [41,42]. YOLOv4 is a state-of-the-art detector, which is faster and more accurate than other available detectors [42]. The training of the red fox posture detector based on YOLOv4 was performed with the parameters from Table 3 and the training set (Table 2).

In this study, we have limited the detectable classes for postures exclusively to three different postures, i.e., 'sitting', 'lying', and 'standing'. A detailed description of the training procedure is given in the Appendix A.

Table 3. Parameters of the YOLOv4 red fox detection model.

Parameter	Value
Input size	416×416
Classes	3
Maxbatches	6000
Filters	24
Steps	4800, 5400
Learning rate	0.001
Batch size	64

2.6. Evaluation of Model Performance

In order to verify the performance of the model, the following indicators were determined:

(i)   Mean average precision (mAP) (Equation (2));
(ii)   Precision (Equation (3));
(iii)   Recall (Equation (4));
(iv)   Detection speed.

Intersection over Union (IoU) was used to determine if a detection was true positive or false positive (see Equation (1)). When IoU ≥ 0.5, it was true positive and false positive if IoU < 0.5. If an image was labeled and the model does not detect anything, it was false negative. The following values were computed:

$$IoU = \frac{area(BB_p) \cap area(BB_{gt})}{area(BB_p) \cup area(BB_{gt})}, \tag{1}$$

$$mAP = \frac{\sum_{c=1}^{C} AP(c)}{C}, \tag{2}$$

$$precision = \frac{TP}{TP + FP}, \tag{3}$$

$$recall = \frac{TP}{TP + FN}, \tag{4}$$

where IoU: Intersection over Union, BB_p: predicted bounding box (BB) from the model, BB_{gt}: ground-truth bounding box (e.g., manually labeled), AP: average precision, C: number of classes, TP: number of true positives, FP: number of false positives, FN: number of false negatives. The AP is a measure for the detection accuracy of the model (for more details see [33,43]). Here we used the 11-point interpolated AP [43]:

$$AP = \frac{1}{11} \sum_{r \in \{0,0.1,...,1\}} p_{interp}(r) \tag{5}$$

with

$$p_{interp}(r) = \max_{\hat{r}:\hat{r} \geq r} p(\hat{r}), \tag{6}$$

where $p(\hat{r})$ is the precision at recall $\hat{r}$. Equation (6) results in a smoothening of the precision-recall curve.

For each image f, the trained object detection algorithm returns whether a red fox is on the image, and if so, the class ('lying', 'sitting, 'standing'), the confidence of the detection, and the bounding box center position (x_f, y_f), and width and height—standardized between 0 and 1, respectively.

2.7. Automatic Evaluation: Activity Analysis

The bounding boxes were used to measure the activity level of the red fox [33]. To this end, the movement of the center of the bounding box between two consecutive frames was determined. This movement of the center of the BB corresponds to the distance covered by the red fox between two consecutive frames. $m_{f,f+1} = \sqrt{(x_{f+1} - x_f)^2 + (y_{f+1} - y_f)^2}$ with (x_f, y_f): coordinates of the center of the BB in frame f and (x_{f+1}, y_{f+1}): coordinates of the center of the BB in frame $f + 1$. For the calculation of the mean vector norm, a sliding window of 5 s with a step size of 1 s was used. $\bar{m}_t = 1/F \sum_{f=1}^{F-1} m_{f,f+1}$ with mean vector norm $\bar{m}_t$, time period t, and number of frames F in t. The maximum of the mean vector norm for different periods and kinds of movement behavior can be used to determine thresholds for different activity levels. Three activity levels were considered:

(i) Highly active: considerable movement of the bounding box (BB), i.e., the localization of the red fox changes, e.g., walking or running;
(ii) Active: slight movement of the BB, i.e., the localization of the red fox does not change, but there is some movement inside the BB, e.g., rotation or minimal movements, such as scratching or stretching;
(iii) Inactive: no movement of the BB, i.e., the red fox does not move, e.g., lying, sitting, or standing still.

To distinguish between the activity levels, the thresholds from Schütz et al. [33] were used (mean norm $\geq$ 0.0094: highly active; 0.0054 to 0.0094: active; 0 to 0.0054: inactive).

For all three postures (lying, sitting, standing), highly active, active, and inactive variants are possible.

2.8. Automatic Evaluation: Behavior Analysis

For behavioral analysis the activity levels were considered along with body posture to draw conclusions about the behavior of the respective red fox. Therefore, a behavior was assigned to each possible combination of a body posture and activity level (see below).

2.9. Workflow for Automated Video Evaluation

For the evaluation of the videos we used the trained red fox detector as described above. The video analysis was implemented as follows:

1. Frame extraction (5 frames per second);
2. Red fox posture detection on each frame;
3. Activity analysis using the BB values for the activity level determination;
4. Behavior analysis using the posture and activity level for the behavior classification.

The evaluation workflow is depicted in Figure 1.

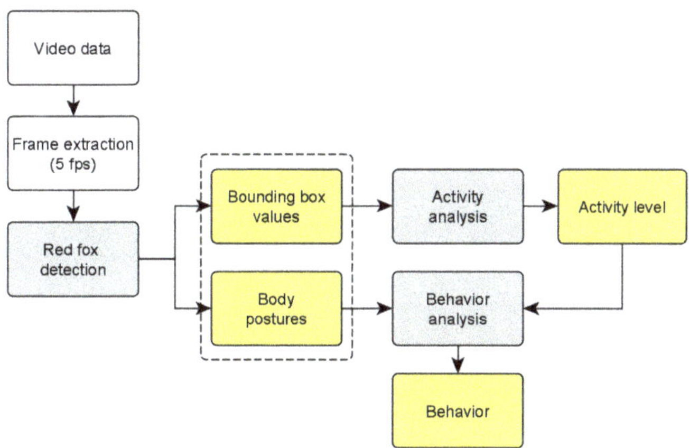

Figure 1. Workflow: video data evaluation. White: raw video material. Gray: detection and analysis. Yellow: results.

For the joint evaluation of both cameras of the same animal, each video was evaluated separately for the same period. It is possible that the evaluation of both cameras differs, e.g., if the fox is not completely visible for camera 1 because the legs are in the blind spot, the fox is classified as lying for camera 1, while camera 2 (seeing the complete fox) classifies it as standing. In case of mismatches between the cameras, the larger vector norm and the body posture with the higher confidence was chosen. To avoid single individual false classifications for the body posture, a sliding window of 5 s duration was used to select the most frequently occurring body posture.

3. Results

3.1. Model Training and Evaluation

The performance of the model was evaluated by comparing the label of manually labeled images (test set) with the results of the automated detection of the trained model. The results for the three classes are shown in Table 4, and the overall performance is shown in Table 5.

Table 4. Performance of the trained model for the classes 'lying', 'sitting', and 'standing'.

Class	Precision	Recall	Average IoU	AP
Sitting	99.24%	99.57%	0.93	99.97%
Lying	96.95%	98.45%	0.90	99.79%
Standing	98.16%	99.16%	0.90	99.96%

Table 5. Overall performance of the trained model.

Precision	Recall	Average IoU	mAP	Detection Speed
98.61%	95.12%	0.91	99.91%	73.31 ms

The *precision* and *recall* of the model are 98.61% and 95.12%, respectively; the *average IoU* is 0.91, the *mAP* is 99.91% and the detection speed reaches 73.31 ms per frame.

Figure 2 shows examples of red fox detection for each of the three postures 'sitting' (Figure 2a–d), 'standing' (Figure 2e–h), and 'lying' (Figure 2i–l) for day and night scenes.

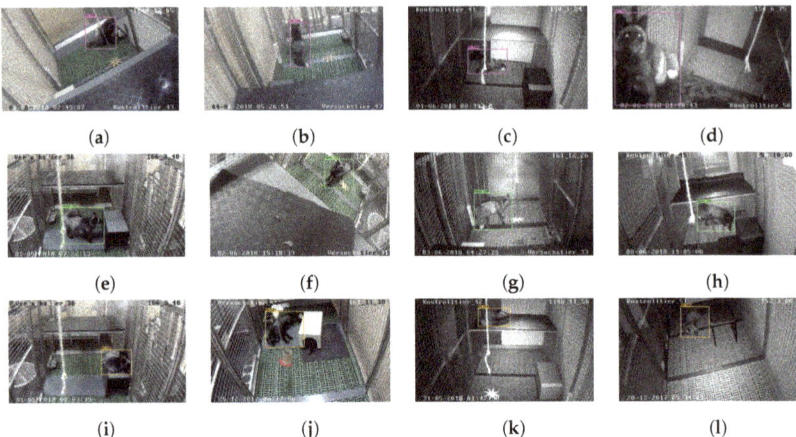

Figure 2. Detection of red foxes in single frames. (**a–d**) detection examples of sitting foxes; (**e–h**), standing foxes; and (**i–l**) lying foxes. Two day scene images (left) and night scene images (right), respectively.

3.1.1. Activity Detection

Figure 3 provides overviews of the activity level for the evaluated days during the 11-day observation period.

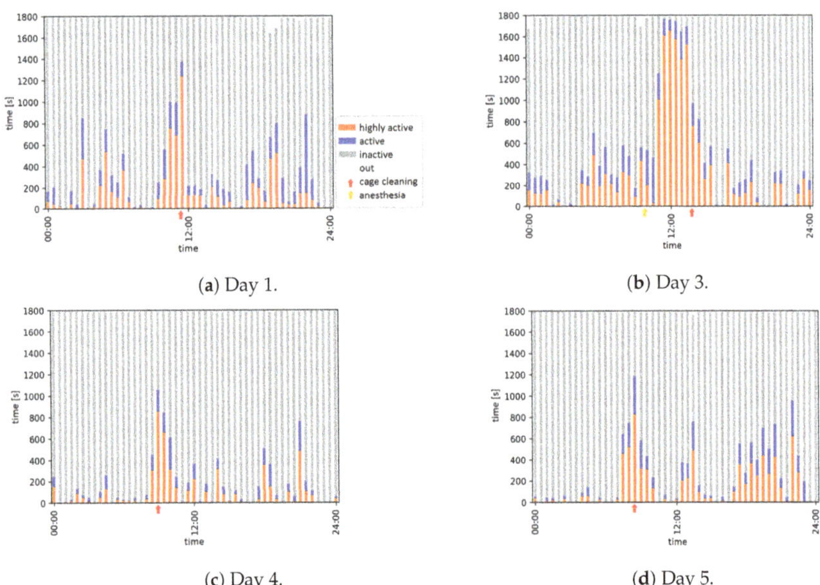

(**a**) Day 1.

(**b**) Day 3.

(**c**) Day 4.

(**d**) Day 5.

Figure 3. *Cont.*

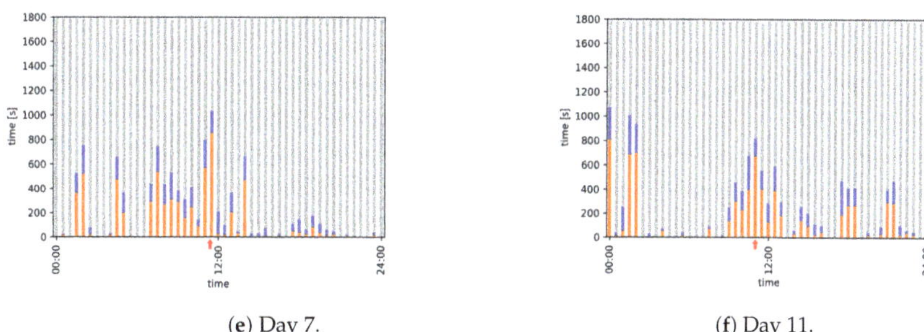

(**e**) Day 7.   (**f**) Day 11.

Figure 3. Activity overview of six days from a time period of 11 days.

For example, day 1 represents the normal activity of the red fox. More active and inactive phases alternate throughout the day. The period during which the animal caretakers entered the room and cleaned the cages related with increased red fox activity. This can be seen particularly well on day 1, from 11:30 to 12:00, where the largest amount of 'highly active' coincides with the presence of an animal caretaker in the room (11:30 to 11:55). The anesthetic phase, i.e., anesthesia and subsequent recovery phase, on day 3 relates with the active phase on that day. Activity increased suddenly with the beginning of the wake-up phase at 11:00. From 11:00 to 14:00, the activity level was almost only 'highly active' and related with walking around in the cage after the wake-up phase. From 14:00 to 16:00, the proportion of 'highly active' decreased and the red fox showed the activity level 'inactive' for a longer time. This is followed by a complete hour (16:00 to 17:00) in the activity level 'inactive', which coincided with the sleeping phase from 16:01 to 17:09.

3.1.2. Posture Detection

The trained model was used to classify the postures, 'sitting', 'lying', and 'standing'. Figure 4 shows the posture overviews for all recorded days during the observation period of 11 days.

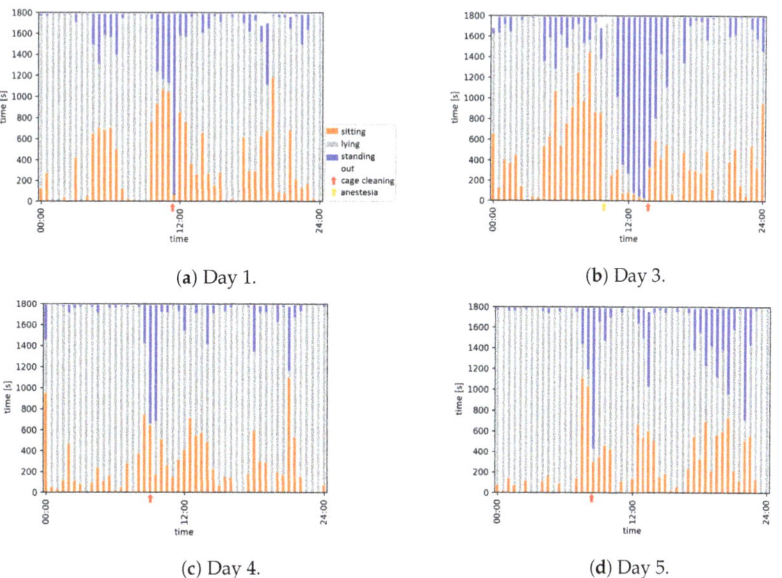

(**a**) Day 1.   (**b**) Day 3.

(**c**) Day 4.   (**d**) Day 5.

Figure 4. *Cont.*

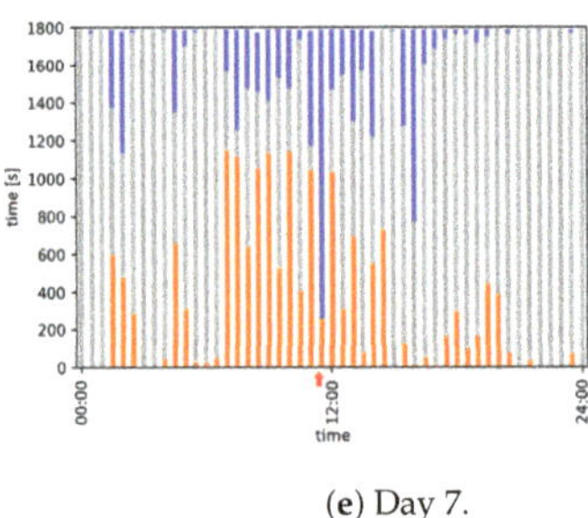 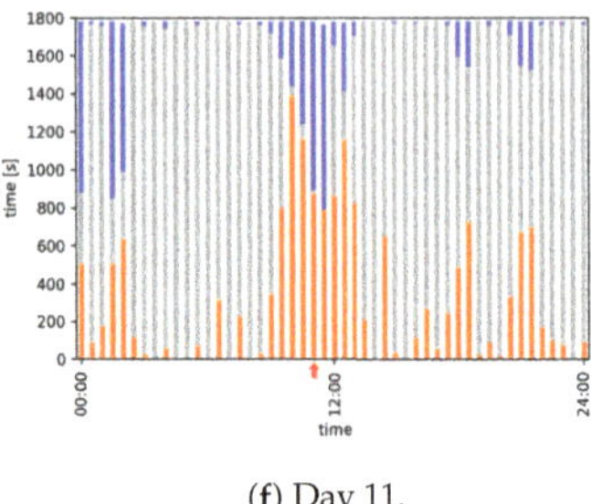

(**e**) Day 7.   (**f**) Day 11.

Figure 4. Posture overview of 6 days from a period of 11 days.

In all plot, periods when an animal caretaker was in the room were very well recognizable, e.g., from 11:30 to 11:55 on day 1, (Figure 4a), and also on day 7 (Figure 4e). The red fox was only standing or sitting, but hardly lying. The anesthesia on day 3 with the subsequent wake-up phase (described in Section 3.1.1) is also reflected in the classified postures. On day 11, there was a phase in which the red fox almost exclusively showed the postures 'sitting' and 'standing', which lasted from about 10:00 to 12:00. The classified postures agree with the manually analyzed video data for this period, i.e., the red fox mostly sat or stood from 10:00 onwards, it laid down at 12:20. Another remarkable half hour was on day 5 from 9:00 p.m. to 9:30 p.m. Here the body posture detection shows that the red fox only used the body postures 'sitting' and 'standing'. The classified postures are consistent with the video.

The classified postures were used to determine the numbers of body posture changes. The numbers of changes for day 1, summed up per half hour, are shown in Figure 5.

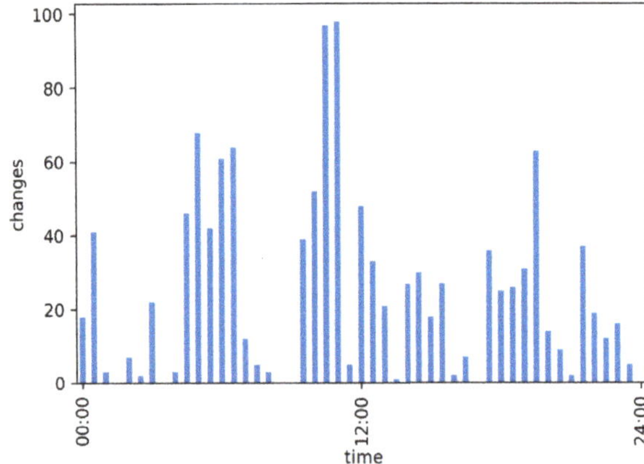

Figure 5. Number of changes on day 1 in half hour steps.

Randomly three arbitrarily selected half-hour videos were manually analyzed and compared with the determined number of changes. For this purpose, one video with many changes (11:00 to 11:30—98 changes), one with few changes (07:30 to 08:00—5 changes), and one with no changes (17:00 to 17:30—zero changes) were randomly selected from all the videos of day 1. In all three videos, the automatic detected numbers of changes matched the numbers seen during manual inspection of the video and, in addition, the time points of the changes matched. Figure 6 shows a timeline for each of the periods with posture changes.

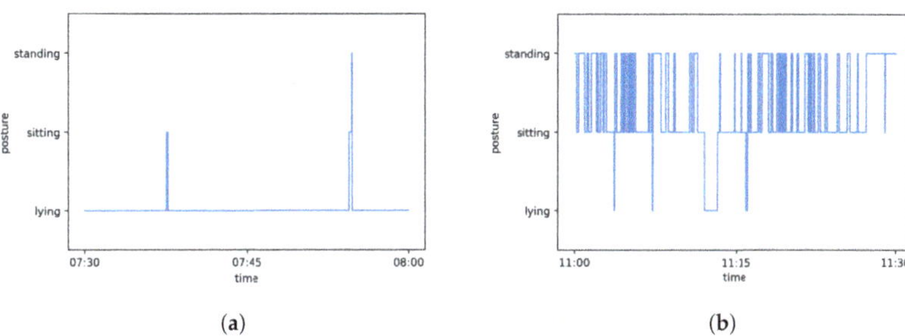

(a)   (b)

Figure 6. Postures of the red fox over 30 min during two different periods. Each vertical line represents a posture change. (**a**) Day 1—07:30 to 08:00; (**b**) Day 1—11:00 to 11:30.

The five changes during the period lasting from 07:30 to 08:00 period are illustrated in Figure 6a, where each vertical line represents a posture change. A second period (Figure 6b) depicts all 98 posture changes that were recorded between 11:00 and 11:30. Most of the changes occurred between the postures 'sitting' and 'standing'.

3.1.3. Behavior Detection

To determine the behavior, all combinations of posture ('lying', 'sitting', and 'standing') and activity level ('inactive', 'active', and 'highly active') were considered and a behavior assigned to each combination, i.e., 'highly active standing', 'active standing', 'standing still', 'active lying', 'lying motionless', 'active sitting', and 'sitting still'. This is shown as a decision tree in Figure 7.

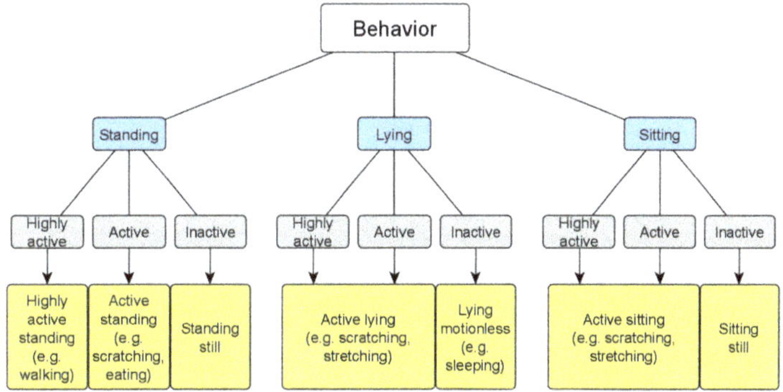

Figure 7. Decision tree: Determination of behavior (yellow) through a combined analysis of body posture (blue) and activity level (gray). For example, if a fox is classified as 'active' and 'standing', it shows the behavior 'active standing'. This refers to a standing fox with an activity level of 'active', and this could be scratching, eating, etc.

With the decision tree (Figure 7) it is possible to generate a continuous behavior overview with a resolution of one value per second (see Section 2.7). For example, if a red fox is classified as 'standing' in an image, the combined view with the determined activity level provides the behavior of the fox. If the activity level is 'highly active', the behavior of 'highly active standing' can be inferred. If the corresponding activity level is 'active', this means that the fox is 'standing active'. The last possible combination is the activity level 'inactive', which means that the behavior of the fox is 'standing still'.

The determined behavior of the red fox is exemplarily represented by two timelines over half an hour each in Figure 8.

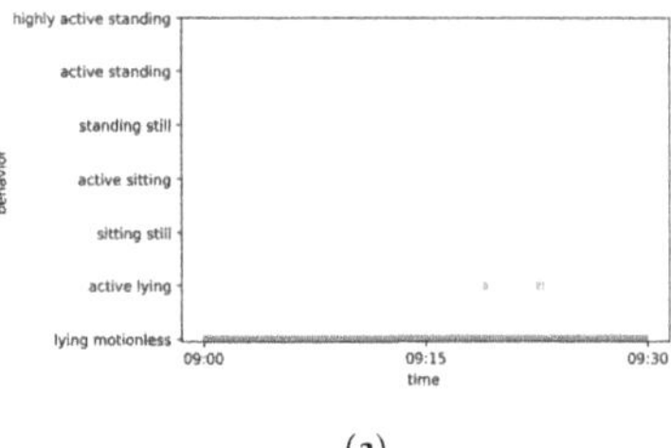
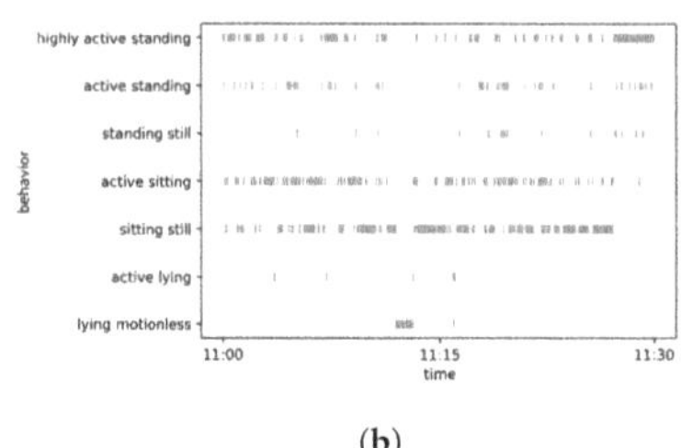

(a)   (b)

Figure 8. Overview of the exhibited behavior of the red fox in timelines of 30 min for two different periods. (**a**) Day 1—09:00 to 09:30; (**b**) Day 1—11:00 to 11:30. Note: The fox does not show two behaviors at the same time at any point, although it may appear like that in the figure due to the high density of data points.

The first half hour (Figure 8a) shows the behavior of a sleeping fox. It can be seen that the animal was 'lying motionless' for almost the entire observation period. Only at two short moments was there movement during sleep. The behavior coincides with the real behavior as determined by visual inspection of the video data, where it became evident that the fox slept curled up and changed its lying position only twice. In the second timeline (Figure 8b), one can see that the red fox showed all the defined behaviors. Here, visual inspection of the video showed that the red fox changed its location very often, and changed frequently between the body postures 'sitting', 'standing', and 'lying'.

4. Discussion

We were able to train a classifier with a very high precision for posture detection, and we can determine animal behavior in detail in a high temporal resolution. Our results suggest that the presented method may be useful for monitoring animals, especially posture, behavior, activity, and additional posture change monitoring. In our setting, the results show that the model achieves high performance for the posture detection. Moreover, the detection speed is sufficient for a real-time detection with 5 fps, which was used in this study. Furthermore, the animals do not need to be equipped with a sensor or collar, as was the case in other studies (e.g., [20,44]). This non-invasive approach is a major advantage of computer vision [45].

Activity determination works in the same way with the new method for the posture detection in our previous study on red fox detection [33]. The activity overview can also be used to detect events like the anesthesia or the presence of an animal caretaker in the room. In particular, increased activity in the presence of humans illustrates their influence on animal behavior and the advantages of video observation in getting an—at least in this respect—unbiased insight. Changes in movement patterns can provide information relevant to animal welfare and health [44,46], especially over long terms these data can be achieved automatically by the proposed approach.

In this study, three major postures of red foxes were investigated; 'lying', 'sitting', and 'standing'. The posture 'lying' consists of all the shown lying postures of the red fox, including 'lying curled up', 'lying on side', 'lying on back', and 'lying prone'. The red fox also shows other body postures like 'standing on the hind legs' or 'standing front end lowered'. For these rare postures, it is difficult to create a large enough image set for training, e.g., only 25 images could be labeled with the posture 'standing front end lowered' from the total extracted image set. The classified postures are also suitable for determining periods with special events. Furthermore, in contrast to pure activity monitoring, periods with an unremarkable activity, but a noticeable ratio of the postures were detected. For example, in the period from 09:00 to 09:30 on day 4, the red fox shows almost only 'sitting' and 'standing' postures (Figure 4c), but no particularly striking values regarding of activity levels. Thus, the different duration of the postures could be used as an indicator of animal behavior. As an example, a reduction of lying time is a typical behavior change before a calving in dairy cows [47]. Similarly, the determination of the number of changes is an indicator for behavioral changes. In dairy cows, an increase of posture changes is another typical behavior change before calving [48].

Analysis of the behavior as a result of the combined evaluation of activity and posture yields a detailed overview of the behaviors under consideration, i.e., 'highly active standing', 'active standing', 'standing still', 'active lying', 'lying motionless', 'active sitting', and 'sitting still'. The use of the developed decision tree provided good insights into detailed behavior. Even though we did not provide a detailed paired analysis here, given the high accuracy of the used algorithm (as discussed in Section 3.1), the method is capable of reproducing the manual results. However, different kinds of behavior that show the same combination of activity level and posture cannot be distinguished. For example, the behaviors 'lying motionless—sleeping' and 'lying motionless—awake' have the same combination and the decision tree provides 'lying motionless'. However, sleeping and resting phases could be used to refine these categories, which may be relevant as indicators of animal welfare [6]. There is a further limitation in assessing welfare based on behavior, e.g., 'highly active standing' could be due to the fox's interaction with enrichment (indicating good welfare) and also pacing (indicating poor welfare). Thus, the number of determinable behaviors is a limitation [45]. This limitation can be minimized, for example, by training a model with more postures.

Our proposed method is based on single snapshots—as opposed to estimating behavior from long-time windows. It shows a high precision and accuracy here (correct detection of behavior compared to manual observation). In order to estimate the accuracy of long-term evaluations, a long-term study using a large amount of video data with this method might be part of future research.

We showed that computer vision systems are useful to generate activity, posture, and behavior overviews and the number of posture changes.

5. Conclusions

The aim of this study was to investigate the potential of computer vision for the detection of red foxes and the classification of their body postures to use the results for activity, posture, and behavior analysis. On the basis of the YOLOv4 algorithm, a model for the detection of red foxes was realized that classifies the body postures 'lying', 'sitting', and 'standing'. Along with the subsequent analysis of the posture changes, activity levels, and behavior of the red foxes, this study provides a method for seamless monitoring.

The generated daily, weekly, or monthly overviews (activity, posture, behavior) can be used to monitor animal activities, posture, and behavior, and may thus help to establish indicators for animal welfare.

Author Contributions: Conceptualization, T.H.-B. and H.H.K.L.; methodology, A.K.S. and H.H.K.L.; software, A.K.S.; validation, A.K.S. and E.T.K.; formal analysis, A.K.S.; investigation, A.K.S.; resources, T.M., C.M.F. and F.J.C.; data curation, A.K.S.; writing—original draft preparation, A.K.S.; writing— review and editing, A.K.S., E.T.K., M.F., T.M., C.M.F., F.J.C., T.H.-B. and H.H.K.L.; visualization, A.K.S.; supervision, T.H.-B. and H.H.K.L.; project administration, T.H.-B. All authors have read and agreed to the published version of the manuscript.

Funding: This study was supported by an intramural collaborative research grant on lyssaviruses at the Friedrich-Loeffler-Institut (A.S.).

Institutional Review Board Statement: The data used in this paper is a subset of data generated in an experimental study with red foxes conducted over 450 days [34]. Animal housing and maintenance were in accordance with national and European legislation and followed the guidelines for the veterinary care of laboratory animals. The experimental study was approved by the local authority in Mecklenburg-Western Pomerania (Landesamt für Landwirtschaft, Lebensmittelsicherheit und Fischerei Mecklenburg-Vorpommern, # FLI-7221.3-1-087/16) and was conducted at the Friedrich-Loeffler-Institut (FLI), Greifswald, Insel Riems, Germany.

Informed Consent Statement: Not applicable.

Data Availability Statement: The data sets during and/or analyzed during the current study are available from the corresponding author on reasonable request.

Acknowledgments: We thank the animal attendants of the Friedrich-Loeffler-Institut for taking care of the red foxes and providing further help in organizing the experiments. We also thank Björn-Uwe Kaeß and Oliver Sanders for providing technical support with the videos and the motion detectors. Thanks to Verena Schöler for the discussions about the foxes. We thank Anna Lea Bühring, Kathrin Körner and Ute Lenert for screening the video material. We are grateful to Thomas C. Mettenleiter and Lars Schrader for their generous support and stimulating discussions.

Conflicts of Interest: The authors declare no conflict of interest. The funders had no role in the design of the study; in the collection, analyses, or interpretation of data; in the writing of the manuscript, or in the decision to publish the results.

Abbreviations

The following abbreviations are used in this manuscript:

AP	average precision
BB	bounding box
CNN	convolutional neural networks
FLI	Friedrich-Loeffler-Institut
FP	false positive
FN	false negative
IoU	intersection over union
mAP	mean average precision
TP	true positive
YOLO	you only look once

Appendix A

In this section, we provide a detailed description on how to train the algorithm. The procedure is very similar to our previous work [33]. The training was implemented following the instructions of the YOLOv4 Github page [49] with pre-trained weights (i.e., a pre-trained weights-file `yolov4.conv.137`, downloaded from GitHub [49]) and the parameters shown in Table 3.

The algorithm was trained as follows:

1. Download and extract YOLOv4 from GitHub [49].
2. Copy the content of `cfg/yolov4-custom.cfg` to the new created file `yolo-obj.cfg` and change the following lines:

    line 3: `batch=64`
    line 4: `subdivisions=1`
    line 8: `width=416`
    line 9: `height=416`
    line 20: `max_batches=6000` ($classes * 2000$)
    line 22: `steps=4800,5400` (80 and 90 % of maxbatches)
    lines 603, 689, 776: `filters=24` (($classes + 5$) $* 3$)
    lines 610, 696, 783: `classes=3`

3. Create a file `obj.names` with the name of each object in separate lines:

    ```
    sit
    lie
    stand
    ```

4. Label each image of the image set, such that for each image there exists a `.txt` file with the following values for every labeled object:

    `<object-class> <BB x_center> <BB y_center> <BB width> <BB hight>`

with `<object-class>` an integer between 0 and number of classes−1, and `<BB x_center>`, `<BB y_center>`, `<BB width>` and `<BB hight>` are float values between $(0, 1]$, relative to the image height and width. Thus, the directory with the images contains a `.txt` file for each image with the same name.

Create the files `train.txt` and `test.txt`. Split the image set into a training and test set and save the file names of the images, with respect to the full path relative to the directory darknet, in the respective file (one file name per line).

5. Create a file `obj.data` containing the number of classes and paths to `train.txt`, `obj.names`, and the backup folder:

```
classes = 3
train = data/train.txt
names = data/obj.names
backup = backup/
```

6. For starting the training, run the code:

```
./darknet detector train obj.data yolo-obj.cfg yolov4.conv.137
```

The training can take several hours. During training the trained weights are saved in the backup/ directory, `yolo-obj_xxxx.weights` every 1000 iterations and `yolo-obj_last.weights` every 100 iterations. After training the final weight, `yolo-obj_final.weights` is also stored there.

7. Evaluate the results for trained weights:

```
./darknet detector map obj.data yolo-obj.cfg
backup/yolo-obj_final.weights
```

8. Using the trained detector:

```
./darknet detector test obj.data yolo-obj.cfg
backup/yolo-obj_final.weights
```

References

1. Farm Animal Welfare Council (FAWC). *Second Report on Priorities for Research and Development in Farm Animal Welfare*; DEFRA: London, UK, 1993.
2. Mellor, D.J. Updating animal welfare thinking: Moving beyond the "Five Freedoms" towards "a Life Worth Living". *Animals* **2016**, *6*, 21. [CrossRef] [PubMed]
3. Webster, J. Animal welfare: Freedoms, dominions and "a life worth living". *Animals* **2016**, *6*, 35. [CrossRef] [PubMed]
4. Mason, G.J.; Mendl, M. Why Is There No Simple Way of Measuring Animal Welfare? *Anim. Welf.* **1993**, *2*, 301–319.
5. Sénèque, E.; Lesimple, C.; Morisset, S.; Hausberger, M. Could posture reflect welfare state? A study using geometric morphometrics in riding school horses. *PLoS ONE* **2019**, *14*, e0211852. [CrossRef]
6. Owczarczak-Garstecka, S.C.; Burman, O.H.P. Can Sleep and Resting Behaviours Be Used as Indicators of Welfare in Shelter Dogs (Canis lupus familiaris)? *PLoS ONE* **2016**, *11*, e0163620.
7. Buller, H.; Blokhuis, H.; Lokhorst, K.; Silberberg, M.; Veissier, I. Animal Welfare Management in a Digital World. *Animals* **2020**, *10*, 1779. [CrossRef]
8. Müller, R.; Schrader, L. A new method to measure behavioural activity levels in dairy cows. *Appl. Anim. Behav. Sci.* **2003**, *83*, 247–258. [CrossRef]
9. White, B.J.; Coetzee, J.F.; Renter, D.G.; Babcock, A.H.; Thomson, D.U.; Andresen, D. Evaluation of two-dimensional accelerometers to monitor behavior of beef calves after castration. *Am. J. Vet. Res.* **2008**, *69*, 1005–1012. [CrossRef]
10. Dutta, R.; Smith, D.; Rawnsley, R.; Bishop-Hurley, G.; Hills, J.; Timms, G.; Henry, D. Dynamic cattle behavioural classification using supervised ensemble classifiers. *Comput. Electron. Agric.* **2015**, *111*, 18–28. [CrossRef]
11. Dawkins, M.S. Behaviour as a tool in the assessment of animal welfare. *Zoology* **2003**, *106*, 383–387. [CrossRef]
12. Matthews, S.G.; Miller, A.L.; Clapp, J.; Plötz, T.; Kyriazakis, I. Early detection of health and welfare compromises through automated detection of behavioural changes in pigs. *Vet. J.* **2016**, *217*, 43–51. [CrossRef] [PubMed]
13. Fureix, C.; Hausberger, M.; Seneque, E.; Morisset, S.; Baylac, M.; Cornette, R.; Biquand, V.; Deleporte, P. Geometric morphometrics as a tool for improving the comparative study of behavioural postures. *Naturwissenschaften* **2011**, *98*, 583–592. [CrossRef] [PubMed]
14. Hosey, G. Hediger revisited: How do zoo animals see us? *J. Appl. Anim. Welf. Sci. JAAWS* **2013**, *16*, 338–359. [CrossRef]

15. Hemsworth, P.H.; Barnett, J.L.; Coleman, G.J. The Human-Animal Relationship in Agriculture and its Consequences for the Animal. *Anim. Welf.* **1993**, *2*, 33–51.

16. Sorge, R.E.; Martin, L.J.; Isbester, K.A.; Sotocinal, S.G.; Rosen, S.; Tuttle, A.H.; Wieskopf, J.S.; Acland, E.L.; Dokova, A.; Kadoura, B.; et al. Olfactory exposure to males, including men, causes stress and related analgesia in rodents. *Nat. Methods* **2014**, *11*, 629–632. [CrossRef] [PubMed]

17. Frost, A.R.; Parsons, D.J.; Stacey, K.F.; Robertson, A.P.; Welch, S.K.; Filmer, D.; Fothergill, A. Progress towards the development of an integrated management system for broiler chicken production. *Comput. Electron. Agric.* **2003**, *39*, 227–240. [CrossRef]

18. Mendl, M.; Burman, O.H.; Parker, R.M.; Paul, E.S. Cognitive bias as an indicator of animal emotion and welfare: Emerging evidence and underlying mechanisms. *Appl. Anim. Behav. Sci.* **2009**, *118*, 161–181. [CrossRef]

19. Oh, J.; Fitch, W.T. CATOS (Computer Aided Training/Observing System): Automating animal observation and training. *Behav. Res. Methods* **2017**, *49*, 13–23. [CrossRef]

20. Robert, B.; White, B.J.; Renter, D.G.; Larson, R.L. Evaluation of three-dimensional accelerometers to monitor and classify behavior patterns in cattle. *Comput. Electron. Agric.* **2009**, *67*, 80–84. [CrossRef]

21. Kaler, J.; Mitsch, J.; Vázquez-Diosdado, J.A.; Bollard, N.; Dottorini, T.; Ellis, K.A. Automated detection of lameness in sheep using machine learning approaches: Novel insights into behavioural differences among lame and non-lame sheep. *R. Soc. Open Sci.* **2020**, *7*, 190824. [CrossRef]

22. Diosdado, J.A.V.; Barker, Z.E.; Hodges, H.R.; Amory, J.R.; Croft, D.P.; Bell, N.J.; Codling, E.A. Classification of behaviour in housed dairy cows using an accelerometer-based activity monitoring system. *Anim. Biotelem.* **2015**, *3*, 15. [CrossRef]

23. Naguib, M.; Krause, E.T. *Methoden der Verhaltensbiologie*, 2nd ed.; Springer Spektrum: Berlin/Heidelberg, Germany, 2020.

24. Iserbyt, A.; Griffioen, M.; Borremans, B.; Eens, M.; Müller, W. How to quantify animal activity from radio-frequency identification (RFID) recordings. *Ecol. Evol.* **2018**, *8*, 10166–10174. [CrossRef] [PubMed]

25. Will, M.K.; Büttner, K.; Kaufholz, T.; Müller-Graf, C.; Selhorst, T.; Krieter, J. Accuracy of a real-time location system in static positions under practical conditions: Prospects to track group-housed sows. *Comput. Electron. Agric.* **2017**, *142*, 473–484. [CrossRef]

26. Nasirahmadi, A.; Edwards, S.A.; Sturm, B. Implementation of machine vision for detecting behaviour of cattle and pigs. *Livest. Sci.* **2017**, *202*, 25–38. [CrossRef]

27. Valletta, J.J.; Torney, C.; Kings, M.; Thornton, A.; Madden, J. Applications of machine learning in animal behaviour studies. *Anim. Behav.* **2017**, *124*, 203–220. [CrossRef]

28. Nasirahmadi, A.; Sturm, B.; Edwards, S.; Jeppsson, K.H.; Olsson, A.C.; Müller, S.; Hensel, O. Deep Learning and Machine Vision Approaches for Posture Detection of Individual Pigs. *Sensors* **2019**, *19*, 3738. [CrossRef] [PubMed]

29. Yang, A.; Huang, H.; Zheng, B.; Li, S.; Gan, H.; Chen, C.; Yang, X.; Xue, Y. An automatic recognition framework for sow daily behaviours based on motion and image analyses. *Biosyst. Eng.* **2020**, *192*, 56–71. [CrossRef]

30. Redmon, J.; Divvala, S.; Girshick, R.; Farhadi, A. *You Only Look Once: Unified, Real-Time Object Detection*; Cornell University: Ithaca, NY, USA, 2016. Available online: https://arxiv.org/abs/1506.02640 (accessed on 10 January 2022) .

31. Wang, J.; Wang, N.; Li, L.; Ren, Z. Real-time behavior detection and judgment of egg breeders based on YOLO v3. *Neural Comput. Appl.* **2020**, *32*, 5471–5481. [CrossRef]

32. Jiang, M.; Rao, Y.; Zhang, J.; Shen, Y. Automatic behavior recognition of group-housed goats using deep learning. *Comput. Electron. Agric.* **2020**, *177*, 105706. [CrossRef]

33. Schütz, A.K.; Schöler, V.; Krause, E.T.; Fischer, M.; Müller, T.; Freuling, C.M.; Conraths, F.J.; Stanke, M.; Homeier-Bachmann, T.; Lentz, H.H.K. Application of YOLOv4 for Detection and Motion Monitoring of Red Foxes. *Animals* **2021**, *11*, 1723. [CrossRef] [PubMed]

34. Freuling, C.M.; Kamp, V.T.; Klein, A.; Günther, M.; Zaeck, L.; Potratz, M.; Eggerbauer, E.; Bobe, K.; Kaiser, C.; Kretzschmar, A.; et al. Long-Term Immunogenicity and Efficacy of the Oral Rabies Virus Vaccine Strain SPBN GASGAS in Foxes. *Viruses* **2019**, *11*, 790. [CrossRef]

35. Kukekova, A.V.; Trut, L.N.; Oskina, I.N.; Johnson, J.L.; Temnykh, S.V.; Kharlamova, A.V.; Shepeleva, D.V.; Gulievich, R.G.; Shikhevich, S.G.; Graphodatsky, A.S.; et al. A meiotic linkage map of the silver fox, aligned and compared to the canine genome. *Genome Res.* **2007**, *17*, 387–399. [CrossRef] [PubMed]

36. Thurmon, J.C.; Tranquilli, W.J.; Benson, G.J.; Lumb, W.V. *Lumb & Jones' Veterinary Anesthesia*, 3rd ed.; Williams & Wilkins: Baltimore, MD, USA, 1996; pp. 241–296.

37. Voipio, H.M.; Baneux, P.; Gomez de Segura, I.A.; Hau, J.; Wolfensohn, S. Guidelines for the veterinary care of laboratory animals: Report of the FELASA/ECLAM/ESLAV Joint Working Group on Veterinary Care. *Lab. Anim.* **2008**, *42*, 1–11.

38. Tzutalin, D. LabelImg: Git Code. 2015. Available online: https://github.com/tzutalin/labelImg (accessed on 10 January 2022).

39. Kluyver, T.; Ragan-Kelley, B.; Pérez, F.; Granger, B.E.; Bussonnier, M.; Frederic, J.; Kelley, K.; Hamrick, J.B.; Grout, J.; Corlay, S.; et al. Jupyter Notebooks—A publishing format for reproducible computational workflows. *Stand Alone* **2016**, *2016*, 87–90.

40. van Rossum, G.; Drake, F.L., Jr. *The Python Language Reference*; Python Software Foundation: Wilmington, DE, USA, 2014.

41. Redmon, J.; Farhadi, A. *YOLOv3: An Incremental Improvement*; Cornell University: Ithaca, NY, USA, 2018. Available online: https://arxiv.org/abs/1804.02767 (accessed on 10 January 2022).

42. Bochkovskiy, A.; Wang, C.Y.; Liao, H.Y.M. *YOLOv4: Optimal Speed and Accuracy of Object Detection*; Cornell University: Ithaca, NY, USA, 2020. Available online: https://arxiv.org/abs/2004.10934 (accessed on 10 January 2022).

43. Everingham, M.; van Gool, L.; Williams, C.K.I.; Winn, J.; Zisserman, A. The Pascal Visual Object Classes (VOC) Challenge. *Int. J. Comput. Vis.* **2010**, *88*, 303–338. [CrossRef]
44. Fernández-Carrión, E.; Barasona, J.Á.; Sánchez, Á.; Jurado, C.; Cadenas-Fernández, E.; Sánchez-Vizcaíno, J.M. Computer Vision Applied to Detect Lethargy through Animal Motion Monitoring: A Trial on African Swine Fever in Wild Boar. *Animals* **2020**, *10*, 2241. [CrossRef] [PubMed]
45. Rushen, J.; Chapinal, N.; De Passille, A. Automated monitoring of behavioural-based animal welfare indicators. *Anim. Welf.-UFAW J.* **2012**, *21*, 339. [CrossRef]
46. Fernández-Carrión, E.; Martínez-Avilés, M.; Ivorra, B.; Martínez-López, B.; Ramos, Á.M.; Sánchez-Vizcaíno, J.M. Motion-based video monitoring for early detection of livestock diseases: The case of African swine fever. *PLoS ONE* **2017**, *12*, e0183793. [CrossRef] [PubMed]
47. Rice, C.A.; Eberhart, N.L.; Krawczel, P.D. Prepartum Lying Behavior of Holstein Dairy Cows Housed on Pasture through Parturition. *Animals* **2017**, *7*, 32. [CrossRef]
48. Speroni, M.; Malacarne, M.; Righi, F.; Franceschi, P.; Summer, A. Increasing of Posture Changes as Indicator of Imminent Calving in Dairy Cows. *Agriculture* **2018**, *8*, 182. [CrossRef]
49. GitHub. AlexeyAB/darknet, 22 November 2021. Available online: https://github.com/AlexeyAB/darknet (accessed on 10 January 2022).

Article

Evaluation of Welfare in Commercial Turkey Flocks of Both Sexes Using the Transect Walk Method

Nina Mlakar Hrženjak [1], Hristo Hristov [2], Alenka Dovč [1], Jana Bergoč Martinjak [3], Manja Zupan Šemrov [4], Zoran Žlabravec [1], Jožko Račnik [1], Uroš Krapež [1], Brigita Slavec [1] and Olga Zorman Rojs [1,*]

[1] Institute of Poultry, Birds, Small Mammals and Reptiles, Faculty of Veterinary Medicine, University of Ljubljana, Gerbičeva Ulica 60, 1000 Ljubljana, Slovenia; nina.mlakarhrzenjak@vf.uni-lj.si (N.M.H.); alenka.dovc@vf.uni-lj.si (A.D.); zoran.zlabravec@vf.uni-lj.si (Z.Ž.); josko.racnik@vf.uni-lj.si (J.R.); uros.krapez@vf.uni-lj.si (U.K.); brigita.slavec@vf.uni-lj.si (B.S.)
[2] Nutrition Institute, Tržaška Cesta 40, 1000 Ljubljana, Slovenia; hristovhristo@outlook.com
[3] Perutnina Ptuj, Perutnina Ptuj d.o.o., Potrčeva 10, 2250 Ptuj, Slovenia; Jana.Brgoc@perutnina.eu
[4] Department of Animal Science, Biotechnical Faculty, University of Ljubljana, Groblje 3, 1230 Domžale, Slovenia; manja.zupansemrov@bf.uni-lj.si
* Correspondence: olga.zorman-rojs@vf.uni-lj.si

Simple Summary: In the last decade, increased attention has been directed toward the welfare of commercial poultry. In current turkey production systems, males and females are typically reared in the same facility until slaughtering the hens. Hens are reared for 12 to 14 weeks, while toms are reared for up to 22 weeks. This study examines farm health and welfare in commercial turkey flocks of both sexes during the fattening cycle using the transect walk method. Flocks, separately for males and females, were assessed at 3 to 4 weeks of age, 1 week before slaughtering the hens and 1 week before slaughtering the toms. We found several differences in the frequency of welfare indicators between different assessments and between male and female populations. The period just before slaughtering the hens was found to be most problematic for both sexes, although several welfare indicators suggested that health problems were mainly already present at 3 to 4 weeks of age and also continued after hen depopulation. Our results show that transect walks used at different ages may provide relevant information on animal health and welfare during the fattening cycle.

Citation: Hrženjak, N.M.; Hristov, H.; Dovč, A.; Martinjak, J.B.; Šemrov, M.Z.; Žlabravec, Z.; Račnik, J.; Krapež, U.; Slavec, B.; Rojs, O.Z. Evaluation of Welfare in Commercial Turkey Flocks of Both Sexes Using the Transect Walk Method. *Animals* **2021**, *11*, 3253. https://doi.org/10.3390/ani11113253

Academic Editors: Melissa Hempstead and Danila Marini

Received: 29 September 2021
Accepted: 11 November 2021
Published: 13 November 2021

Publisher's Note: MDPI stays neutral with regard to jurisdictional claims in published maps and institutional affiliations.

Abstract: The study was conducted between March and September 2019 in six meat-type turkey flocks with similar management standard procedures using the transect walk method. The concept of the method is based on visual observation of the birds while slowly walking across the entire farm in predetermined transects. Each flock was evaluated at three different times during the fattening cycle: at 3 to 4, 12 to 13, and 19 to 20 weeks of age, and total number of males and females that were immobile or lame, had visible head, vent, or back wounds, were small, featherless, dirty, or sick, had pendulous crop, or showed aggression toward birds or humans were recorded. At each visit, NH_3 and CO_2 were measured within the facilities. In the first assessment, the most frequently observed welfare indicators were small size (0.87%) and immobility (0.08%). Males showed a significantly higher prevalence of small size ($p < 0.01$), sickness ($p < 0.05$), and dirtiness ($p < 0.1$) compared to females. In the second assessment, the most common findings in both sexes were dirtiness (1.65%) and poor feather condition (1.06%), followed by immobility (0.28%). Males were significantly dirtier ($p < 0.001$), had more immobile birds ($p < 0.01$) and birds with vent wounds ($p < 0.1$), but had fewer sick birds ($p < 0.05$). In the last assessment, an increase in immobile, lame, sick, and dead birds was recorded, indicating an increase in health problems. Higher CO_2 (3000 and 4433 ppm) and NH_3 (40 and 27.6 ppm) values were noted only at the first assessment in two facilities. Further analyses showed that slightly elevated NH_3 and CO_2 levels did not influence the occurrence of welfare indicators. This study is the first description of the welfare of commercial turkey flocks in Slovenia.

Keywords: welfare; mixed commercial turkey flocks; on-farm assessment

1. Introduction

In the last decade, increased attention has been directed toward the welfare of commercial poultry and its assessment. Not only the awareness of the public, but also of farmers and stakeholders, is increasing due to the obvious impact of animal welfare issues on production [1–4].

Several on-farm animal-based assessment methods have been introduced to evaluate the welfare of poultry. In 2009, the Welfare Quality Assessment Protocol was introduced for layers and broilers [5], but it proved to be complex and time consuming [6], and it could even pose a risk to birds or handlers when used for other poultry species because it requires handling the birds. On the other hand, the transect walk method has proven to be practical, reliable, and efficient without the need to catch birds or subject the flock to excessive stress. The method has been used successfully in large flocks of broilers [1–3]. and in turkeys [7–9]. The concept of the method is based on visual observation of the birds while slowly walking across the entire farm in predetermined transects, as is usually done during routine inspection by farmers. While walking, the observer records every bird affected by health and welfare indicators such as small size, dirty or featherless birds, birds with head, back, or tail wounds, immobile, lame, sick, terminally ill, or dead birds, and birds showing aggressive behavior toward other birds or humans [10]. The transect walk can be used at different ages in turkeys, provides relevant information on animal welfare during the rearing period, and allows farmers to make changes and improve welfare for current flocks [11]. The health and welfare of commercially farmed turkeys supposed to be attributed to the high growth potential of the commonly used commercial hybrids [12,13], and also depend on environmental factors such as air quality [14,15], ambient temperature [16,17], light intensity and duration of day length [18,19], and stocking density [2,20]. All these factors, if not within the recommended limits, can cause significant physical distress to the animals [16,17] and consequently have a negative impact on animal performance [2,20], and post-slaughter product quality [21–23].

In Slovenia, commercial turkeys are reared in a conventional housing system and, as in many other countries, birds of both sexes are kept in the same facilities separated by the wire mash until slaughtering the hens. Hens are slaughtered when the birds reached an average body weight of 9 kg at around 14 weeks, while toms are reared until 21 to 22 weeks of age. There are limited field studies on welfare in such mixed flocks during the fattening period because most previous studies have been carried out before slaughter [7,8,11]. The aim of this study was to identify transect-based on-farm welfare indicators of commercial turkeys of both sexes at three different points in time during the fattening cycle; 3 to 4 weeks after placement and before slaughtering the hens and toms. We hypothesized that the welfare problems identified would differ at different ages and between males and females. Namely, the first weeks after placement of turkey chicks on farms are critical due to health problems caused by infectious diseases such as colibacillosis and aspergillosis [24,25] and specific behavior and environmental requirements of poults [1]. The second and third time points were chosen because in these two periods the facilities are at their maximum capacity regarding stocking density and ventilation, which may influence animal welfare [9,26]. In addition, we investigated the importance of selected climate conditions on the occurrence of animal welfare indicators.

2. Materials and Methods

2.1. Meat-Type Turkey Flocks

The study was conducted between March and September 2019 on six meat-type turkey flocks of both sexes (hens and toms) of two hybrids (Converter and British United Turkeys (B.U.T.) Big 6). All flocks were kept on farms in the central region of Slovenia. The owners of the farms were subcontractors of one poultry producer; therefore, it was expected that the management practices would be similar. The size of the houses varied between 900 and 1440 m^2, and the number of birds housed ranged from 4300 to 7300. All birds were beak trimmed in the hatchery. Males and females were housed separately, but in the same house, with the toms occupying about 60% of the area until the hens were slaughtered, after which the toms were placed throughout the entire area. From flocks 1 and 2, about one-third of the animals were removed at the age of 38 days and placed on another farm. Birds from all flocks were vaccinated against Newcastle disease and hemorrhagic enteritis. The number of birds found dead or that were culled was recorded daily. When health problems were noticed by the poultry farmers, a field veterinarian inspected the flock. Based on the results of the clinical observations and pathological findings, a decision on treatment was made where appropriate. The flock's information is summarized in Table 1.

All facilities were fully enclosed and insulated. They had a concrete floor and were equipped with either manually or automatically controlled ventilation systems, automatic drinkers, and automatic feeders. The birds were reared on wood shavings. The natural light entering the house through the windows was supplemented by artificial lighting for a total of 23 to 24 h of light per day during the whole rearing period. The light intensity varied from 3 to 27 lux. Birds did not have any environmental enrichment and access to elevated areas. All flocks had the same feed supplier and were slaughtered in the same slaughterhouse. The hens were slaughtered at around 14 to 15 weeks of age, when the birds reached an average body weight (BW) of 9 kg. Toms were slaughtered when the animals reached an average BW of 20 kg at 21 to 22 weeks of age. The flocks' information is summarized in Table 1.

2.2. Evaluation of Animal Welfare

Each flock was visited at three different times during the fattening cycle; the first visit took place at 3 to 4 weeks of age, the second visit took place approximately 1 week before slaughtering the hens (i.e., at 12 to 13 weeks of age), and the last assessment was conducted before slaughtering the toms at 19 to 20 weeks of age (Table 1). All visits were carried out by the same observers. At each assessment, information on cumulative mortality was collected from farm records, and stocking density was calculated for each flock, for males and females separately (Table 2).

The transect walk approach methodology developed by Marchewka et al. (2015) [7] was used to assess the welfare of commercial turkeys. At the first and second visits, toms and hens were assessed on the same day. Because both male and female animals were housed in the same house, each part (male and female) was divided into three to four longitudinal transects, depending on the size of the building. This approach was used in the first two evaluations. In the last assessment only toms were present, and therefore the entire barn was divided into three to four longitudinal sections. The assessor walked through the transect parts, from the entrance wall to the wire mesh or to the opposite wall (third assessment). The observer moved slowly to minimize disturbance to the birds during the assessment and recorded all observed occurrences of birds that fell into any of the predefined animal welfare indicator categories shown in Table 3.

2.3. Environmental Parameters

Inside temperature, ammonia (NH_3), and CO_2, were checked using Dräger X-am 1/2/5000 (Dräger, Lübeck, Germany) in each facility. All measurements were performed at animal level at six different locations: left and right at the entrance, in the middle, and at the end of the facility. Average values were calculated for each parameter.

Table 1. Information relating to flocks of turkeys: number of birds and hybrids, age of birds at each assessment, veterinary treatment, age, and average live weight at slaughter.

Flock	Barn Size (m²)	Number of Birds Placed		Hybrid	Age of Birds at Each Assessment (Days)			Veterinary Intervention	Age at Slaughter (Days)		Average Live Weight (kg)	
		Toms	Hens		I	II	III		Toms	Hens	Toms	Hens
1 [1]	1100	3900	3700	Converter	22	91	132	Yes [3]	143	97	20.6	9.25
2 [2]	1100	3900	3700	Converter	22	91	132	Yes [3]	143	97	20.4	9.23
3	900	2300	2000	Converter	21	84	125	Yes [4]	143	98	20.28	9.47
4	1380	4000	3300	BUT 6	22	85	126	Yes [3]	147	102	20.36	8.98
5	960	2700	2200	BUT 6	28	92	135	Yes [3]	147	104	19.67	9.28
6	1440	3400	3300	BUT 6	27	91	131	Yes [3]	150	101	19.27	9.19

[1] At the age of 38 days, a total of 1305 males and 1245 females were taken out and placed on another farm. [2] At the age of 39 days, a total of 1260 males and 1170 females were taken out and placed on another farm. [3] Birds were treated because of necrotic enteritis at 5 to 6 weeks of age. [4] Birds were treated due to *E. coli* infection at 1 and 5 weeks of age.

Table 2. Number of birds examined, cumulative mortality, and stocking density at each assessment.

Flock	Sex	First Assessment			Second Assessment			Third Assessment		
		Birds Exam.	Mortality (%)	Stock. Density (Birds/m²)	Birds Exam.	Mortality (%)	Stock. Density (Birds/m²)	Birds Exam.	Mortality (%)	Stock. Density (Birds/m²)
1	M	3818	2.11	5.78	2435	3.73	3.85	2360	7.66	1.86
	F	3543	0.95	8.32	2390	1.81	5.43			
2	M	3844	1.44	5.82	2430	2.66	3.85	2335	5.49	2.26
	F	3665	1.25	8.08	2344	1.77	5.35			
3	M	2221	3.43	8.23	2168	5.74	4.01	2088	8.21	2.32
	F	1936	3.20	10.76	1903	4.85	5.28			
4	M	3956	1.10	9.15	3846	3.85	4.45	3790	5.25	2.63
	F	3262	1.15	11.32	3242	1.75	5.62			
5	M	2653	1.74	9.21	2556	5.33	3.90	2481	8.11	2.58
	F	2173	1.23	11.31	2149	2.32	5.59			
6	M	3270	3.82	5.66	3203	5.79	4.67	3168	6.82	2.77
	F	3195	3.18	8.31	3166	4.06	6.94			

Table 3. Description of the birds' behavior and appearance in each of the welfare indicator categories.

Indicator	Description
Immobile	Birds not moving when approached or, after being gently touched, only able to move by propping themselves up on their wings
Lame	Birds walking with obvious difficulties; one of the legs not placed on the ground, bird moving away from the observer but stopping after two to three paces to rest
Head wounds	Visible alterations on the head, snood, beak, or neck related to fresh or older wounds
Back wounds	Bird with visible fresh or older wounds on the back, wings, or legs
Vent wounds	Visible wounds around tail, or on its sides, including fresh, older, or bleeding wounds
Small size	Birds that are approximately 50% the size of an average bird in the flock
Featherless	Missing or damaged feathers on the majority of the back area, including the wings and tail
Dirty	Very clear and dark staining of the back, wing, and/or tail feathers of the bird, covering at least 50% of the body area
Sick	Bird showing mild to severe clinical signs of impaired health; pale comb and eyes, watery discharge, and swollen sinuses, visibly breathing
Dead	Dead birds found during the assessment
Pendulous crop	Birds with a pendulous crop hanging in front of the breast
Aggression toward birds	Bird chases or pecks, hits, flies into, or leaps onto another bird
Aggression toward humans	Bird perceptibly hits human with the wings, or runs into, jumps onto, or pecks the human

2.4. Statistical Analyses

Incidence of welfare indicators were calculated for each flock, for males and females separately, and therefore the analyses were conducted with the flock and sex as experimental unit. Data were analyzed using STATA version 15.1 (StataCorp LLC, Lakeway Dr, College Station, TX, USA). A Z-test for difference in proportions was used to analyze differences between male and female populations in the occurrence of different welfare indicators. Logistic regression analyses were used to determine the effect of CO_2 and NH_3 on animal welfare represented by the presence of indicators that describe the level of animal welfare, including mortality.

3. Results

3.1. Evaluation of Animal Welfare

The mean values of each welfare indicator recorded at each assessment in male and female turkeys are presented in Table 4 and Figure 1.

In the first assessment, the most frequently observed welfare indicators in both males and females were small size and immobility. Overall, 0.997% of the males and 0.721% of the females were half the size of the other birds. Immobility was observed in 0.076% of male and 0.078% of female birds. All other indicators were rarely or never observed. No birds with pendulous crop or aggressive behavior were noted at this age. There were significantly smaller ($p < 0.01$), sick ($p < 0.05$), and dirty birds ($p < 0.1$) among males compared to females.

In the second assessment, the most common findings in both sexes were dirtiness and poor feather condition, followed by immobility. Males were significantly dirtier ($p < 0.001$), and there were more immobile birds ($p < 0.002$) and birds with vent wounds ($p < 0.100$), but fewer sick birds ($p < 0.048$) compared to females. At this assessment, pendulous crop was observed in both sexes, but the difference was not significant. No aggressive behavior was found.

The most common welfare indicators found in males before slaughter were immobility and dirtiness (0.53%), followed by poor feather condition (0.302%), lameness (0.197%), small size (0.129%), and sick birds (0.105%). At this age, aggression toward humans and other birds was also observed, but in very few birds (0.012%).

3.2. Effects of Selected Environmental Parameters on the Presence of Welfare Indicators

The average CO_2 and NH_3 values and temperature within each turkey facility obtained at each assessment are presented in Table 5.

Table 4. Welfare indicator mean values for male and female turkeys expressed as percentage at each assessment.

Indicator	Sex	I (3–4 w) n (%)	z	p-Value	II (13–14 w) n (%)	z	p-Value	III (17–19 w) n (%)
Immobile	M	15 (0.076)	−0.06	ns	62 (0.372)	3.16	0.002 ***	86 (0.530)
	F	14 (0.078)			28 (0.184)			
Lame	M	3 (0.015)	−0.14	ns	11 (0.066)	−0.83	ns	32 (0.197)
	F	3 (0.017)			14 (0.092)			
Head wounds	M	2 (0.010)	−0.09	ns	8 (0.048)	−1.10	ns	2 (0.012)
	F	2 (0.011)			12 (0.079)			
Back wounds	M	4 (0.020)	−0.12	ns	12 (0.072)	1.00	ns	5 (0.031)
	F	4 (0.022)			11 (0.072)			
Vent wounds	M	3 (0.015)	−0.14	ns	11 (0.066)	1.65	0.100 *	0 (0)
	F	3 (0.017)			4 (0.026)			
Small size	M	197 (0.997)	2.66	0.008 ***	21 (0.126)	1.09	ns	21 (0.129)
	F	128 (0.721)			13 (0.086)			
Featherless	M	6 (0.030)	0.75	ns	164 (0.986)	−1.38	ns	49 (0.302)
	F	3 (0.017)			174 (1.145)			
Dirty	M	4 (0.020)	1.74	0.081 *	343 (2.062)	6.00	0.001 ***	86 (0.530)
	F	0 (0)			183 (1.204)			
Sick	M	6 (0.030)	2.14	0.033 **	4 (0.024)	−1.98	0.048 **	17 (0.105)
	F	0 (0)			11 (0.072)			
Dead	M	0 (0)			2 (0.012)	1.35	ns	12 (0.074)
	F	0 (0)			0 (0)			
Pendulous crop	M	0 (0)			10 (0.060)	−1.04	ns	9 (0.055)
	F	0 (0)			14 (0.092)			
Agression towards birds	M	0 (0)			0 (0)			2 (0.012)
	F	0 (0)			0 (0)			
Agression towards humans	M	0 (0)			0 (0)			2 (0.012)
	F	0 (0)			0 (0)			

Note: Population of birds examined in each visit: first N_{male} = 19,762, N_{female} = 17,754; second N_{male} = 16,638, N_{female} = 15,194; third N_{male} = 16,222. Significance: * $p < 0.1$; ** $p < 0.05$; *** $p < 0.01$.

The average CO_2 values ranged from 850 to 4433 ppm. The values above the recommended (CO_2 < 2500 ppm) were recorded at the first assessment in flocks 1 and 2. At that time point, the NH_3 level in both facilities exceeded the anticipated level of 20 ppm. At the second and the third assessment, NH_3 values were low in all flocks and did not exceed 4 ppm. The CO_2 values were also low and ranged from 850 to 2200 ppm. Inside temperatures recorded at all three visits were slightly higher than recommended for the specific age of turkeys [27].

Table 6 shows the influence of different CO_2 and NH_3 levels on occurrences of welfare indicators and mortality.

For NH_3, the probability of occurrence of welfare indicators was lower at concentrations of NH_3 > 0 ppm compared to NH_3 = 0 ppm; for CO_2, the possibility of the development of welfare indicators was higher at CO_2 < 1600 ppm compared to CO_2 between 1600 and 3000 ppm and more than 3000 ppm.

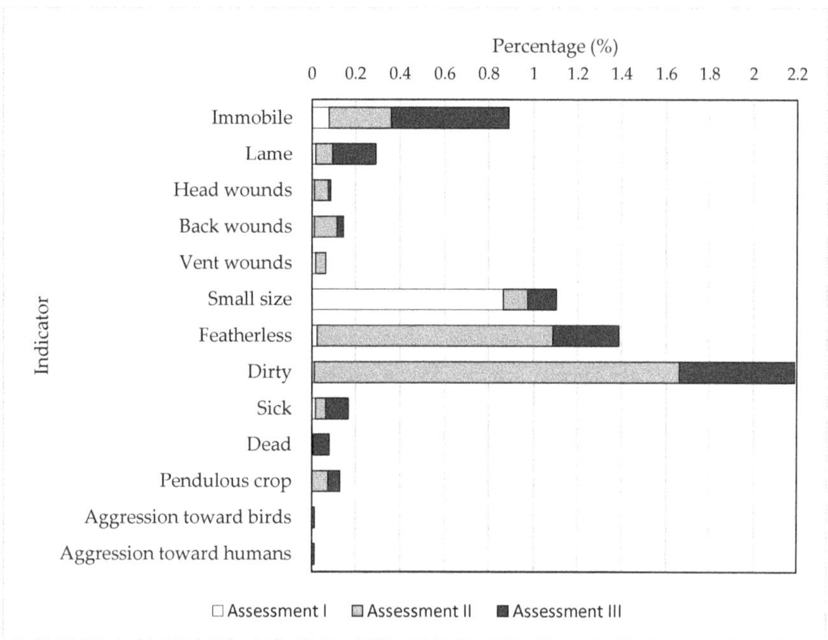

Figure 1. Mean values of welfare indicators recorded in meat-type turkeys at 3 to 4 (Assessment I), 13 to 14 (Assessment II), and 17 to 19 weeks of age (Assessment III).

Table 5. Mean temperature and CO_2 and NH_3 values in turkey facilities at each assessment.

Parameter	Flock					
	1	2	3	4	5	6
Assessment I						
Inside temp. (°C)	26.5	26.6	26.7	28.9	26.8	27.0
NH_3 (ppm)	40	27.6	4	0	0	0
CO_2 (ppm)	4433	3000	1630	1200	1050	975
Assessment II						
Inside temp. (°C)	21.5	21	19.5	26.0	26.5	27.6
NH_3 (ppm)	3	2.3	4	0	0	0
CO_2 (ppm)	1600	1217	1750	1200	950	1025
Assessment III						
Inside temp. (°C)	25	28	23.2	21	23	20.5
NH_3 (ppm)	0	0	0	0	2.3	0
CO_2 (ppm)	1000	850	867	1000	2200	1000

Note: Recommended values according to Aviagen: ambient temperature at 3–4 weeks = 23–25 °C, at more than 9 weeks = 16–17 °C; NH_3: <20 ppm; CO_2: <2500 ppm [27].

Table 6. Welfare indicators' prevalence (including mortality) by different CO_2 and NH_3 levels.

Parameter	Observations *n* (%)	Prevalence *n* (%)	Odds Ratio (CI)	SE	*p*-Value
CO					
<1600 ppm	247 (63.33)	106 (42.91)	1		
1600–3000 ppm	91 (23.33)	32 (35.16)	0.20 (0.07–0.54)	0.103	0.002 ***
>3000 ppm	52 (13.33)	10 (19.23)	0.15 (0.04–0.55)	0.100	0.004 ***
NH3					
0 ppm	221 (56.67)	86 (38.91)	1		
>0 ppm	169 (43.33)	62 (36.69)	0.35 (0.12–1.02)	0.192	0.056 *

Note: * Significance at $p < 0.1$; *** $p < 0.01$.

4. Discussion

In Slovenia, turkeys represent less than 2% of poultry meat production and, until this study, no data were available on the welfare profile of this species. The aim of this study was to identify transect based on-farm health and welfare indicators in commercial turkey flocks of both sexes during the fattening cycle. Flocks, separately for males and females, were assessed at 3 to 4 weeks of age, 1 week before slaughtering the hens, and 1 week before slaughtering the toms. In brief, we found several differences in the frequency of welfare indicators between different assessments and between male and female populations.

To date, no data on the welfare of turkeys based on the transect walk approach at the age of 3 to 4 weeks are available. The results of our study showed that small birds were identified as the main welfare problem at this age. Such birds were found in both sexes, although significantly more small birds were observed in the male population. The second most common indicator was immobility. A high incidence of smaller birds and immobility may indicate compromised health and welfare on a farm due to either general housing or bird health problems. Infections caused by *E. coli* are commonly present in young chicks. Localized infections such as omphalitis and yolk sac infection or systemic colibacillosis generally resulted in higher mortality in the first weeks after placement. Affected birds are usually undersized, because they may have difficulties in walking, which alters weight and leads to weakness [25]. Of the flocks included in the study, colibacillosis was diagnosed in one flock and the birds were treated with antibiotics. Although no veterinary intervention was required in the other flocks, the cumulative mortality indicated that health problems due to *E. coli* or other unidentified infections were present in at least one other flock.

Excessive NH_3 and CO_2 levels may also have negative impact on birds' health and metabolism in young turkeys [14,28]. It was shown that young poults exposed to 4000 ppm CO_2 had suppressed body weight gain compared to those exposed to 2000 ppm [14]; in addition, NH_3 levels greater than 10 ppm can also reduce feed intake with effects on body weight [28]. In our study, higher NH_3 (40 ppm and 27.6 ppm) and CO_2 (4433 ppm and 3000 ppm) values were detected in two facilities. Elevated levels of both gasses directly correlate with reduced ventilation. Under field conditions, such a situation is often seen in the first weeks after placement of turkeys due to reduced heating costs [29]. Nevertheless, the higher mortality as well as the significantly higher incidence of sick birds found in males indicate that health problems seem to play a more important role than environmental conditions, which were equal for males and females.

As shown in other studies [9], the most problematic period for both sexes in mixed commercial turkey flocks seems to be the time before slaughtering the hens. Indeed, more indicators were present at the second assessment than the first, and the overall prevalence of altered birds was higher compared to the first and third assessments. The most common findings were dirtiness and poor feather condition, followed by immobility. Dirty feathers were observed in more than 2% of males and in 1.204% of females. Dirtiness before slaughtering the hens has also been reported by other authors, although not with such a high prevalence as in our study. In a study conducted in Norway, an average of 0.36% dirty males and 0.15% females were observed at 11 weeks [9,11,30]. In Italian commercial turkey flocks, dirtiness was recorded in 0.022% of females. Unfortunately, males of this age were not scored [8], and so no direct comparison was possible. Previous studies have shown that poor litter quality, dust, and high stocking density can significantly impact dirtiness [1]. Unfortunately, litter quality was not scored in our study. CO_2 and NH_3 values that reflect inadequate ventilation and poor litter management [31] were low in all facilities, but the measurements were performed only during assessments so no relevant conclusions could be made on the significance of poor litter quality on such a high prevalence of dirtiness. Recently, dirtiness in turkeys was found to be correlated with immobility and lameness more than with poor litter quality [9]. In our study, immobility was the third most frequently observed indicator at this age. Similar to previously reported findings [8,32], this occurred significantly more frequently in males than in females, which could explain why males were significantly dirtier. Dirty feathers have been suggested as an indicator of health problems of the digestive system [33], but it is unlikely that necrotic enteritis diagnosed at 5 weeks of age in five of six flocks contributed to such a high percentage of dirty birds observed 7 weeks later. Necrotic enteritis is an acute disease caused by toxins produced by *Clostridium perfringens*. The course of the disease is usually short, and birds respond very well to antibiotic treatment if it is given immediately after the onset of the disease [34,35].

Featherless condition was the second most frequently observed welfare indicator during this period. Missing or damaged feathers, particularly in the tail region, were observed in 0.986% of males and in 1.145% of females. The etiology is not entirely known, but it is likely that poor plumage is the consequence of feather pecking. In turkeys, mild feather pecking can be a form of social or investigative behavior [36,37]. When pecking becomes more severe, it may result in severely damaged feathers and feather loss, or even cannibalism [38]. The incidence of feather pecking is known to increase with age, although damaging pecking can occur as early as the 1st or 2nd week of age [39]. Under field conditions, high stocking density, inappropriate lighting, feed deficiency, breed, and sex are considered to influence injurious feather pecking [2,8,40]. Such aggressive pecking often results in wounds seen on the head, around the tail, on the wings, and in the back region. In our study, injuries were rarely found and were observed in both sexes, although vent wounds were found significantly more frequently in males. This is consistent with the results from commercial turkey flocks in Norway [9] and Italy [8]. To prevent feather pecking and aggressiveness, beak trimming is still common practice in commercial turkeys [40], but it seems that beak trimming does not play an essential role in preventing poor feather condition. The incidences of featherless birds observed in our study and in Italian commercial flocks [8]—both included beak-trimmed birds—were higher compared to the study performed in non-trimmed commercial flocks in Norway [9].

The percentage of sick birds was low, but the difference between the sexes was significant. For classifying a bird as sick, other studies also included birds with pendulous crop [7–9,11,30]. In our study, birds with pendulous crop were recorded separately, and so no direct comparison could be made. The etiology of pendulous crop is yet not fully understood, but hereditary predisposition, dietary influence, increased liquid intake in hot weather, and the effect of lighting period have been suggested. Unfortunately, no treatment is available, and the carcasses of affected birds are usually condemned at processing [41–43]. In our study, birds with pendulous crop were observed in both sexes, although at much lower frequencies compared to the study performed by Vermette et al. [32]. In their clinical study, pendulous crop was found to be the second major reason for morbidity and mortality in turkeys, and females were significantly more affected than males. In comparison to the first assessment, the size of birds was more uniform. This supports previous suggestions that the number of small birds decreases with age [30].

At the last visit, only male turkeys were assessed. Compared to the second assessment, the prevalence of immobile as well as lame, sick, and dead birds increased, indicating health problems, most likely caused by poorer leg health. Lameness and immobility as its consequence are important welfare and health issues in commercial turkey flocks, especially in males [8,32]. Due to their heavy weight and longer production cycle, males' legs are exposed to more stressors, resulting in chronic pain and movement difficulties [23]. In addition to degenerative and development disorders, bacterial and viral infections such as *Staphylococcus aureus*, mycoplasmas, and reovirus are involved in skeletal and joint lesions, causing acute or chronical local inflammations or even systemic septicemia, resulting in higher mortality [44]. Because no further investigations were performed, we do not know the exact causes of immobility. Dirty and featherless animals were frequently observed, although the prevalence was much lower than in the second assessment. This is consistent with recent findings that poor feather condition decreases with lower stocking density [20,26], but may still persist within the flock due to leg problems [9]. The incidence of head, back, and vent wounds also decreased to less than half after depopulation of females. These results are in agreement with previous findings that injurious feather pecking and wounds may be a consequence of behavioral disturbances due to high stocking density [1]. In the EU, specific minimum stocking density requirements have been established for broilers [45], but these do not directly apply to turkeys. In some countries, such as Norway, specific regulations depending on the live weight of turkeys have been adopted [9] or recommendations for turkeys have been published [46]. Unfortunately, no such documents are available in Slovenia.

During the production cycle, NH_3 and CO_2 are the prevalent gases in turkey facilities [47]. NH_3 in high concentrations can have severe adverse health effects, causing lesions of the upper respiratory tract and inflammation of the cornea and conjunctive [15,48]. High concentration of CO_2 can be harmful to turkeys due to hypoxia [14], which may lead to dilatated cardiomyopathy [49]. Although no precise concentration limits have been established for turkeys, Directive 98/58/EC [50] requires that gas concentrations in turkey facilities be kept within safe limits. For broilers, EU regulation established the maximum NH_3 concentration inside a poultry barn at 20 ppm, and for CO_2 the concentration should not exceed 3000 ppm measured at the level of the chickens' heads. If higher levels are detected, corrective actions must be taken [45]. In Slovenia, most commercial turkey farms are equipped with sensors for temperature and humidity, but gas concentrations are not measured on a daily basis. Apart from higher CO_2 and NH_3 values recorded at two facilities in the first assessment, the levels were low in all facilities in the next two assessments. Similar results were also obtained in the study performed in commercial turkey flocks in Poland, where a significant decreasing trend was observed during the production cycle [47]. Although accurate limits should be defined for turkeys, our results indicate that NH_3 levels of 40 ppm or less did not influence the occurrence of welfare indicators. Moreover, a higher probability could be expected at NH3 = 0 ppm compared to more than 0 ppm. For CO_2 the probability of occurrence of animal welfare indicators was higher at levels below 1600 ppm than at levels between 1600 and 3000 ppm or above 3000 ppm. The reason for this could be that birds exposed to slightly elevated NH_3 or CO_2 concentrations are likely to be less active. Similar results were recently obtained by Candido et al. [14], who found that poults housed at a lower CO_2 level (2000 ppm) showed reduced movement compared to those exposed to higher CO_2 concentrations. However, further studies should be performed to obtain a balance between welfare and the optimal production of turkeys.

5. Conclusions

Compared to other poultry species, there is a lack of field studies on welfare problems in commercial turkey flocks. In this study, we investigated some aspects of health and welfare in commercial turkey flocks of both sexes in Slovenia. We cannot assume that our limited sample of flocks is representative of the commercial turkey industry in Slovenia, but it provides an estimation of problems that may exist during the production cycle and emphasizing the importance of setting specific standards and regulations regarding levels of harmful gases and stocking density for commercial turkeys. Our study confirmed that assessing welfare using transect walk approach performed in different times during the fattening cycle provides important information on animal health and welfare and could help farmers improve welfare in commercial turkey flocks.

Author Contributions: Conceptualization, O.Z.R., M.Z.Š. and A.D., methodology, O.Z.R.; software, H.H.; formal analysis, H.H.; investigation, O.Z.R., J.B.M. and A.D.; resources, M.Z.Š.; data curation, H.H.; writing—original draft preparation, N.M.H., H.H. and O.Z.R.; writing—review and editing, M.Z.Š., O.Z.R., A.D., Z.Ž., J.R., B.S., J.B.M. and U.K.; supervision, M.Z.Š.; funding acquisition, M.Z.Š. and O.Z.R. All authors have read and agreed to the published version of the manuscript.

Funding: This research was funded by the Slovenian Ministry of Agriculture, Forestry, and Food and the Slovenian Research Agency, project no. V-4 1817, and by the Slovenian Research Agency, research core funding no. P4-0092.

Institutional Review Board Statement: The national Animal Protection Act (Official Gazette of RS 38/2013; 21/2018; 92/2020 and 159/2021) defines in 21 a. article that non-invasive and non-experimental clinical veterinary practices are not considered as a procedure on animals for scientific purposes.

Informed Consent Statement: Not applicable.

Data Availability Statement: Not applicable.

Acknowledgments: The authors thank Perutnina Ptuj d.o.o. for its support as well as all farmers for their kindness and willingness to help.

Conflicts of Interest: The authors declare no conflict of interest.

References

1. Marchewka, J.; Watanabe, T.; Ferrante, V.; Estevez, I. Review of the social and environmental factors affecting the behavior and welfare of turkeys (*Meleagris gallopavo*). *Poult. Sci.* **2013**, *92*, 1467–1473. [CrossRef]
2. Erasmus, M.A. A review of the effects of stocking density on turkey behavior, welfare, and productivity. *Poult. Sci.* **2017**, *96*, 2540–2545. [CrossRef]
3. BenSassi, N.; Averós, X.; Estevez, I. Broiler Chickens On-Farm Welfare Assessment: Estimating the Robustness of the Transect Sampling Method. *Front. Veter.-Sci.* **2019**, *6*, 236. [CrossRef]
4. Costa, E.D.; Tranquillo, V.; Dai, F.; Minero, M.; Battini, M.; Mattiello, S.; Barbieri, S.; Ferrante, V.; Ferrari, L.; Zanella, A.; et al. Text Mining Analysis to Evaluate Stakeholders' Perception Regarding Welfare of Equines, Small Ruminants, and Turkeys. *Animals* **2019**, *9*, 225. [CrossRef]
5. Butterworth, A.; van Niekerk, T.G.C.M.; Veissier, I.; Keeling, L.J. *Welfare Quality®Assessment Protocols for Poultry (Broilers, Laying Hens) Welfare Quality Consortium*; Welfare Quality®Consortium: Lelystad, The Netherlands, 2009.
6. de Jong, I.; Hindle, V.A.; Butterworth, A.; Engel, B.; Ferrari, P.; Gunnink, H.; Moya, T.P.; Tuyttens, F.A.M.; van Reenen, C.G. Simplifying the Welfare Quality®assessment protocol for broiler chicken welfare. *Animal* **2016**, *10*, 117–127. [CrossRef]
7. Marchewka, J.; Estevez, I.; Vezzoli, G.; Ferrante, V.; Makagon, M.M. The transect method: A novel approach to on-farm welfare assessment of commercial turkeys. *Poult. Sci.* **2015**, *94*, 7–16. [CrossRef]
8. Ferrante, V.; Lolli, S.; Ferrari, L.; Watanabe, T.T.N.; Tremolada, C.; Marchewka, J.; Estevez, I. Differences in prevalence of welfare indicators in male and female turkey flocks (*Meleagris gallopavo*). *Poult. Sci.* **2019**, *98*, 1568–1574. [CrossRef]
9. Marchewka, J.; Vasdal, G.; Moe, R.O. Identifying welfare issues in turkey hen and tom flocks applying the transect walk method. *Poult. Sci.* **2019**, *98*, 3391–3399. [CrossRef]
10. AWIN Welfare Assessment Protocol for Turkeys. Available online: https://www.researchgate.net/publication/279953184_AWIN_Welfare_assessment_protocol_for_Turkeys#fullTextFileContent (accessed on 21 September 2021).
11. Marchewka, J.; Vasdal, G.; Moe, R.O. Associations between on-farm welfare measures and slaughterhouse data in commercial flocks of turkey hens (*Meleagris gallopavo*). *Poult. Sci.* **2020**, *99*, 4123–4131. [CrossRef]
12. Martrenchar, A. Animal welfare and intensive production of turkey broilers. *World's Poult. Sci. J.* **1999**, *55*, 143–152. [CrossRef]
13. Olschewsky, A.; Riehn, K.; Knierim, U. Suitability of Slower Growing Commercial Turkey Strains for Organic Husbandry in Terms of Animal Welfare and Performance. *Front. Veter. -Sci.* **2021**, *7*. [CrossRef]
14. Cândido, M.; Xiong, Y.; Gates, R.; Tinôco, I.; Koelkebeck, K. Effects of carbon dioxide on turkey poult performance and behavior. *Poult. Sci.* **2018**, *97*, 2768–2774. [CrossRef]
15. Kristensen, H.H.; Wathes, C.M. Ammonia and poultry welfare: A review. *World's Poult. Sci. J.* **2000**, *56*, 235–245. [CrossRef]
16. Yahav, S.; Straschnow, A.; Plavnik, I.; Hurwitz, S. Blood system response of chickens to changes in environmental temperature. *Poult. Sci.* **1997**, *76*, 627–633. [CrossRef]
17. Mendes, A.S.; Moura, D.J.; Morello, G.M.; Carvalho, T.M.R.; Sikorski, R.R. Turkey wattle temperature response to distinct envi-ronmental factors. *Braz. J. Poult. Sci.* **2015**, *17*, 439–444. [CrossRef]
18. Vermette, C.; Schwean-Lardner, K.; Gomis, S.; Crowe, T.G.; Classen, H.L. The impact of graded levels of daylength on turkey productivity to eighteen weeks of age. *Poult. Sci.* **2016**, *95*, 985–996. [CrossRef]
19. Lewis, P.D.; Perry, G.C.; Sherwin, C.M. Effect of photoperiod and light intensity on the performance of intact male turkeys. *Anim. Sci.* **1998**, *66*, 759–767. [CrossRef]
20. Beaulac, K.; Schwean-Lardner, K. Assessing the Effects of Stocking Density on Turkey Tom Health and Welfare to 16 Weeks of Age. *Front. Veter.-Sci.* **2018**, *5*, 213. [CrossRef]
21. Mitterer-Istyagin, H.; Ludewig, M.; Bartels, T.; Krautwald-Junghanns, M.E.; Ellerich, R.; Schuster, E.; Berk, J.; Petermann, S.; Fehlhaber, K. Examinations on the prevalence of footpad lesions and breast skin lesions in B.U.T. Big 6 fattening turkeys in Germany. Part II: Prevalence of breast skin lesions (breast buttons and breast blisters). *Poult. Sci.* **2011**, *90*, 775–780. [CrossRef]
22. Krautwald-Junghanns, M.-E.; Ellerich, R.; Mitterer-Istyagin, H.; Ludewig, M.; Fehlhaber, K.; Schuster, E.; Berk, J.; Petermann, S.; Bartels, T. Examinations on the prevalence of footpad lesions and breast skin lesions in British United Turkeys Big 6 fattening turkeys in Germany. Part I: Prevalence of footpad lesions. *Poult. Sci.* **2011**, *90*, 555–560. [CrossRef]
23. Erasmus, M.A. Welfare issues in turkey production. In *Advances in Poultry Welfare*; Mench, J.A., Ed.; Woodhead Publishing: Cambridge, MA, USA, 2018; pp. 265–282.

24. Arné, P.; Thierry, S.; Wang, D.; Deville, M.; Loc'h, L.; Desoutter, A.; Féménia, F.; Nieguitsila, A.; Huang, W.; Chermette, R.; et al. Aspergillus fumigatusin Poultry. *Int. J. Microbiol.* **2011**, *2011*, 1–14. [CrossRef] [PubMed]
25. Nolan, L.K.; Vaillan, J.P.; Barbieri, N.L.; Logue, C.M. Colibacillosis. In *Diseases of Poultry*, 14th ed.; Swayne, D.E., Boulianne, M., Logue, C.M., McDougald, L.R., Nair, V., Suarez, D.L., Eds.; Wiley-Blackwell: Hoboken, NJ, USA, 2020; Volume 1, pp. 770–830.
26. Martrenchar, A.; Huonnic, D.; Cotte, J.P.; Boilletot, E.; Morisse, J.P. Influence of stocking density on behavioural, health and productivity traits of turkeys in large flocks. *Br. Poult. Sci.* **1999**, *40*, 323–331. [CrossRef]
27. Aviagen®Turkeys. Available online: https://www.aviagenturkeys.com/en-gb/products/b-u-t-6?disableredirect=true (accessed on 21 September 2021).
28. Carlile, F.S. Ammonia in Poultry Houses: A Literature Review. *World's Poult. Sci. J.* **1984**, *40*, 99–113. [CrossRef]
29. Noll, S.L.; Janni, A.K.; Halvorson, A.D.; Clanton, C.J. Market turkey performance, air quality, and energy consumption affected by partial slotted flooring. *Poult. Sci.* **1997**, *76*, 271–279. [CrossRef] [PubMed]
30. Vasdal, G.; Marchewka, J.; Moe, R.O. Associations between animal-based measures at 11 wk and slaughter data at 20 wk in turkey toms (*Meleagris gallopavo*). *Poult. Sci.* **2020**, *100*, 412–419. [CrossRef]
31. Glatz, P.; Rodda, B. Turkey farming: Welfare and husbandry issues. *Afr. J. Agric. Res.* **2013**, *8*, 6149–6163. [CrossRef]
32. Vermette, C.; Schwean-Lardner, K.; Gomis, S.; Grahn, B.H.; Crowe, T.G.; Classen, H.L. The impact of graded levels of day length on turkey health and behavior to 18 weeks of age. *Poult. Sci.* **2016**, *95*, 1223–1237. [CrossRef]
33. de Jong, I.C.; Gunnink, H.; van Harn, J. Wet litter not only induces footpad dermatitis but also reduces overall welfare, technical performance, and carcass yield in broiler chickens. *J. Appl. Poult. Res.* **2014**, *23*, 51–58. [CrossRef]
34. Lyhs, U.; Perko-Mäkelä, P.; Kallio, H.; Brockmann, A.; Heinikainen, S.; Tuuri, H.; Pedersen, K. Characterization of Clostridium perfringens isolates from healthy turkeys and from turkeys with necrotic enteritis. *Poult. Sci.* **2013**, *92*, 1750–1757. [CrossRef] [PubMed]
35. Openrart, K.; Baulianne, M. Necrotic enteritis. In *Diseases of Poultry*, 14th ed.; Swayne, D.E., Boulianne, M., Logue, C.M., McDou-gald, L.R., Nair, V., Suarez, D.L., Eds.; Wiley-Blackwell: Hoboken, NJ, USA, 2020; Volume 1, pp. 972–976.
36. Busayi, R.; Channing, C.; Hocking, P. Comparisons of damaging feather pecking and time budgets in male and female turkeys of a traditional breed and a genetically selected male line. *Appl. Anim. Behav. Sci.* **2006**, *96*, 281–292. [CrossRef]
37. Marchewka, J.; Watanabe, T.T.N.; Ferrante, V.; Estevez, I. Welfare assessment in broiler farms: Transect walks versus individual scoring. *Poult. Sci.* **2013**, *92*, 2588–2599. [CrossRef]
38. Duggan, G.; Widowski, T.; Quinton, M.; Torrey, S. The development of injurious pecking in a commercial turkey facility. *J. Appl. Poult. Res.* **2014**, *23*, 280–290. [CrossRef]
39. Moinard, C.; Lewis, P.D.; Perry, G.C.; Sherwin, C.M. The effects of light intensity and light source on injuries due to pecking of male domestic turkeys (*Meleagris Gallopavo*). *Anim. Welf.* **2001**, *10*, 131–139.
40. Dalton, H.; Wood, B.; Torrey, S. Injurious pecking in domestic turkeys: Development, causes, and potential solutions. *World's Poult. Sci. J.* **2013**, *69*, 865–876. [CrossRef]
41. Almeida, E.A.; Silva, F.H.A.; Crowe, T.G.; Macari, M.; Furlan, R.L. Influence of rearing temperature and feed format in the development of the pendulous crop in broilers. *Poult. Sci.* **2018**, *97*, 3556–3563. [CrossRef]
42. Crespo, R.; Shivaprasad, H.L. Developmental, metabolic, and other noninfectious disorders. In *Diseases of Poultry*, 13th ed.; Swayne, D.E., Glisson, J.R., Eds.; Wiley-Blackwell: Ames, Iowa, 2013; pp. 1233–1270.
43. Willems, O.; Buddiger, N.J.H.; Wood, B.; Miller, S. The genetic and phenotypic relationship between feed efficiency and pendulous crop in the turkey (*Meleagris gallopavo*). In Proceedings of the 10th World Congress on Genetics Applied to Livestock Production, Vancouver, BC, Canada, 17–22 August 2014. [CrossRef]
44. Julian, R.J. Production and growth related disorders and other metabolic diseases of poultry—A review. *Veter.-J.* **2005**, *169*, 350–369. [CrossRef]
45. European Union. Council Directive 2007/43/EC of 28 June 2007 Laying down Minimum Rules for the protection of Chickens Kept for Meat Production. *OJEC* **2007**, 19–28. Available online: https://eur-lex.europa.eu/legal-content/en/ALL/?uri=CELEX%3A32007L0043 (accessed on 21 September 2021).
46. Government of United Kingdom—Department for Environment Food & Rural Affairs. Turkeys: Welfare Recommendations: Guidance; UK. 2019. Available online: https://www.gov.uk/government/publications/poultry-on-farm-welfare/turkeys-welfare-recommendations (accessed on 28 September 2021).
47. Witkowska, D. Volatile gas concentrations in turkey houses estimated by Fourier Transform Infrared Spectroscopy (FTIR). *Br. Poult. Sci.* **2013**, *54*, 1–9. [CrossRef]
48. Wathes, C.M.; Jones, E.K.M.; Kristensen, H.H.; Mckeegan, D.E.F. Ammonia and poultry production: Biological responses, wel-fare and environmental impact. In Proceedings of the 22th World's Poultry Congress, Istanbul, Turkey, 8–13 June 2004; pp. 1–9.
49. Julian, R.J.; Mirsalimi, S.M.; Bagley, L.G.; Squires, E.J. Effect of Hypoxia and Diet on Spontaneous Turkey Cardiomyopathy (Round-Heart Disease). *Avian Dis.* **1992**, *36*, 1043. [CrossRef] [PubMed]
50. European Union. Council Directive 98/58/EC of 20 July 1998 Concerning the Protection of Animals Kept for Farming Purposes. *OJEC* **1998**, 23–27. Available online: https://eur-lex.europa.eu/LexUriServ/LexUriServ.do?uri=OJ:L:2010:276:0033:0079:en:PDF (accessed on 21 September 2021).

Article

Recumbency as an Equine Welfare Indicator in Geriatric Horses and Horses with Chronic Orthopaedic Disease

Zsofia Kelemen [1], Herwig Grimm [2], Mariessa Long [2], Ulrike Auer [3,*] and Florien Jenner [1,*]

[1] Equine Surgery Unit, University Equine Hospital, Department of Companion Animals and Horses, University of Veterinary Medicine Vienna, Veterinaerplatz 1, 1210 Vienna, Austria; Zsofia.Kelemen@vetmeduni.ac.at

[2] Unit of Ethics and Human-Animal-Studies, Messerli Research Institute, University of Veterinary Medicine Vienna, Medical University of Vienna, University of Vienna, Veterinaerplatz 1, 1210 Vienna, Austria; Herwig.Grimm@vetmeduni.ac.at (H.G.); Mariessa.Long@vetmeduni.ac.at (M.L.)

[3] Anaesthesiology and Perioperative Intensive Care Medicine Unit, Department of Companion Animals and Horses, University of Veterinary Medicine Vienna, Veterinaerplatz 1, 1210 Vienna, Austria

* Correspondence: Ulrike.Auer@vetmeduni.ac.at (U.A.); Florien.Jenner@vetmeduni.ac.at (F.J.)

Citation: Kelemen, Z.; Grimm, H.; Long, M.; Auer, U.; Jenner, F. Recumbency as an Equine Welfare Indicator in Geriatric Horses and Horses with Chronic Orthopaedic Disease. *Animals* **2021**, *11*, 3189. https://doi.org/10.3390/ani11113189

Academic Editors: Melissa Hempstead and Danila Marini

Received: 30 September 2021
Accepted: 4 November 2021
Published: 8 November 2021

Publisher's Note: MDPI stays neutral with regard to jurisdictional claims in published maps and institutional affiliations.

Simple Summary: Horses have to lie down to achieve rapid eye movement (REM) sleep. Horses that do not lie down for environmental reasons or pain suffer from an REM sleep deficiency that negatively affects their welfare and health. The present study aimed to assess the influence of chronic orthopedic disease and old age on the time horses lie down. Wearable automated sensor technology was used to monitor the time 83 old and young adult horses with or without chronic lameness spent lying down, moving, or standing. Interestingly, neither age nor lameness due to chronic orthopedic disease significantly influenced the time spent lying down. Horses showing symptoms of REM sleep deficiency had shorter lying times and reduced times spent moving, indicating a general compromise of their well-being. The study shows that wearable sensor technology can be used to identify horses with short recumbency times at risk for REM sleep deficiency. Furthermore, the technology can be used to assess and monitor equine welfare objectively and optimize husbandry conditions so that old horses and horses suffering from chronic orthopedic conditions can achieve lying-down times comparable to younger, healthy horses.

Abstract: Recumbency is a prerequisite for horses achieving rapid eye movement (REM) sleep and completing a full sleep cycle. An inability to lie down due to environmental insecurities or pain results in REM sleep deficiency, which can cause substantial impairment of welfare and health. Therefore, the present study used wearable automated sensor technology on 83 horses housed in an animal sanctuary to measure and compare the recumbency, locomotion, and standing time budgets of geriatric horses with and without chronic lameness to younger adult sound and lame horses. Recumbency times ranged from 0 to 319 min per day with an overall mean of 67.4 ($\pm$61.9) minutes; the time budget for locomotion was 19.1% ($\pm$11.2% s.d.) and for standing 75.6% ($\pm$13.1 s.d.). Interestingly, neither age nor lameness due to chronic orthopedic disease had a significant influence on recumbency times in this study. Eight horses showed symptoms of REM deficit. These horses had significantly shorter lying times (7.99 $\pm$ 11.4 min) and smaller locomotion time budgets than the other horses enrolled in this study (73.8 $\pm$ 61.8 min), indicating a general compromise of well-being. Thus, wearable sensor technology can be used to identify horses with low recumbency times at risk for REM sleep deficiency and to assess and monitor equine welfare objectively.

Keywords: welfare; horse; equine; sleep; lying; time budget; locomotion; geriatric; orthopedic; recumbency

1. Introduction

Recumbency is a prerequisite for horses achieving rapid eye movement (REM) sleep and complete a full sleep cycle [1–4]. While adult horses sleep only 2.5–5 h/day, 80%

of which is in a standing position, they need a minimum of 30 min of recumbency per day to achieve 3.5–4.5 min of REM sleep and avoid REM sleep deprivation with excessive secondary drowsiness and collapse [1–16]. However, as a prey species, horses only lie down when they feel comfortable to do so [2,4,5,14,17–23]. Hence, measuring lying behavior is an essential component of equine welfare assessment [12,13].

Adult (>4 years) horses, both domestic and (semi-)feral, spend 3–15% (43–216 min) of their total daily time budget (=percentage of time spent on specific activities) in recumbency, of which about 20 min or 15% are spent in lateral and the rest in sternal recumbency [1,7,12,16,20,21,24–33]. The consistency of lying times between the vastly different domestic and free-ranging living conditions emphasizes the importance of recumbency as basic maintenance behavior. In comparison, the time budgets for eating, resting standing, and locomotion vary greatly between adult domestic and free-ranging horses, with the latter spending 50.82–66.6% foraging, 12.9–29.3% of their day standing, and 4.3–13.4% in movement (excl. grazing), while domestic horses divide their time between 10–64% eating, 15.6–68% standing, and 2.5–19.3% locomotion.

Sleep is a basic maintenance behavior, essential for physiological and cognitive function. Horses sleep in a polyphasic pattern, distributed over 5–7 episodes, with most sleep occurring between midnight and 4:00 am [2–5,8,10,12,14,32–37]. Based on postural and behavioral indicators and specific cortical electronic activity, four sleep–wakefulness states are differentiated: wakefulness (18 h/d [3]), drowsiness (2 h/d [3]), slow-wave sleep (SWS, 3 h/d [3]), and paradoxical or rapid eye movement (REM) sleep (<1 h/d [3]) with only the latter two counting toward the total sleep time budget of typically 2.9–3.5 h/d (10–21% of the total daily time budget) [3,5–8,12,16,31–33,38–40]. REM sleep represents the smallest proportion (10–15%) of the total sleep time, while SWS, at 65% takes up most sleep time [4,5,7]. While sleep in horses, in contrast to most other species, is not uniquely associated with recumbency, as horses can go through SWS in both standing and recumbent positions, the muscle atonia associated with REM sleep requires sternal or lateral recumbency [1–5,7,11,14,16,31]. Indeed, since horses usually fall asleep shortly after lying down, recumbency can be used as an inferred measure of sleep [7,11,12,16]. Polysomnographic studies demonstrated that horses were in REM sleep 29.7% of the time spent sleeping in sternal and 33.7% of the time in lateral recumbency, with SWS accounting for the remaining time [16,39]. As a reduction in recumbent sleep states cannot be compensated for by increased sleep time standing, the reluctance of a horse to enter a recumbent position causes REM sleep deficiency and can have substantial effects on health and quality of life [14].

The duration of lying, and with it the quality and length of sleep, is affected by various environmental influences, including the availability of a suitable lying area, space allowance, the presence and type of bedding, and lighting conditions [11,14,19,20,22,23,36,39,41–47]. Also, age influences the lying times with foals (up to 53.1% in domestic foals [10] and up to 15% in semi-feral foals [48]) and young horses (<2 years; up to 27% in domestic horses raised for meat production [49,50]; up to 8% in semi-feral horses [27]), who spend more time in recumbency than adults (3–15% in domestic and semi-feral horses [1,7,12,16,20,21,24–33,51,52]); however, the influence of old age on lying times has not yet been reported. In addition, painful conditions can modify lying times [4,18,53–57]. While recumbency is increased in acute pain due to colic or acute laminitis, it has recently been reported to decrease in horses suffering from angular limb deformities; analgesia administration resulted in a return to regular lying times [4,18,53–57]. However, the effect of other chronic orthopedic diseases such as osteoarthritis, tendinopathy, or chronic laminitis on recumbency has not been evaluated yet. The paucity of data on equine recumbency times is mainly due to the time and resource requirements for measuring this predominantly nocturnal behavior by direct or video observation without affecting the behavior studied [58,59]. The individual variation of the equine behavioral circadian rhythm requires detailed surveillance over several successive days [12,60–63]. However, to date, only a few studies, four in (semi-)feral and five in domesticated horses, measured recumbency times for a minimum of 24 continuous hours [12,24–29,48–50]. Recent advances in biotelemetry, and biologging, using wearable automated tracking equipment, provide increased objectivity and new opportunities to remotely quantify behavior at scales and temporal resolutions that were not previously possible [12,64]. These new technologies facilitate accurate time budget analysis over several successive days as an objective, quantitative measure of behavior, and have the potential to become a reliable tool for on-farm assessment of equine welfare.

Given the aging equine population and the prevalence of musculoskeletal problems in horses [65–74], the present study aimed to assess the influence of chronic musculoskeletal disease and old age on the lying time budgets of horses using wearable sensor technology. We hypothesize that geriatric horses and horses suffering from chronic orthopedic discomfort spend less time recumbent than healthy adult control horses.

2. Materials and Methods

2.1. Horses, Housing and Management Conditions

This prospective, observational cohort study was carried out in 83 horses, 39 warmbloods, 17 draft horses, and 27 horses of other breeds (Supplementary Table S1), owned by an animal sanctuary. Horses were housed in familiar environments, in individual box stalls (16 m^2, n= 55) or group housing (2–10 horses/group, $\geq$11 m^2/horse, n = 27), and had daily paddock or pasture turn-out (season and weather-dependent) in groups that had been stable for at least 6 months. Lying surfaces were bedded with straw (n = 64) or shavings (n = 18). Horses had ad libitum access to water and were fed a predominantly grass and hay-based diet ad libitum or rationed, depending on their nutritional requirements.

Prior to inclusion in the study and every three months for the duration of the study, horses were examined by the same veterinarian, and their physical health and body condition score (BCS, range 1 (=extremely emaciated) to 9 (=extremely fat) [75]) were recorded. Based on their age, and physical and orthopedic exams, horses were assigned to one of four health/age groups: (1) horses younger than 20 years with chronic orthopedic diseases (chronic lameness > 1 (on the American Association of Equine Practitioner (AAEP) scale), n = 31); (2) geriatric horses ($\geq$20 years) with chronic orthopedic disease (n = 40); (3) sound (lameness $\leq$ 1) geriatric horses (n = 7); and (4) sound horses younger than 20 years (control group, n = 5). Horses with cardiovascular, respiratory, or abdominal disease or acute onset or exacerbation of lameness were excluded from the study. Horses that were observed to collapse or that exhibited the associated pathognomonic skin lesions on the dorsal aspect of their carpi and fore fetlocks (Figure 1, Supplementary Videos S1 and S2) were considered REM-sleep deprived. Horses that were included in the study but developed additional health problems or an acute exacerbation of their musculoskeletal disease after the first tracking round was completed did not participate in additional tracking rounds to avoid masking the effects of the chronic conditions that were the focus of this study with acute disease.

2.2. Automated Equine Monitoring

Using the Trackener$^{\circledR}$ (London, UK) automated equine monitoring system [76], horses were tracked 1–3 times within 15 months for a minimum of 60 continuous hours (max of 360 h), each with horses that showed abnormal recumbency patterns being tracked longer and repeatedly. The Trackener$^{\circledR}$ system measures the horse's body position (standing, sternal, or lateral recumbency), gait, speed, rein (left, straight, right), and location [76]. The wearable horse unit (140 $\times$ 50 $\times$ 30 mm, weight: 190 g), containing a MEMS 3-axis accelerometer, a gyroscope, a barometer, a temperature sensor, and a GPS, is carried within a special lycra bib (Figure 1).

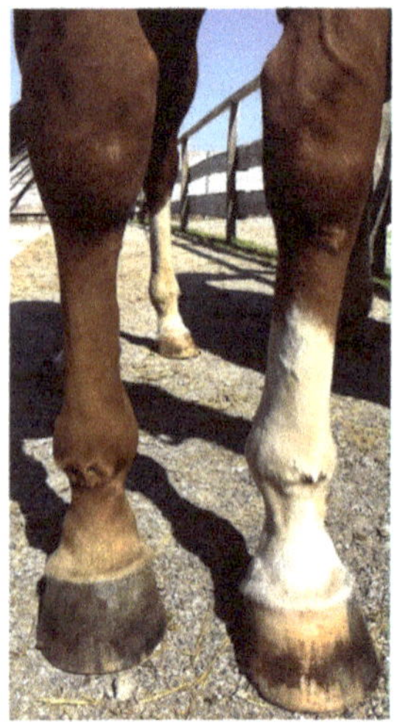 

Figure 1. The photograph on the left shows the pathognomonic skin lesions over the dorsal aspect of both metacarpophalangeal (fetlock) joints and carpi of a horse with chronic REM deficit. The photograph on the right displays a horse equipped with the automated equine monitoring system (Trackener®) in the lycra bib.

The wearable horse units send the data via GSM communication (3G network) to the cloud. An artificial intelligence algorithm analyses the data and displays the amount of time the horse spent resting (detailed by body position into standing, sternal, or lateral recumbency) and active (divided into walk, trot, and canter). The activity data measured by the Trackener® device have been previously validated in an equine hospital setting against direct observation of horses' activities recorded manually based on CCTV recordings, which yielded a mean agreement of 95.7% [76].

Time budgets are presented as minutes per hour and day in the corresponding Trackener® app for each behavior category. In addition, data differentiated between recumbency versus upright body position and between standing and movement are provided in csv format for further analysis. Recorded lying times of less than 1 min were considered artifacts and not included in the recumbency time budget. To facilitate reading, recumbency times are indicated in minutes rather than the percentage of the 24 h time budget used for movement and standing, as the small lying time budgets expressed in percentages were too cumbersome to interpret in the context of horses' daily behavioral routines.

2.3. Statistical Analysis

Continuous variables were expressed as mean ± standard deviation (s.d.), and categorical variables were expressed as percentages. The effect of time and horse on time budgets was analyzed using a Brown–Forsythe ANOVA to account for the difference in group size. A generalized linear model using time budgets as target variables and age, lameness (yes/no), REM sleep deficiency (yes/no), BCS, sex (gelding/mare), breed group (warmblood, draft, or other), and pasture (yes/no) as explanatory variables was calculated. Statistical analyses were carried out using Graphpad Prism version 9 (Graphpad Software, San Diego, CA, USA) [77]. A p-value < 0.05 was considered significant.

2.4. Ethics Statement

This study was non-invasive and entailed only monitoring the horses under their current conditions of life. No specific veterinary treatments or interventions were carried out for the purpose of this study. The study was thus reviewed by the Institutional Ethics Committee of the University of Veterinary Medicine Vienna (ETK-152/09/2019) in accordance with the "Good Scientific Practice. Ethics in Science and Research" guidelines implemented at the University of Veterinary Medicine, Vienna and national legislation, and ethical approval was waived.

3. Results

3.1. Horses and Tracking

Of the 83 horses included in this study, 38 were mares and 45 geldings. Their age ranged from 2 to 32 years (20.7 ± 6.2 years), and their body condition score from 3 to 7 (5.7 ± 1.1) (Supplementary Table S1). In health/age group 1 (lame, <20 y, $n = 31$), horses' mean age was 15.7 (±4.2 s.d.); in group 2 (lame, ≥20 y, $n = 40$), 24.9 (±3.5 s.d.); in group 3 (sound, ≥20 y, $n = 7$), 25 (±3.5 s.d.); and in group 4 (sound, <20 y, $n = 5$), 12 (±5.8 s.d.). Eight horses (age: 24.5 ± 5.5), all of whom suffered from chronic orthopedic disease) showed signs of REM sleep deficit (collapse or pathognomonic skin lesions observed). Four of the eight REM-sleep-deprived horses had single stalls with straw bedding, 2 had single stalls with shavings, and 2 were in small group housing with straw bedding.

BCS was significantly affected by breed ($p = 0.0048$, F(Dfn, DFd) = 5.598 (2, 115)) and chronic lameness ($p = 0.002$, F(Dfn, DFd) = 9.974 (1, 115)), but not by age ($p = 0.4029$), REM sleep deficiency ($p = 0.0686$), sex ($p = 0.1784$), pasture access ($p = 0.6774$), bedding ($p = 0.5547$), or housing ($p = 0.1634$) conditions.

All horses tolerated the wearable horse unit well, and no dermal irritations were observed. Data collection and transfer functioned well in 131 tracking cycles, and no technical problems were encountered.

3.2. Time Budgets for Lying and the Influence of Age, Lameness, Presence of REM Deficit Symptoms, Sex, BCS, Breed, and Pasture Access on Recumbency

The overall mean duration of recumbency was 67.4 min (±61.9 s.d. range: 0–319 min) per day. Young, lame horses were recumbent for 85 min (±70.3 s.d.); old, lame horses for 59.7 min (±58.5 s.d.); old, sound horses for 36.5 min (±26.5 s.d.); and young, sound horses for 64.1 min ((±50.7 s.d.) (Figure 2, Table 1, Supplementary Tables S1 and S2). The effects of age ($p = 0.7408$, F(Dfn, DFd) = 0.11 (1, 115)) and lameness ($p = 0.3072$, F(Dfn, DFd) = 1.052 (1, 115)) were not statistically significant.

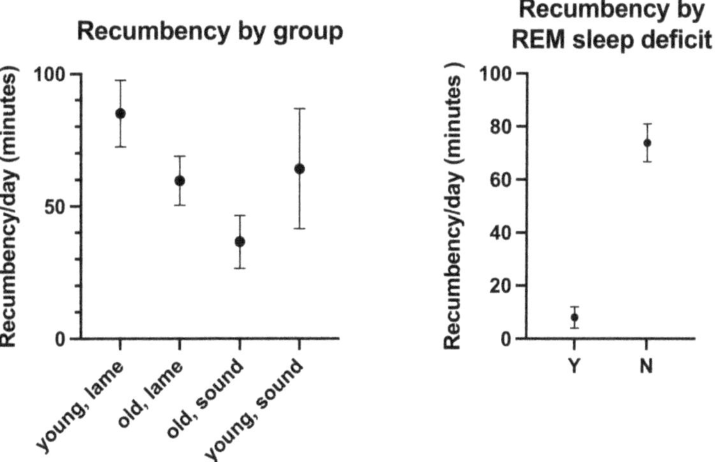

Figure 2. Comparative recumbency times (mean +/−SEM minutes per day) of the horses enrolled in this study by age/lameness group and by the diagnosis of REM deficit (Yes (Y) and No (N)).

Table 1. The recumbency duration is detailed by group (young lame, old lame, old sound, young sound) and by the presence of REM deficit symptoms. In addition, the number of horses per recumbency duration period is listed.

	n	Recumbency: Minutes/Day (Mean)	Recumbency: Minutes/Day (s.d.)	No Recumbency	Recumbency <10 min	Recumbency 10–30 min	Recumbency 30–60 min	Recumbency 60–120 min	Recumbency 120–180 min	Recumbency >180 min
overall	83	67.4	61.9	13	7	9	23	19	10	1
young, lame	31	85.0	70.3	7	1	2	6	9	5	0
old, lame	40	59.7	58.5	6	4	3	23	8	4	1
old, sound	7	36.5	26.5	0	1	3	2	1	0	
young, sound	5	64.1	50.7	0	1		2	1	1	
REM deficit	8	7.99	11.4	6	1	1	0			
No REM deficit	75	73.8	61.8	5	6	8	23	19	10	1

Horses with clinically established REM sleep deficit had significantly ($p = 0.0003$, F(Dfn, DFd) = 14.25 (1, 115)) shorter lying times (7.99 min $\pm$ 11.4 s.d.) than other horses (73.8 min $\pm$ 61.8 s.d., Table 1, Figure 2). Also, breed had a statistically significant effect on recumbency times ($p = 0.0052$, F(Dfn, DFd) = 5.513 (2, 115)), with draft horses (88 $\pm$ 84 min/d) lying significantly more than warmbloods (58 $\pm$ 59 min/d) or horses of other breeds (69 $\pm$ 67 min/d). Furthermore, BCS ($p = 0.0263$, F(Dfn, DFd) = 2.867 (4, 115)) significantly influenced lying times. Horses with a BCS of 4 had longer recumbency times (122 $\pm$ 122 min/d) than horses with lower (BCS 3: 60 $\pm$ 55 min/d) or higher BCS (BCS 5: 50 $\pm$ 47, BCS 6: 73 $\pm$ 63, BCS 7: 51 $\pm$ 52). However, bedding (straw versus shavings, $p = 0.9427$, F(Dfn, DFd) = 0.0052 (1, 115)), housing (single box stall versus group housing, $p = 0.735$, F(DFn, DFd) = 0.1151 (1, 115)), sex ($p = 0.3094$, F(Dfn, DFd) = 1.042 (1, 115)), and pasture access ($p = 0.2901$, F(Dfn, DFd) = 1.13 (1, 115)) did not significantly affect lying times.

Time of day and horse had a significant influence on lying time budgets ($p < 0.0001$, Figure 3, Supplementary Figure S1). Eight horses slept only at night (0:00–4:00), one only during the day (4:00–20:00), the others distributed over the 24 h day. Between 0:00 and 4:00, lying bouts lasted 2 to 57 min (16 $\pm$ 15 s.d.); in the time between 4:00–20:00, 7–12 min (10 $\pm$ 7 s.d.); and between 20:00–0:00, 7–16 min (10 $\pm$ 12 s.d.).

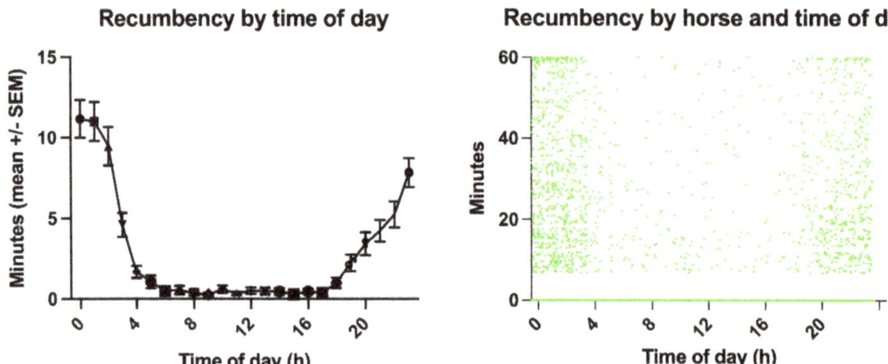

Figure 3. The left graph illustrates the recumbency times throughout the day, showing the primarily nocturnal recumbency distribution with a peak between midnight and 4:00 a.m. The right graph, the scatter plot of the recumbency minutes per horse, demonstrates the individual variation of the circadian sleep rhythm of the horses enrolled in this study.

Twenty-nine horses were recumbent for less than 30 min per day throughout the entire study period, 13 of which all suffered from chronic orthopedic disease and did not lie down at all for more than 72 h (Figure 4); five of these horses lay down during subsequent tracking periods for 2–48 min. Thirty horses were consistently recumbent for more than one hour per day (Figure 4).

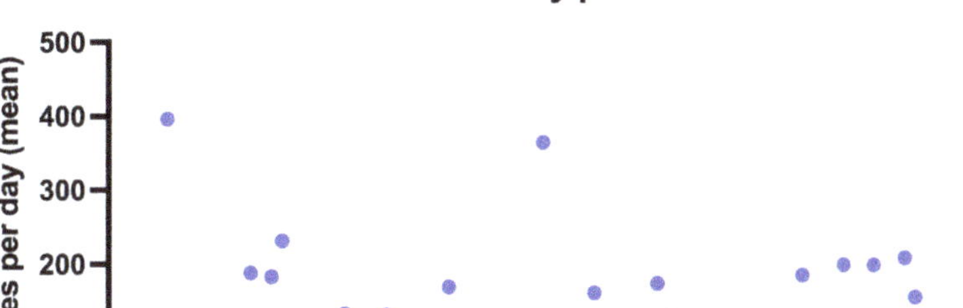

Recumbency per horse

Figure 4. Recumbency time budget (mean %) per horse (horses are distributed alphabetically along the x-axis). Horses with established REM deficit are highlighted in red, turquoise indicates horses with less than 30 min sleep per day throughout the study period, and horses that lay down more than average are marked in blue.

The coefficient of variation was, on average, 370% (range 0–1076%) for evidently REM-deficient horses and 396% (range 0–990%) for other horses (Figure 5). Horses with REM deficit symptoms notably either had a coefficient of variation of 0% because they did not lie down at all, or above 500% because they lay down only rarely. In total, 19 horses had a coefficient of variation above 500%, and 6 of 0%, all of whom had recumbency times well below the population average.

Recumbency Coefficient of Variation per Horse

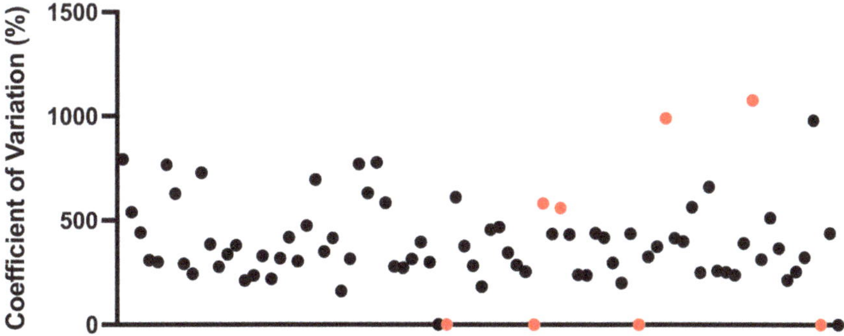

Figure 5. Coefficient of Variation (%) of recumbency times. Horses (distributed alphabetically along the x-axis) with established REM deficiency symptoms are indicated in red.

3.3. Time Budgets for Locomotion and the Influence of Age, Lameness, Presence of REM Deficit Symptoms, Sex, BCS, Breed, and Pasture Access on Movement

The overall mean time budget for locomotion was 19.1% (±11.2% s.d., range: 4.09–55.8%, Supplementary Tables S1 and S3). Young, lame horses moved for 20.79% (±15.13% s.d.) of their day, old, lame horses for 17.23% (±12.05% s.d.), old, sound horses for 15.92% (±6.8% s.d.), and young, sound horses for 31.7% (±11.85% s.d.) (Figure 6, Supplementary Tables S1 and S3). The effect of age (p = 0.0302, F(Dfn, DFd) = 4.82 (1, 114)) but not of lameness (p = 0.6867, F(Dfn, DFd) = 0.1635 (1, 114)) on movement time budgets was statistically significant, with young horses moving more (22.39% ± 15.15%) than old horses (17.05% ± 11.46%). Horses with evident REM deficits moved significantly less

(16.5% ± 5.42%) than other horses (19.3% ± 7.07%, *p*= 0.0383, F(Dfn, DFd) = 4.392 (1, 114)). Unsurprisingly, horses with access to pasture moved significantly more (33.31% ± 12.74% of their time budget) than horses with more restricted turn-out (12.56% ± 7–08%, *p* < 0.0001, F(Dfn, DFd) = 119.7 (1, 114)). Bedding (*p* = 0.5511, F(Dfn, DFd) = 0.3575 (1, 114)), housing (*p* = 0.066, F(Dfn, DFd) = 3.446 (1, 114)), sex (*p* = 0.5827, F(Dfn, DFd) = 0.3036 (1,114)), breed (*p* = 0.1492, F(Dfn, DFd) = 1.935 (2, 114)), and BCS (*p* = 0.8084, F(Dfn, DFd) = 0.3999 (4, 114)) had no significant effect on movement. Time of day had a significant influence on the time budgets for locomotion (*p* < 0.0001, Figure 7), but horse did not (*p* = 0.7124).

3.4. Time Budget for Standing and the Influence of Age, Lameness, Presence of REM Deficit Symptoms, Sex, BCS, Breed, and Pasture Access on Standing Times

The overall mean time budget for standing was 75.6% (±13.1 s.d., range: 32.2–95.9%, Supplementary Tables S1 and S4). Young, lame horses were standing for 72.43% (±15.99% s.d.) of their day, old, lame horses for 78.08% (±13.74% s.d.), old, sound horses for 81.45% (±7.03% s.d.), and young, sound horses for 63.05% (±15.7% s.d.) (Figure 8, Supplementary Tables S1 and S2). The effect of age (*p* = 0.0489, F(Dfn, DFd) = 3.965 (1, 114)) but not lameness (*p* = 0.7067, F(Dfn, DFd) = 0.1424 (1, 114)) was statistically significant, with old horses standing more (78.53 ± 13.06) than young horses (71.1 ± 16.23). REM sleep deficit (*p* = 0.6012, F(Dfn, DFd) = 0.2747 (1, 114)), BCS (*p* = 0.9484, F(Dfn, DFd) = 0.1799 (4, 114)), bedding (*p* = 0.5069, F(Dfn, DFd) = 0.4433 (1, 114)), housing (*p* = 0.126, F(Dfn, DFd) = 2.376 (1, 114)), sex (*p* = 0.4451, F(Dfn, DFd) = 0.5872 (1, 114)), and breed (*p* = 0.9769, F(Dfn, DFd) = 0.0234 (2, 114)) had no statistically significant effect on standing times, but pasture access (*p* < 0.0001, F(Dfn, DFd) = 96.36 (1, 114)) did. Horses on pasture stood significantly less (60.4% ± 14.13% of their time budget) than those with more restricted turn-out (82.38% ± 8.77%).

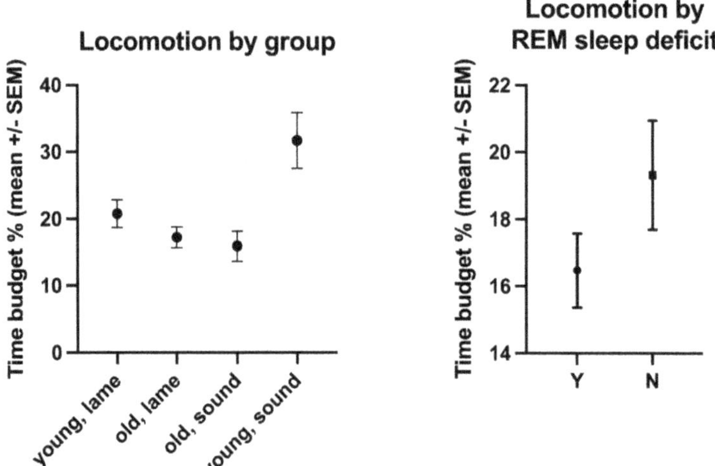

Figure 6. Comparative locomotion time budgets (mean % of 24 h +/− SEM) of the horses enrolled in this study by age/lameness group and by the diagnosis of REM deficit (Yes (Y) and No (N)).

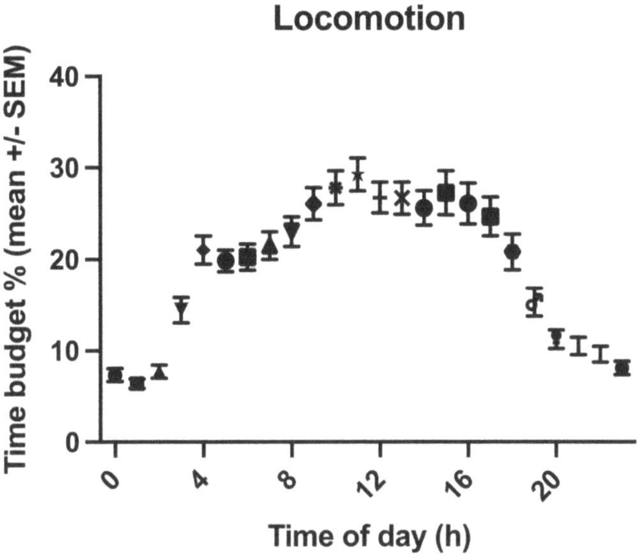

Figure 7. Graph of the locomotion time budget (mean % of 24 h +/−SEM) throughout the day, showing the movement peak during daytime hours and less movement at night.

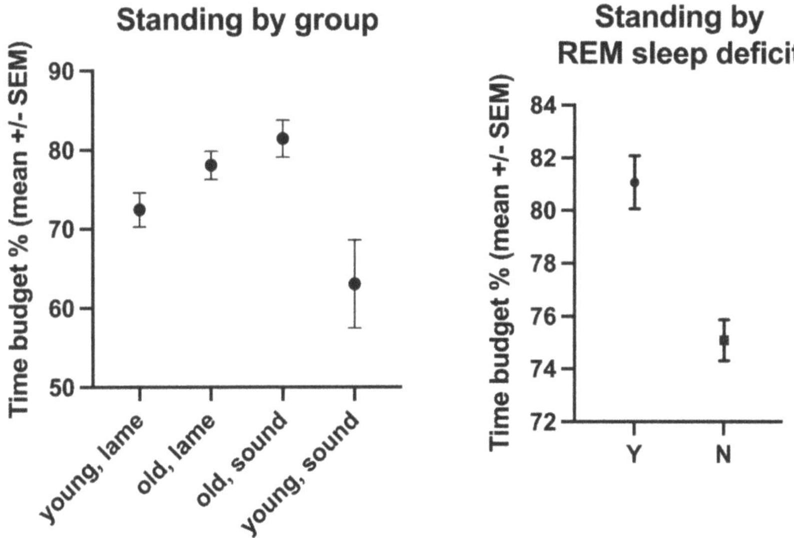

Figure 8. Comparative standing time budgets (mean % of 24 h +/−SEM) of the horses enrolled in this study by age/lameness group and by the diagnosis of REM deficit (Yes (Y) and No (N)).

Time of day significantly influenced standing time budgets ($p < 0.0001$, Figure 9), but horse did not $p = 0.4778$).

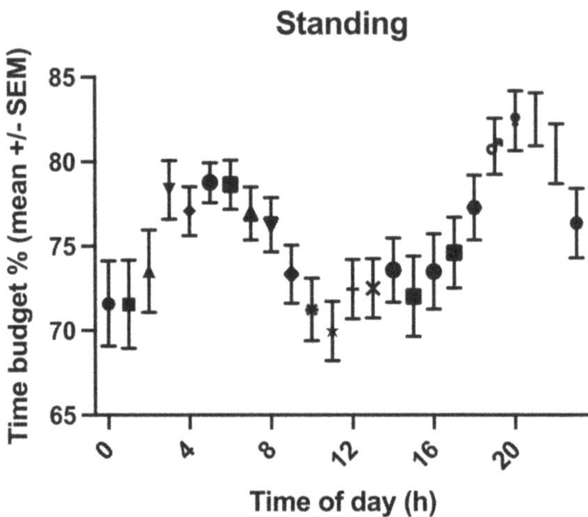

Figure 9. Graph of the standing time budget (mean % of 24 h +/−SEM) throughout the day with a peak in standing times in the morning and evening.

4. Discussion

Recumbency times of the primarily geriatric and lame population of horses in this study ranged from 0 to 319 min per day, with a mean of 67.4 min. Interestingly, neither age nor lameness due to chronic orthopedic disease significantly influenced recumbency times in this study. Young, lame horses lay down for 85, old, lame horses for 59.7, old, sound horses for 36.5, and young, sound horses for 64.1 min per day. As sleep duration is inversely proportional to the risk of predation, horses sleep only for short periods, typically for 2–15 min, and rarely remain recumbent for longer than 30 min at a time [1,7,12,16,20,21,24–33]. The occurrence and duration of recumbency depend on the horse's ability to find a comfortable and safe place to lie [3–5,47]. Correspondingly, decreased lying time budgets are associated with unsuitable environmental conditions, stress, social insecurity, and pain [18,24,51,56,78]. Adaptions to a horse's environmental conditions should therefore be considered if a horse shows insufficient lying times.

Intriguingly, while (semi-)feral horses were observed to prefer open spaces for recumbency [2,7,11,27,28,34], in domesticated horses, recumbency times are longer in box stalls than in free-stall housing or on pasture [16,19,23]. Also, the effect of an individual's hierarchical status on recumbency times depends on space availability [7,11,20,27,28,79,80]. In free-ranging horses, social rank has no effect on lying times, since the lack of spatial limitations under natural conditions seems to allow each individual within a group to satisfy their demand for recumbency [7,11,20,27,28,79,80]. In contrast, in group-housed domesticated horses, for whom a suitable lying area represents a potentially limited resource, larger lying surfaces increase the duration of recumbency and decrease the proportion of forcedly terminated lying bouts in low-ranking horses [7,11,20,27,28,79,80]. In the current study, neither pasture access nor housing conditions (single box stable versus group housing) significantly affected recumbency times, which may be due to the stable group composition and availability of adequate lying surfaces in the equine sanctuary. The effect of hierarchy on equine recumbency times was not assessed in this study, but further studies looking at the effect of social rank on recumbency times of group-housed horses are needed to establish evidence-based husbandry recommendations to improve equine welfare.

Both (semi-)feral and domestic horses prefer dry, clean, and soft lying surfaces [11,14,19–23,31,36,39,41–47,80]. While straw bedding has been reported to increase recumbency times compared to shavings or rubber mats [19–22,41–45,47], there was no difference in lying times between horses with straw versus shavings bedding in this study.

Visual and auditory stimuli have also been observed to influence equine recumbency and sleep times, with absent stimuli increasing SWS, but also especially REM sleep [2,3,5,14,32,38,39,41]. Unsurprisingly, artificial light overnight or during late-night checks, which may affect melatonin cycles and subsequently sleep patterns, decreases recumbency [14,38,39]. In contrast, music reduced alertness and increased recumbency periods, possibly by masking the occurrence of trivial novel environmental auditory stimuli [38,81,82].

In addition to environmental factors, personal factors also influence sleep. While age and sex had no impact on recumbency duration in this study, BCS and breed did significantly affect lying times, with draft horses (88 ± 84 min/d) lying down more than warmbloods (58 ± 59 min/d) or other breeds (69 ± 67 min/d). Surprisingly, moderately thin (BCS 4) horses had significantly longer recumbency times than horses in a thin (BCS 3) or moderate (BCS 5) to fleshy (BCS 7) condition. While in humans, sleep deprivation is associated with weight gain [83], the BCS of REM-sleep-deprived horses (5.38 ± 0.96) was similar to the other horses (5.71 ± 1.06). The BCS in this study (5.7 ± 1.1) was slightly above the midpoint of the scale, which is consistent with results obtained from other equine populations [84] and the emergence of equine obesity as one of the most important equine welfare issues in the western world [85,86]. However, the known association between breed type and BCS, with draft horses having the highest BCS [87], requires caution when interpreting results, and further studies, with selected breeds and a wider spread of BCS, to independently assess the influence of these variables on recumbency times.

In this study, 8 out of 83 horses showed symptoms of REM deficit. These horses had significantly shorter lying times (7.99 min ± 11.4 s.d.) and locomotion time budgets (16.5% ± 5.42%) than the other horses enrolled in this study (73.8 min ± 61.8 s.d. lying, 19.3% ± 7.07% locomotion). To identify potential REM-deficient horses, it proved essential to track them for more extended periods, to determine whether they do not lie down at all, or only do so when exhausted or under specific environmental conditions. For example, one horse with physical problems lying down due to severe osteoarthritis in both carpi did lie down on an incline in the pasture, which made it easier for the horse to get up.

REM-sleep deficiency due to recumbent sleep deprivation caused by illness, ethological deficits, or husbandry shortcomings typically manifests in excessive drowsiness and horses' literally falling asleep while standing and partially collapsing before suddenly waking again, resulting in pathognomonic skin lesions on the dorsal aspect of the carpi and front fetlocks (Figure 1) [11]. The collapse is commonly incorrectly diagnosed as narcolepsy, a rare neurological sleep disorder characterized by daytime sleepiness, cataplexy, and sleep paralysis [9]. Recumbent sleep deprivation may have psychological or physical causes. Environmental-insecurity-associated excessive drowsiness arises when horses are not feeling comfortable to lie down due to social insecurity or fear of predation [2–5,7,9,10,14,15,17]. Also, stereotypical behaviors are associated with both suboptimal environmental conditions and decreased REM sleep [56,88–92]. Horses suffering from chronic musculoskeletal disease may be hesitant to assume recumbency due to pain or mechanical difficulties during lying down or rising. In contrast, administering analgesics to horses suffering from orthopedic pain increased lying times [12,18,51,56]. As REM-sleep deprivation is associated with hyperalgesia and persistence of chronic pain in other species [12,18,93,94], reduced recumbent sleep due to chronic pain may intensify pain sensation, further contributing to the problem [18]. REM sleep deficiency thus impairs welfare and health, necessitating further studies to investigate methods for early diagnosis and management as an essential basis to adjust husbandry and welfare decisions accordingly.

Interestingly, age, but not lameness significantly affected the time budget for locomotion and standing in this study, with old horses standing significantly more than younger horses. The standing time budget measured with the wearable tracking device in this study encompasses a broad repertoire of behaviors, including standing while eating or resting. The poor resolution of the standing behavior is one of the limitations of this study and would require the addition of other wearable equipment to remedy. Another limitation of this study is the low number of young and healthy horses. As the population of horses in animal shelters tend to include a large proportion of geriatric and lame horses, a more even distribution of groups was not possible within the framework of this study. To address this limitation, we also compared our results to time budgets reported for healthy adult horses in the literature.

Recumbency times in this study were lower than the 203 +/− 46.5 min reported in the literature. However, the recumbency duration of adult horses has to date only been quantified in observational studies using scan or focal sampling, which may yield less detailed and accurate measurements or in polysomnographic analyses of equine sleep phases over 24 h only, which is insufficient time to measure cyclic recumbency patterns. The wearable automated tracking equipment used in this study enables the continuous measurement of recumbency, locomotion, and standing times over several days with a temporal resolution of one second. The ease of use, excellent tolerance of the sensor-bib by the horses, and automated app-based data analysis facilitate its use on-farm to identify horses with inadequate recumbency times or problems with locomotion, for welfare assessment and monitoring of the success of interventions. However, although the horses were housed in an animal sanctuary under appropriate management conditions, as is shown by the homogenous distribution of recumbency throughout the age and lameness groups, further optimization of environmental conditions, with individual adaptions to accommodate the unique population of horses living in an animal sanctuary, may increase recumbency times.

5. Conclusions

Recumbency is a prerequisite for horses achieving rapid eye movement (REM) sleep and completing a full sleep cycle. Hence, measuring lying behavior is an essential component of equine welfare assessment. Wearable sensor technology can identify horses with low recumbency times and at risk for a REM-sleep deficit. Horses with REM deficit symptoms have not only lower recumbency but also decreased locomotion time budgets, indicating a general compromise of well-being. Interestingly, neither age nor lameness due to chronic orthopedic disease significantly influenced recumbency times in this study. Thus, geriatric horses and horses suffering from chronic orthopedic conditions can achieve recumbency times comparable to younger, healthy horses, but may require optimized husbandry conditions.

Supplementary Materials: The following are available online at https://www.mdpi.com/article/10.3390/ani11113189/s1, Supplementary Table S1 details the age, sex, breed, group, the presence of lameness or REM deficit symptoms, BCS, access to pasture, and the time budgets (mean, s.d.) for lying, locomotion, and standing for all horses enrolled in this study. Supplementary Video S1: Horse collapsing due to REM-sleep deficit. Supplementary Video S2: Horse with REM sleep deficit falling asleep standing as it cannot lie down due to severe carpal osteoarthritis. Supplementary Table S2 details the recumbency time budgets per hour of the day for each horse enrolled in this study. Supplementary Figure S1: The distribution of recumbency times of three exemplary horses is shown for all 7–11 days they were tracked to demonstrate the strong individual variation. Horse A had long recumbency times and a regular daily rhythm, while horse B (the red line is not visible as it is continuously on the zero-line) never lay down (in 7 days), and horse C lay down only once in 11 days. Horses B and C showed REM deficit symptoms. Supplementary Table S3 details the locomotion time budgets per hour of the day for each horse enrolled in this study. Supplementary Table S4 details the standing time budgets per hour of the day for each horse enrolled in this study.

Author Contributions: Conceptualization: U.A. and F.J.; Data collection: Z.K., U.A. and F.J., Data analysis: Z.K., U.A. and F.J.; writing—original draft preparation: U.A. and F.J.; writing—review and editing: Z.K., H.G., M.L., U.A. and F.J. All authors have read and agreed to the published version of the manuscript.

Funding: This research was funded by Gut Aiderbichl.

Institutional Review Board Statement: This study was non-invasive and entailed only monitoring the horses under their current conditions of life. No specific veterinary treatments or interventions were carried out for the purpose of this study. The study was thus reviewed by the Institutional Ethics Committee of the University of Veterinary Medicine Vienna (ETK-152/09/2019) in accordance with the "Good Scientific Practice. Ethics in Science and Research" guidelines implemented at the University of Veterinary Medicine Vienna and national legislation; ethical approval was waived.

Informed Consent Statement: Not applicable.

Data Availability Statement: All pertinent data is included in the manuscript and Supplementary Materials.

Acknowledgments: The authors thank Gut Aiderbichl and its team for the financial and technical support.

Conflicts of Interest: The authors declare no conflict of interest.

References

1. Littlejohn, A.; Munro, R. Equine Recumbency. *Vet. Rec.* **1972**, *90*, 83. [CrossRef]
2. Houpt, K. The Characteristics of Equine Sleep. *Equine Pract.* **1980**, *2*, 8–17.
3. Dallaire, A. Rest Behavior. *Vet. Clin. N. Am. Equine Pract.* **1986**, *2*, 591–607. [CrossRef]
4. Aleman, M.; Willams, C.; Holliday, T. Sleep and Sleep Disorders in Horses. *AAEP Proc.* **2008**, *54*, 180–185.
5. Belling, T. Sleep Patterns in the Horse. *Equine Pract.* **1990**, *12*, 22–27.
6. Dallaire, A.; Rucklebusch, Y. Sleep and Wakefulness in the Housed Pony under Different Dietary Conditions. *Can. J. Comp. Med.* **1974**, *38*, 65–71.
7. Wöhr, A.; Kalus, M.; Reese, S.; Fuchs, C.; Erhard, M. Equine Sleep Behaviour and Physiology Based on Polysomnographic Examinations. *Equine Vet. J.* **2016**, *48*, 9. [CrossRef]
8. Williams, D.C.; Aleman, M.; Holliday, T.A.; Fletcher, D.J.; Tharp, B.; Kass, P.H.; Steffey, E.P.; LeCouteur, R.A. Qualitative and Quantitative Characteristics of the Electroencephalogram in Normal Horses during Spontaneous Drowsiness and Sleep. *J. Vet. Intern. Med.* **2008**, *22*, 630–638. [CrossRef] [PubMed]
9. Fuchs, C.; Kiefner, C.; Reese, S.; Erhard, M.; Wöhr, A. Narcolepsy: Do Adult Horses Really Suffer from a Neurological Disorder or Rather from a Recumbent Sleep Deprivation/Rapid Eye Movement (REM)-Sleep Deficiency? *Equine Vet. J.* **2016**, *48*, 9. [CrossRef]
10. Zanker, A.; Wöhr, A.-C.; Reese, S.; Erhard, M. Qualitative and Quantitative Analyses of Polysomnographic Measurements in Foals. *Sci. Rep.* **2021**, *11*, 16288. [CrossRef] [PubMed]
11. Burla, J.-B.; Rufener, C.; Bachmann, I.; Gygax, L.; Patt, A.; Hillmann, E. Space Allowance of the Littered Area Affects Lying Behavior in Group-Housed Horses. *Front. Vet. Sci.* **2017**, *4*, 23. [CrossRef] [PubMed]
12. Auer, U.; Kelemen, Z.; Engl, V.; Jenner, F. Activity Time Budgets—A Potential Tool to Monitor Equine Welfare? *Animals* **2021**, *11*, 850. [CrossRef] [PubMed]
13. Kelemen, Z.; Grimm, H.; Vogl, C.; Long, M.; Cavalleri, J.M.V.; Auer, U.; Jenner, F. Equine Activity Time Budgets: The Effect of Housing and Management Conditions on Geriatric Horses and Horses with Chronic Orthopaedic Disease. *Animals* **2021**, *11*, 1867. [CrossRef]
14. Greening, L.; Downing, J.; Amiouny, D.; Lekang, L.; McBride, S. The Effect of Altering Routine Husbandry Factors on Sleep Duration and Memory Consolidation in the Horse. *Appl. Anim. Behav. Sci.* **2021**, *236*, 105229. [CrossRef]
15. Fuchs, C. Narkolepsie Oder REM-Schlafmangel? Ph.D. Thesis, LMU, München, Germany, 2017.
16. Kalus, M. Schlafverhalten Und Physiologie Des Schlafes Beim Pferd Auf Der Basis Polysomnographischer Untersuchungen. Ph.D. Thesis, LMU, München, Germany, 2014.
17. Bertone, J.J. Excessive Drowsiness Secondary to Recumbent Sleep Deprivation in Two Horses. *Vet. Clin. N. Am. Equine Pract.* **2006**, *22*, 157–162. [CrossRef]
18. Clothier, J.; Small, A.; Hinch, G.; Barwick, J.; Brown, W.Y. Using Movement Sensors to Assess Lying Time in Horses with and Without Angular Limb Deformities. *J. Equine Vet. Sci.* **2019**, *75*, 55–59. [CrossRef] [PubMed]
19. Köster, J.; Hoffmann, G.; Bockisch, F.-J.; Kreimeier, P.; Köster, J.R.; Feige, K. Lying Behaviour of Horses Depending on the Bedding Material in Individual Housing in Boxes with or without Adjacent Pen. *Pferdeheilkunde Equine Med.* **2017**, *33*, 43–51. [CrossRef]
20. Baumgartner, M.; Zeitler-Feicht, M.H.; Wöhr, A.-C.; Wöhling, H.; Erhard, M.H. Lying Behaviour of Group-Housed Horses in Different Designed Areas with Rubber Mats, Shavings and Sand Bedding. *Pferdeheilkunde Equine Med.* **2015**, *31*, 211–220. [CrossRef]
21. Chaplin, S.; Gret, L. Effect of Housing Conditions on Activity and Lying Behaviour of Horses. *Animal* **2010**, *4*, 792–795. [CrossRef]
22. Werhahn, H.; Hessel, E.F.; Bachhausen, I.; Van den Weghe, H.F.A. Effects of Different Bedding Materials on the Behavior of Horses Housed in Single Stalls. *J. Equine Vet. Sci.* **2010**, *30*, 425–431. [CrossRef]
23. Raabymagle, P.; Ladewig, J. Lying Behavior in Horses in Relation to Box Size. *J. Equine Vet. Sci.* **2006**, *26*, 11–17. [CrossRef]
24. Price, J.; Catriona, S.; Welsh, E.M.; Waran, N.K. Preliminary Evaluation of a Behaviour-based System for Assessment of Postoperative Pain in Horses Following Arthroscopic Surgery. *Vet. Anaesth. Analg.* **2003**, *30*, 124–137. [CrossRef]
25. Aristizabal, F.; Nieto, J.; Yamout, S.; Snyder, J. The Effect of a Hay Grid Feeder on Feed Consumption and Measurement of the Gastric PH Using an Intragastric Electrode Device in Horses: A Preliminary Report. *Equine Vet. J.* **2014**, *46*, 484–487. [CrossRef]

26. Correa, M.G.; e Silva, C.F.R.; Dias, L.A.; Junior, S.D.S.R.; Thomes, F.R.; do Lago, L.A.; de Mattos Carvalho, A.; Faleiros, R.R. Welfare Benefits after the Implementation of Slow-Feeder Hay Bags for Stabled Horses. *J. Vet. Behav.* **2020**, *38*, 61–66. [CrossRef]
27. Duncan, P. Time-Budgets of Camargue Horses II. Time-Budgets of Adult Horses and Weaned Sub-Adults. *Behaviour* **1980**, *72*, 26–48. [CrossRef]
28. Duncan, P. Time-Budgets of Camargue Horses III. Environmental Influences. *Behaviour* **1985**, *92*, 188–208. [CrossRef]
29. Boyd, L.E. Time Budgets of Adult Przewalski Horses: Effects of Sex, Reproductive Status and Enclosure. *Appl. Anim. Behav. Sci.* **1988**, *21*, 19–39. [CrossRef]
30. Kiley-Worthington, M. The Behavior of Horses in Relation to Management and Training—Towards Ethologically Sound Environments. *J. Equine Vet. Sci.* **1990**, *10*, 62–75. [CrossRef]
31. Carson, K.; Wood-Gush, D.G.M. Equine Behaviour: II. A Review of the Literature on Feeding, Eliminative and Resting Behaviour. *Appl. Anim. Ethol.* **1983**, *10*, 179–190. [CrossRef]
32. Dallaire, A.; Ruckebusch, Y. Sleep Patterns in the Pony with Observations on Partial Perceptual Deprivation. *Physiol. Behav.* **1974**, *12*, 789–796. [CrossRef]
33. Ruckebusch, Y. The Relevance of Drowsiness in the Circadian Cycle of Farm Animals. *Anim. Behav.* **1972**, *20*, 637–643. [CrossRef]
34. Boyd, L.E.; Carbonaro, D.A.; Houpt, K.A. The 24-Hour Time Budget of Przewalski Horses. *Appl. Anim. Behav. Sci.* **1988**, *21*, 5–17. [CrossRef]
35. Maisonpierre, I.N.; Sutton, M.A.; Harris, P.; Menzies-Gow, N.; Weller, R.; Pfau, T. Accelerometer Activity Tracking in Horses and the Effect of Pasture Management on Time Budget. *Equine Vet. J.* **2019**, *51*, 840–845. [CrossRef] [PubMed]
36. Raspa, F.; Tarantola, M.; Bergero, D.; Bellino, C.; Mastrazzo, C.M.; Visconti, A.; Valvassori, E.; Vervuert, I.; Valle, E. Stocking Density Affects Welfare Indicators in Horses Reared for Meat Production. *Animals* **2020**, *10*, 1103. [CrossRef] [PubMed]
37. Ogilvie-Graham, T. Time Budget Studies in Stalled Horses. Ph.D. Thesis, The University of Edinburgh, Edinburgh, UK, 1994.
38. Greening, L.; Hartman, N. A Preliminary Study Investigating the Influence of Auditory Stimulation on the Occurrence of Nocturnal Equine Sleep Related Behaviour in Stabled Horses. *J. Equine Vet. Sci.* **2019**, *82*, 102782. [CrossRef]
39. Amiouny, D. The Effects of Night Light and Bedding Depth on Equine Sleep Duration and Memory Consolidation. Master's Thesis, Aberystwyth University, Penglais, UK, 2020.
40. Güntner, K. Polysomnographische Untersuchung Zum Schlafverhalten des Pferdes. Ph.D. Thesis, LMU, München, Germany, 2010.
41. Greening, L.; Shenton, V.; Wilcockson, K.; Swanson, J. Investigating Duration of Nocturnal Ingestive and Sleep Behaviors of Horses Bedded on Straw versus Shavings. *J. Vet. Behav. Clin. Appl. Res.* **2013**, *8*, 82–86. [CrossRef]
42. Ninomiya, S.; Aoyama, M.; Ujiie, Y.; Kusunose, R.; Kuwano, A. Effects of Bedding Material on the Lying Behavior in Stabled Horses. *J. Equine Sci.* **2008**, *19*, 53–56. [CrossRef]
43. Aoyama, M.; Yoshimura, N.; Sugita, S.; Kusunose, R. Effects of Used Bedding Straw and Drying It in Sunshine on Lying Behavior in Stable Horses. *J. Equine Sci.* **2004**, *15*, 67–73. [CrossRef]
44. Kwiatkowska-Stenzel, A.; Sowińska, J.; Witkowska, D. The Effect of Different Bedding Materials Used in Stable on Horses Behavior. *J. Equine Vet. Sci.* **2016**, *42*, 57–66. [CrossRef]
45. Pedersen, G.R.; Søndergaard, E.; Ladewig, J. The Influence of Bedding on the Time Horses Spend Recumbent. *J. Equine Vet. Sci.* **2004**, *24*, 153–158. [CrossRef]
46. Mills, D.S.; Eckley, S.; Cooper, J.J. Thoroughbred Bedding Preferences, Associated Behaviour Differences and Their Implications for Equine Welfare. *Anim. Sci.* **2000**, *70*, 95–106. [CrossRef]
47. Daniel, J.A.; Groux, R.; Wilson, J.A.; Krawczel, P.D.; Lee, A.R.; Whitlock, B.K. Short Communication: Trimming and Re-Shoeing Results in More Steps per Day and More Time Spent Lying per Day. *J. Equine Vet. Sci.* **2020**, *88*, 102947. [CrossRef]
48. Boy, V.; Duncan, P. Time-Budgets of Camargue Horses I. Developmental Changes in the Time-Budgets of Foals. *Behaviour* **1979**, *71*, 187–201. [CrossRef]
49. Raspa, F.; Tarantola, M.; Bergero, D.; Nery, J.; Visconti, A.; Mastrazzo, C.M.; Cavallini, D.; Valvassori, E.; Valle, E. Time-Budget of Horses Reared for Meat Production: Influence of Stocking Density on Behavioural Activities and Subsequent Welfare. *Animals* **2020**, *10*, 1334. [CrossRef] [PubMed]
50. Sartori, C.; Guzzo, N.; Normando, S.; Bailoni, L.; Mantovani, R. Evaluation of Behaviour in Stabled Draught Horse Foals Fed Diets with Two Protein Levels. *Animal* **2017**, *11*, 147–155. [CrossRef]
51. DuBois, C.; Zakrajsek, E.; Haley, D.B.; Merkies, K. Validation of Triaxial Accelerometers to Measure the Lying Behaviour of Adult Domestic Horses. *Animal* **2015**, *9*, 110–114. [CrossRef]
52. Chung, E.L.T.; Khairuddin, N.H.; Azizan, T.R.P.T.; Adamu, L. Sleeping Pattern of Horses in Selected Local Horse Stables in Malaysia. *J. Vet. Behav.* **2018**, *26*, 1–4. [CrossRef]
53. Pritchett, L.C.; Ulibarri, C.; Roberts, M.C.; Schneider, R.K.; Sellon, D.C. Identification of Potential Physiological and Behavioral Indicators of Postoperative Pain in Horses after Exploratory Celiotomy for Colic. *Appl. Anim. Behav. Sci.* **2003**, *80*, 31–43. [CrossRef]
54. Sutton, G.A.; Dahan, R.; Turner, D.; Paltiel, O. A Behaviour-Based Pain Scale for Horses with Acute Colic: Scale Construction. *Vet. J.* **2013**, *196*, 394–401. [CrossRef]
55. Rietmann, T.R.; Stauffacher, M.; Bernasconi, P.; Auer, J.A.; Weishaupt, M.A. The Association between Heart Rate, Heart Rate Variability, Endocrine and Behavioural Pain Measures in Horses Suffering from Laminitis. *J. Vet. Med. Ser. A* **2004**, *51*, 218–225. [CrossRef]

56. Lesimple, C. Indicators of Horse Welfare: State-of-the-Art. *Animals* **2020**, *10*, 294. [CrossRef] [PubMed]
57. Hausberger, M.; Fureix, C.; Lesimple, C. Detecting Horses' Sickness: In Search of Visible Signs. *Appl. Anim. Behav. Sci.* **2016**, *175*, 41–49. [CrossRef]
58. Leruste, H.; Bokkers, E.A.M.; Sergent, O.; Wolthuis-Fillerup, M.; van Reenen, C.G.; Lensink, B.J. Effects of the Observation Method (Direct v. from Video) and of the Presence of an Observer on Behavioural Results in Veal Calves. *Animal* **2013**, *7*, 1858–1864. [CrossRef] [PubMed]
59. Rattenborg, N.C.; de la Iglesia, H.O.; Kempenaers, B.; Lesku, J.A.; Meerlo, P.; Scriba, M.F. Sleep Research Goes Wild: New Methods and Approaches to Investigate the Ecology, Evolution and Functions of Sleep. *Philos. Trans. R. Soc. B Biol. Sci.* **2017**, *372*, 20160251. [CrossRef] [PubMed]
60. Wolfensohn, S. Too Cute to Kill? The Need for Objective Measurements of Quality of Life. *Animals* **2020**, *10*, 1054. [CrossRef]
61. Murphy, B.A. Circadian and Circannual Regulation in the Horse: Internal Timing in an Elite Athlete. *J. Equine Vet. Sci.* **2019**, *76*, 14–24. [CrossRef]
62. Berger, A.; Scheibe, K.-M.; Eichhorn, K.; Scheibe, A.; Streich, J. Diurnal and Ultradian Rhythms of Behaviour in a Mare Group of Przewalski Horse (*Equus Ferus Przewalskii*), Measured through One Year under Semi-Reserve Conditions. *Appl. Anim. Behav. Sci.* **1999**, *64*, 1–17. [CrossRef]
63. Martin, A.-M.; Elliott, J.A.; Duffy, P.; Blake, C.M.; Attia, S.B.; Katz, L.M.; Browne, J.A.; Gath, V.; McGivney, B.A.; Hill, E.W.; et al. Circadian Regulation of Locomotor Activity and Skeletal Muscle Gene Expression in the Horse. *J. Appl. Physiol.* **2010**, *109*, 1328–1336. [CrossRef]
64. Smith, J.E.; Pinter-Wollman, N. Observing the Unwatchable: Integrating Automated Sensing, Naturalistic Observations and Animal Social Network Analysis in the Age of Big Data. *J. Anim. Ecol.* **2021**, *90*, 62–75. [CrossRef]
65. Ireland, J.L. Demographics, Management, Preventive Health Care and Disease in Aged Horses. *Vet. Clin. N. Am. Equine Pract.* **2016**, *32*, 195–214. [CrossRef]
66. Ireland, J.L.; McGowan, C.M.; Clegg, P.D.; Chandler, K.J.; Pinchbeck, G.L. A Survey of Health Care and Disease in Geriatric Horses Aged 30years or Older. *Vet. J.* **2012**, *192*, 57–64. [CrossRef]
67. Ireland, J.L.; Clegg, P.D.; McGowan, C.M.; McKane, S.A.; Chandler, K.J.; Pinchbeck, G.L. Disease Prevalence in Geriatric Horses in the United Kingdom: Veterinary Clinical Assessment of 200 Cases. *Equine Vet. J.* **2012**, *44*, 101–106. [CrossRef] [PubMed]
68. Ireland, J.L.; Clegg, P.D.; McGowan, C.M.; McKane, S.A.; Chandler, K.J.; Pinchbeck, G.L. Comparison of Owner-Reported Health Problems with Veterinary Assessment of Geriatric Horses in the United Kingdom. *Equine Vet. J.* **2011**, *44*, 94–100. [CrossRef] [PubMed]
69. Ireland, J.L.; Clegg, P.D.; McGowan, C.M.; McKane, S.A.; Pinchbeck, G.L. A Cross-Sectional Study of Geriatric Horses in the United Kingdom. Part 2: Health Care and Disease. *Equine Vet. J.* **2010**, *43*, 37–44. [CrossRef] [PubMed]
70. Ireland, J.L.; Clegg, P.D.; McGowan, C.M.; McKane, S.A.; Pinchbeck, G.L. A Cross-Sectional Study of Geriatric Horses in the United Kingdom. Part 1: Demographics and Management Practices. *Equine Vet. J.* **2010**, *43*, 30–36. [CrossRef] [PubMed]
71. Van Weeren, P.R.; Back, W. Musculoskeletal Disease in Aged Horses and Its Management. *Vet. Clin. N. Am. Equine Pract.* **2016**, *32*, 229–247. [CrossRef]
72. Clegg, P.D. Musculoskeletal Disease and Injury, Now and in the Future. Part 2: Tendon and Ligament Injuries. *Equine Vet. J.* **2012**, *44*, 371–375. [CrossRef]
73. United States Department of Agriculture; Animal and Plant Health Inspection Service; Veterinary Services; National Animal Health Monitoring System. *Changes in the U.S. Equine Industry, 1998–2015*; United States Department of Agriculture: Fort Collins, CO, USA, 2015.
74. Jarvis, N. Clinical Care of the Geriatric Horse. *Practice* **2021**, *43*, 35–44. [CrossRef]
75. Henneke, D.R.; Potter, G.D.; Kreider, J.L.; Yeates, B.F. Relationship between Condition Score, Physical Measurements and Body Fat Percentage in Mares. *Equine Vet. J.* **1983**, *15*, 371–372. [CrossRef]
76. Sinovich, M.; Hewetson, M.; Lombardi, M.; Issard, P.; Archer, D. Hospital-Based Validation of a Commercially Available Remote Activity Tracker for Determining Behaviour Changes, Including Colic, in Horses. *Equine Vet. Educ.* **2021**, *33*, 22. [CrossRef]
77. The R Development Core Team. *R: A Language and Environment for Statistical Computing*; R Foundation for Statistical Computing: Vienna, Austria, 2017.
78. Egan, S.; Kearney, C.M.; Brama, P.A.J.; Parnell, A.C.; McGrath, D. Exploring Stable-Based Behaviour and Behaviour Switching for the Detection of Bilateral Pain in Equines. *Appl. Anim. Behav. Sci.* **2021**, *235*, 105214. [CrossRef]
79. Zeitler-Feicht, M.H.; Prantner, V. Liegeverhalten von Pferden in Gruppenauslaufhaltung. *Arch. Anim. Breed.* **2000**, *43*, 327–336. [CrossRef]
80. Fader, C.; Sambraus, H. The Resting Behaviour of Horses in Loose Housing Systems. *Tierärztliche Umschau* **2004**, *59*, 320–327.
81. Kędzierski, W.; Janczarek, I.; Stachurska, A.; Wilk, I. Comparison of Effects of Different Relaxing Massage Frequencies and Different Music Hours on Reducing Stress Level in Race Horses. *J. Equine Vet. Sci.* **2017**, *53*, 100–107. [CrossRef]
82. Stachurska, A.; Janczarek, I.; Wilk, I.; Kędzierski, W. Does Music Influence Emotional State in Race Horses? *J. Equine Vet. Sci.* **2015**, *35*, 650–656. [CrossRef]
83. Cooper, C.B.; Neufeld, E.V.; Dolezal, B.A.; Martin, J.L. Sleep Deprivation and Obesity in Adults: A Brief Narrative Review. *BMJ Open Sport Exerc. Med.* **2018**, *4*, e000392. [CrossRef] [PubMed]

84. Christie, J.L.; Hewson, C.J.; Riley, C.B.; McNiven, M.A.; Dohoo, I.R.; Bate, L.A. Management Factors Affecting Stereotypies and Body Condition Score in Nonracing Horses in Prince Edward Island. *Can. Vet. J.* **2006**, *47*, 136–143.
85. Jensen, R.B.; Danielsen, S.H.; Tauson, A.-H. Body Condition Score, Morphometric Measurements and Estimation of Body Weight in Mature Icelandic Horses in Denmark. *Acta Vet. Scand.* **2016**, *58*, 59. [CrossRef] [PubMed]
86. Owers, R.; Chubbock, S. Fight the Fat! *Equine Vet. J.* **2013**, *45*, 5. [CrossRef]
87. Schanz, L.; Krueger, K.; Hintze, S. Sex and Age Don't Matter, but Breed Type Does—Factors Influencing Eye Wrinkle Expression in Horses. *Frontiers Vet. Sci.* **2019**, *6*, 154. [CrossRef] [PubMed]
88. Schedlbauer, M. Webbasierte Datenerhebung Und Elektroenzephalographische Messungen Bei Pferden Mit Verhaltensauffälligkeiten. Ph.D. Thesis, LMU, München, Germany, 2021.
89. Marliani, G.; Sprocatti, I.; Schiavoni, G.; Bellodi, A.; Accorsi, P.A. Evaluation of Horses' Daytime Activity Budget in a Model of Ethological Stable: A Case Study in Italy. *J. Appl. Anim. Welf. Sci.* **2020**, *24*, 200–213. [CrossRef] [PubMed]
90. Roberts, K.; Hemmings, A.J.; McBride, S.D.; Parker, M.O. Causal Factors of Oral versus Locomotor Stereotypy in the Horse. *J. Vet. Behav. Clin. Appl. Res.* **2017**, *20*, 37–43. [CrossRef]
91. Sarrafchi, A.; Blokhuis, H.J. Equine Stereotypic Behaviors: Causation, Occurrence, and Prevention. *J. Vet. Behav. Clin. Appl. Res.* **2013**, *8*, 386–394. [CrossRef]
92. Hothersall, B.; Casey, R. Undesired Behaviour in Horses: A Review of Their Development, Prevention, Management and Association with Welfare. *Equine Vet. Educ.* **2012**, *24*, 479–485. [CrossRef]
93. Wei, H.; Zhao, W.; Wang, Y.-X.; Pertovaara, A. Pain-Related Behavior Following REM Sleep Deprivation in the Rat: Influence of Peripheral Nerve Injury, Spinal Glutamatergic Receptors and Nitric Oxide. *Brain Res.* **2007**, *1148*, 105–112. [CrossRef]
94. May, M.E.; Harvey, M.T.; Valdovinos, M.G.; Kline, R.H.; Wiley, R.G.; Kennedy, C.H. Nociceptor and Age Specific Effects of REM Sleep Deprivation Induced Hyperalgesia. *Behav. Brain Res.* **2005**, *159*, 89–94. [CrossRef] [PubMed]

Article

Telomere Length and Regulatory Genes as Novel Stress Biomarkers and Their Diversities in Broiler Chickens (*Gallus gallus domesticus*) Subjected to Corticosterone Feeding

Kazeem Ajasa Badmus [1,2], Zulkifli Idrus [1,2], Goh Yong Meng [2,3], Awis Qurni Sazili [1,2] and Kamalludin Mamat-Hamidi [1,2,*]

[1] Department of Animal Science, Universiti Putra Malaysia, Seri Kembangan 43400, Selangor, Malaysia; alqasiminternational@yahoo.com (K.A.B.); zulidrus@upm.edu.my (Z.I.); awis@upm.edu.my (A.Q.S.)
[2] Institute of Tropical Agriculture and Food Security, Universiti Putra Malaysia, Seri Kembangan 43400, Selangor, Malaysia; ymgoh@upm.edu.my
[3] Department of Veterinary Pre-Clinical Science, Universiti Putra Malaysia, Seri Kembangan 43400, Selangor, Malaysia
* Correspondence: mamath@upm.edu.my

Simple Summary: Assessment of poultry welfare is very crucial for sustainable production in the tropics. There is a demand for alternatives to plasma corticosterone levels as they have received much criticism as an unsuitable predictor of animal welfare due to inconsistency. In this study, we noticed no effect of age on plasma corticosterone (CORT) although it was altered by CORT treatment. However, growth performances and organ weight were affected by CORT treatment and age. The broad sense evaluation of telomere length in this study revealed that telomere length in the blood, muscle, liver and heart was shortened by chronic stress induced by corticosterone administration. The expression profile of the telomere regulatory genes was altered by chronic stress. This study informed us of the potential of telomere length and its regulatory genes in the assessment of animal welfare in the poultry sector for sustainable production.

Citation: Badmus, K.A.; Idrus, Z.; Meng, G.Y.; Sazili, A.Q.; Mamat-Hamidi, K. Telomere Length and Regulatory Genes as Novel Stress Biomarkers and Their Diversities in Broiler Chickens (*Gallus gallus domesticus*) Subjected to Corticosterone Feeding. *Animals* **2021**, *11*, 2759. https://doi.org/10.3390/ani11102759

Academic Editors: Melissa Hempstead and Danila Marini

Received: 22 July 2021
Accepted: 23 August 2021
Published: 22 September 2021

Publisher's Note: MDPI stays neutral with regard to jurisdictional claims in published maps and institutional affiliations.

Abstract: This study was designed to characterize telomere length and its regulatory genes and to evaluate their potential as well-being biomarkers. Chickens were fed a diet containing corticosterone (CORT) for 4 weeks and performances, organ weight, plasma CORT levels, telomere lengths and regulatory genes were measured and recorded. Body weights of CORT-fed chickens were significantly suppressed ($p < 0.05$), and organ weights and circulating CORT plasma levels ($p < 0.05$) were altered. Interaction effect of CORT and duration was significant ($p < 0.05$) on heart and liver telomere length. CORT significantly ($p < 0.05$) shortened the telomere length of the whole blood, muscle, liver and heart. The *TRF1*, *chTERT*, *TELO2* and *HSF1* were significantly ($p < 0.05$) upregulated in the liver and heart at week 4 although these genes and *TERRA* were downregulated in the muscles at weeks 2 and 4. Therefore, telomere lengths and their regulators are associated and diverse, so they can be used as novel biomarkers of stress in broiler chickens fed with CORT.

Keywords: broiler; corticosterone; performance; telomeres; telomere regulators; stress biomarkers

1. Introduction

Broilers are prone to many welfare issues related to genetic differences and environmental challenges. Monitoring animal genetics and physiological data play a pivotal role in the assessment of their welfare [1]. Thus, there is an increasing demand for reliable biomarkers to monitor the health and well-being of poultry in relation to different environmental stresses. The common conventional well-being biomarkers such as hematological values and plasma corticosterone (CORT) levels are not reliable due to inconsistencies in their results [1,2]. The chronic stressor, CORT, has been observed as the product of the hypothalamic–pituitary–adrenal axis response to stress in birds [3–5]. It generates reactive oxygen species (ROS), causing oxidative damage [6]. Many reports show that the elevation of CORT due to stress, be it from plasma or serum, is short-lived due to the biological clock phenomenon and it is known to revert to the normal level after some time [6,7]. However, telomere length and its regulatory genes such as telomerase (*chTERT*), telomeric repeat transcriptional factor 1 (*TRF1*), telomeric repeat-containing RNA (*TERRA*) and heat shock factor 1 (*HSF1*) have recently been shown to be consistently correlated with stress responses [8,9]. Telomeres are nucleoprotein (TTAGGG repeats) structures located at the ends of chromosomes and are usually eroded when exposed to stresses [10,11]. This could indicate the ability to cope with stressful situations, i.e., adaptations [12] and could predict longevity, reproductive capacity or fitness to survive in birds [13,14]. Telomeres are mainly used to safeguard chromosomes and to protect genomic stability by stopping continuous recombination of cells [15]. Several factors, including oxidative stress, feed restriction, antioxidant, breed, sex, age and stocking density could influence the rate of telomere shortening in chicken [16,17]. Chronic stress and reactive oxygen species (ROS) production induced due to CORT administration have been reported to cause drastic reductions in telomere lengths in wild birds [18,19]. In addition, uncovering the role of the sizes of organs on the telomeric DNA becomes important. Organs with high metabolic rate such as liver, heart, brain, and kidney have been reported to be higher in size at an early age [20]. Implication of increase in sizes of these organs is that high metabolic activities will be triggered and will consequently increase the ROS because of increased resting energy expenditure (REE). Metabolic REE was observed to be higher in children per kilogram body weight and reduces steadily during growth [20]. It has been observed that in the first year of life in humans, organs grow in proportion to body weight, but later organ growth decelerates [21]. Decline in the proportional weight of metabolically active organs results in decrease in REE [22]. Association of organ sizes with telomere length can be a good stress indicator for animal welfare assessment.

Telomere length maintenance and restoration due to stress can be assessed via the expression level of the regulatory genes. Shelterin complex that consists of several subunit proteins is crucial for telomere maintenance and genome integrity. A component of shelterin complex, *TRF1*, facilitates and supports the recovery of the telomeric DNA during T-loop formation [23–25]. In addition, telomerase remains essential in telomere integrity as it attaches nucleotides synthesized from the shelterin genes to the telomeric end to maintain the telomere [26,27]. Telomere lengths were maintained in cancer and stem cells of the germline that expressed high levels of telomerase. However, reduced amounts of telomerase were reported in the somatic cell during stress, leading to progressive shortening of the telomere length [28,29]. The telomere maintenance gene 2 (*TELO2*) is another gene that exerts its action via telomerase. Though its expression in chicken fed with CORT is not yet known, it has been predicted to perform a vital role in the telomeric DNA-binding activity [30]. Furthermore, the RNA molecule called telomeric repeat-containing RNA (*TERRA*) was noted as ensuring that very short (or damaged) telomeres are regenerated in humans [31]. This mechanism allows *TERRA* to repair eroded telomeres so that cells can continue to live and keep regenerating. How *HSF1* controls telomerase is unknown to us but it has been reported that human fibroblast showing deficiency in *HSF1* experienced telomeric DNA damage [32]. This shows that *HSF1* is vital for *TERRA* and hence telomerase elevation in a cell under influence of stress. Several studies have recently reported on the role of heat shock in stimulating the upsurge of *TERRA* [33,34].

The reports on how plasma CORT level could be used to predict animal physiology and how it affects traits of economic importance in chicken have been widely criticized [1,7], thus necessitating the current study. We intended to test the hypothesis that CORT administration elevated circulating plasma CORT level and altered telomere length. This study was therefore designed to examine the effects of CORT as a stressor on telomere length, its association with the regulatory genes and how they could be used as novel stress biomarkers in broiler chickens fed with CORT.

2. Materials and Methods

2.1. Animal Management, Housing and Experimental Design

This study was conducted in compliance with the Animal Utilization Protocol approved by the Institutional Animal Care and Use Committee (IACUC) of Universiti Putra Malaysia (approval number: UPM/IACUC/AUP-R019/2018). A total of one hundred (100) male day-old Cobb500 (Leong Hup Farm, Kuala Lumpur, Malaysia) chicks were used for this study and were subjected to 2 × 2 completely randomized factorial design. Chicks were randomly assigned by treatment in groups of 10 into 10 battery cages and subsequently weighed and wing banded. All the cages were placed in a single environmentally controlled chamber. The area of the cages measured 122 cm in length, 91 cm in width and 61 cm in height. The experiment began with initial temperature fixed at 32 °C on day 1 and gradually lowered until it reached day 21. Relative humidity ranged from 70% to 80%. The chickens were vaccinated with infectious bronchitis disease virus and Newcastle disease virus vaccines at day 7 and 14, respectively.

2.2. Diets and Corticosterone Challenge

The diets for the experiment consisted of starter (crumble form) and finisher (pellet form) which were provided ad libitum. The crude protein (CP) of the diet was 21.0% in the starter phase and 19.0% in the finisher phase. Chickens were raised for the first 14 days without CORT feeding (week 2) and then subjected to 30 mg/kg diet CORT feeding [35] on day 15 of age (beginning of week 3) for another 28 days (week 6). The control group consisted of 50 chickens that were given a commercial diet free from CORT (Table 1). The CORT (Abcam, Cambridge, UK) used was an endogenous steroid hormone with an apoptotic-inducing property.

Table 1. Nutrient compositions of commercial broiler starter and finisher diets.

Composition	Starter [1]	Finisher [2]
Crude protein (%)	23.00	19.00
Crude fiber (%)	5.00	5.00
Crude fat (%)	5.00	5.00
Moisture (%)	13.00	13.00
Ash (%)	8.00	8.00
Calcium (%)	0.80	0.80
Phosphorous (%)	0.40	0.40

[1] Commercial starter diet for broiler in crumble form. [2] Commercial finisher diet for broiler in pellet form.

Five hundred mg of CORT was dissolved in 20 mL ethanol for proper solubility before mixing with feed. The CORT mixture was then thoroughly mixed with 16.67 kg feed, producing 30 mg CORT per one kilogram feed. Mixings were repeated at intervals as the diets were being consumed throughout the experiment. Both the CORT and the control group consisted of 5 replicates with 10 chickens each in a cage. Only one level of CORT feeding was used in this study [35].

2.3. Growth Rate Parameters and Animal Sampling

Body weights and feed intakes (FI) of all the chickens were measured weekly using a weighing scale (CAS Corporation, Seoul, Korea). The FI was adjusted for mortality which was registered upon occurrence. The feed conversion ratio (FCR) was calculated as feed/body weight gain. At the end of week 4 and 6 of age, equivalent to 2 and 4 weeks of CORT administration, two (2) chickens were sampled at random from each cage from CORT and control groups for slaughtering. The slaughtering was humanely performed by severing the jugular veins, carotid arteries, trachea, and esophagus with a sharp knife by a single swipe. This sampling was carried out for organ weight measurement, telomeric DNA determination, telomere regulator gene expression and plasma CORT level at week 4 and week 6 (end of second and fourth weeks of CORT administration). The blood samples due to exsanguination were collected into EDTA tubes and stored in ice. The blood samples were stored at $-20\ ^\circ$C prior to DNA extraction. Tissue samples (muscle, liver, and heart) were collected, immediately frozen in liquid nitrogen and then transferred into $-80\ ^\circ$C until use. The wet weight of organs (heart, gizzard, adipose tissue, liver, and small intestine) was measured using a sensitive weighing scale as absolute weight during each sampling period per group. Their relative weight was calculated as percentage organ weight per body weight accordingly. Tissues from muscle, liver and heart were collected and immediately stored in liquid nitrogen and then transferred into $-80\ ^\circ$C freezer.

2.4. Plasma Corticosterone Level Determination

To ensure undisturbed CORT plasma level, randomly selected chickens were carefully captured, weighed, and slaughtered within 3 min before weighing the rest of the chickens. Five (5) mL blood samples were placed into EDTA tubes. The plasma and erythrocyte were separated by centrifuging at 3000 rpm for 30 min at 25 °C and then stored at $-80\ ^\circ$C until they were used for hormone assay. Blank, standard and test sample wells were set and run in duplicate, respectively. The enzyme-linked immunosorbent assay (ELISA) protocol described by the manufacturer (Qayee Biotechnology Co. Ltd., Shanghai, China) was employed for this assay. The final measurement was determined by using a spectrophotometer (Multiskan Go, Thermo Scientific, Waltham, MA, USA). Standard concentration and corresponding OD values were used to calculate the samples' corticosterone concentrations.

2.5. Determination of Telomere Length Using Real-Time Quantitative PCR (qRT-PCR) Analysis

DNA was extracted from the whole blood and tissue samples (muscle, liver and heart) using the blood and tissue DNA innuPREP Mini Kit (Analytik Jena, Jena, Germany) following the recommendations of the manufacturer. The qualities of the DNA extracts were tested using gel electrophoresis and nanodrop (Multiskan Go by Thermo Scientific, USA). Samples with 1.8–2.0 (260/280 ratio) values were stored in a $-20\ ^\circ$C freezer prior to the telomeric length determination analysis. For the discovery of the telomere length, the primers were adopted from an available report [36] (Table 2). The housekeeping gene, glyceraldehyde-3-phosphatase (*GAPDH*) [37] primer sequences were designed and sequenced using information contained in Genbank (NCBI) specific to chicken (Table 2). The primers were tested with DNA amplified using MyTaq Red Mix (Bioline, London, UK) with the aid of the polymerase chain reaction (PTC-100TM, Marshall Scientific, NH, USA) before being run for the qPCR analysis. Twenty nanograms of DNA template was used for both the telomere and the *GAPDH* reactions. The forward and reverse primer concentrations for both telomere and *GAPDH* were 2 μM of each. The primers were mixed with 10 μL SensiFAST SYBR No-ROX qPCR master mix (Bioline, London, UK) for total volume of 20 μL. Ten-fold serial dilutions were performed to obtain standard curves for both the telomere and the housekeeping gene. The samples were arranged accordingly in the PCR machine with identifiers including the non-template control (NTC). The cycling conditions for both telomere and the single copy gene (SCG), *GAPDH* were: 10 min at 95 °C, followed by 40 cycles of 95 °C for 15 s, 60 °C for 1 min, followed by a dissociation (or melt) curve using CFX96 Real-Time PCR System (Bio-Rad, Hercules, CA, USA). Any cycle threshold (Ct) value of standard deviation above one was not used for these analyses. The amplification results (Ct values) were subjected to a Microsoft Excel program designed from standard curves generated to obtain copy numbers in kilobase per reaction (Kb/reaction) of both the telomere and the SCG. The kb/reaction values were then used to calculate the total telomere length in kb per chicken diploid genome according to available information [38].

2.6. Gene Expression Analysis of Telomere Length Regulatory Genes

RNA samples were extracted from the muscle, liver, heart and hypothalamus according to the manufacturer's protocols (InnuPREP RNA Mini kit 2.0, Analytik Jena AG, Germany). The qualities of the RNA were tested using Nanodrop (Multiskan Go by Thermo

Scientific, USA) and RNA with 1.8–2.0 (260/280) values was stored at −80 °C for further analysis. The cDNA was synthesized from the extracted RNA (1 µg) samples using the SensiFast cDNA synthesis kit according to the manufacturer's protocol (Bioline USA Inc., Memphis, TN, USA) with a PCR machine (PTC-100TM, MJ Research Inc., Quebec, Canada) and stored at −20 °C. The cycling conditions for the reverse transcription were: 10 min at 25 °C for annealing, followed by 15 min at 45 °C of reverse transcription, 5 min at 85 °C of inactivation and infinity at 4 °C. The gene expression analysis of chicken telomerase reverse transcriptase (*chTERT*), telomere maintenance gene (*TELO2*), telomeric repeat-containing RNA (*TERRA*), heat shock transcriptional factor 1 (*HSF1*), telomeric repeat transcriptional factor 1 (*TRF1*) and *GAPDH* as the housekeeping gene [37] was obtained by qRT-PCR using the cDNA synthesized from the RNA extracted from the liver, muscle and heart tissues. The primers of these genes specific to chicken were designed and sequenced using information contained in Genbank (NCBI) (Table 2 above). Both the target and the housekeeping genes were simultaneously run in a 96-well plate in duplicates (BIORAD CFX-96, Bio-Rad, CA, USA). The concentrations of the primers for the target and the housekeeping genes were determined by titration; 2 forward and 2 reverse primers were used. The SensiFast Sybr No-ROX kit (Bioline, USA) was used for the amplification of the cDNA. The cycling conditions for the target genes and housekeeping gene were: 10 min at 95 °C, followed by 40 cycles of 95 °C for 15 s, 60 °C for 1 min, followed by a dissociation (or melt) curve using CFX96 Real-Time PCR System (BIO-RAD, USA). Any Ct value of standard deviation above one was not used for further analyses. The mean fold change in the expressions of the target genes at each time was calculated with $2^{-\Delta\Delta CT}$ where $\Delta\Delta Ct$ = (Ct, Target, test–Ct, reference, test)–(Ct, target, Control–Ct, reference, control) [39].

Table 2. Primer sequences for telomere and telomere regulatory genes in chicken.

No	Gene	Primers' Sequence	Accession Numbers	Amplicon Sizes (bp)
1	Telomere	F—GGTTTTTGAGGGTGAGGGTGAGGGTGAGGGTGAGGGT R—TCCCGACTATCCCTATCCCTATCCCTATCCCTATCCCTA	NA	79
2	*TRF1*	F—GGAGGAACGGTTTCCCTAAG R—CTGATGCTGCCCACAGTAGA	NC_006089.5	178
3	*TERRA*	F—GGCCACTGTAAATGGCTGTT R—GTTTGCACAAGGGTCTCCAT	NC_006127.5	219
4	*HSF1*	F—TCTCTGGGTGTCCTTCTGCT R—CTCCTTCCACAGAGCACCTC	NC_006089.5	151
5	*TELO2*	F—GGATGACCCTCAGAGATGGA R—ATTGGTGTGACCAGGAAAGCe	NC_006101.5	249
6	*chTERT*	F—AGGTGCCCAAAACTGAACAC R—CTTCCAAGGGAGACTTGCAG	NC_006089.5	184
7	*GAPDH*	F—ACTATGCGGTTCCCAGTGTC R—TGCCACCATCAGAAAAATGA	NC_006088.5	215

NA = not available.

2.7. Statistical Analysis

Data on body performance, feed consumption, weight gain and FCR were analyses using repeated measure of ANOVA with SAS 9.4 software [40]. Absolute and relative weight of organs and telomere length data were analyzed using general linear model of SAS 9.4 and 2 × 2 factorial analysis. Means were separated using the Duncan Multiple Range test. Comparison between ages and telomere regulatory genes of the tested chickens and the control was subjected to the *t*-test procedure of the general linear model using SAS 9.4 software. Mortality was analyzed using chi-square test of SAS. All statistical tests were conducted at 95% confidence level.

3. Result

3.1. Growth Performance and Mortality

Throughout the CORT administration period, CORT treatment significantly ($p < 0.05$) suppressed body weight, feed consumption and weight gain (Table 3). During the trial phases, CORT treatment led to significantly ($p < 0.05$) higher FCR and higher mortality rate in the CORT-fed chicken than the control. No significant difference was observed in body weight, feed consumption, weight gained and FCR between the two groups before the commencement of CORT administration (week 2).

Table 3. Effects of corticosterone administration on body weight, feed consumption and feed conversion ratio (FCR) of broiler chicken.

Age/Traits	CTRL	CORT-Fed Chicken	SEM	*p*-Values
Body weight (g)				
Week 2	541.88	539.76	8.76	0.8585
Week 4	1509.00	1054.21	23.48	0.001
Week 6	2363.55	1479.18	34.62	0.001
Feed consumption (g/bird/week)				
Week 2	422.00	439.80	8.22	0.1730
Week 4	990.00	843.30	17.65	0.0004
Week 6	1057.70	822.30	22.57	0.0001
Weight gain (g/bird/week)				
Week 2	347.56	344.83	5.41	0.735
Week 4	496.42	259.55	17.39	0.0001
Week 6	1056.67	822.33	22.57	0.0001
FCR (feed/gain)				
Week 2	1.21	1.27	0.02	0.0557
Week 4	2.00	3.27	0.12	0.0001
Week 6	2.27	3.60	0.11	0.0001
Mortality (Week 0–Week 6)	1.00	7.00	0.09	0.024

CORT = corticosterone; Week 2 = period of no CORT administration; Week 4 = 2 weeks of CORT administration; Week 6 = 4 weeks of CORT administration. There were 50 observations per treatment.

3.2. Absolute Weight of Organs

Significant ($p < 0.05$) interaction between age and CORT treatment was noted for absolute weight of the small intestine but not for the heart, liver, abdominal fat, and gizzard (Table 4). After week 4, there was significant ($p < 0.05$) reduction in absolute weight for the small intestine in CORT-fed chicken compared to the control. Significant ($p < 0.05$) effect of age was noted on both control and the CORT-fed chicken. Effect of treatment was significant ($p < 0.05$) on abdominal fat and gizzard sizes but not on heart size. Age significantly ($p < 0.05$) affected heart, liver and abdominal fat sizes but did not affect gizzard size.

3.3. Relative Weight of Organs

Significant ($p < 0.05$) effect of treatment was noted for heart, liver, small intestine, abdominal fat, and gizzard relative weights (Table 5). CORT significantly ($p < 0.05$) led to an increase in relative weights of organs at both weeks 4 and 6. No significant interaction was noted between the treatment and age for all the organs' relative weight. Effect of age was noted on relative weight of liver, abdominal fat and gizzard but was not noted on heart and small intestine relative weights.

3.4. Plasma Level of Corticosterone

No significant interaction between age and the CORT treatment was noted for the plasma CORT level in this study (Table 6). However, CORT administration significantly

($p < 0.05$) elevated circulating plasma CORT level in the CORT-fed chicken. No effect of age was noted for plasma CORT level in both the control and the CORT-fed chicken.

Table 4. Effect of corticosterone feeding and age on absolute organ weight.

Age	Treatment	Heart (g)	Liver (g)	Smallint (g)	AbdFat (g)	Gizzard (g)
Week 4	CTRL	7.85 [b]	36.27 [c]	76.55 [c]	17.87 [d]	29.36 [b]
	CORT	7.22 [b]	55.21 [b]	66.09 [c]	27.24 [c]	36.13 [ab]
Week 6	CTRL	11.24 [a]	49.17 [b]	123.56 [a]	52.03 [b]	29.36 [b]
	CORT	11.74 [a]	75.02 [a]	93.63 [b]	63.12 [a]	42.42 [a]
	SEM	0.54	3.42	2.48	2.48	2.95
p values	Age	<0.0001	<0.0001	<0.0001	<0.0001	0.359
	Treatment	0.907	<0.0001	<0.0001	<0.0001	0.004
	Age × Treatment	0.319	0.363	0.041	0.74	0.287

[a,b,c,d] Means within a column subgroup with no common superscripts are significantly different at $p < 0.05$. SEM = standard error of the mean for main effects (n = 20). Smallint = small intestine; AbdFat = abdominal fat; CTRL = control group; CORT = corticosterone-treated group; Week 4 = 4 weeks of age (2 weeks of CORT administration); Week 6 = 6 weeks of age (4 weeks of CORT administration).

Table 5. Effect of corticosterone feeding and age on relative organ weights.

Age	Treatment	Heart (%)	Liver (%)	Smallint (%)	AbdFat (%)	Gizzard (%)
Week 4	CTRL	0.50 [b]	2.31 [b]	4.88 [b]	1.14 [d]	1.90 [c]
	CORT	0.68 [a]	5.15 [a]	6.21 [a]	2.55 [b]	3.35 [a]
Week 6	CTRL	0.44 [b]	1.93 [b]	4.87 [b]	2.06 [c]	1.17 [d]
	CORT	0.74 [a]	4.59 [a]	5.72 [a]	3.83 [a]	2.61 [b]
	SEM	0.03	0.20	0.22	0.14	0.20
p values	Age	0.756	0.041	0.268	<0.0001	0.0016
	Treatment	<0.0001	<0.0001	<0.0001	<0.0001	<0.0001
	Age × Treatment	0.212	0.667	0.282	0.213	0.999

[a,b,c,d] Means within a column subgroup with no common superscripts are significantly different at $p < 0.05$. SEM = standard error of the mean for main effects (*n* = 20). Smallint = small intestine; AbdFat = abdominal fat; CTRL = control group; CORT = corticosterone-treated group; Week 4 = 4 weeks of age (2 weeks of CORT administration); Week 6 = 6 weeks of age (4 weeks of CORT administration).

Table 6. Effect of corticosterone administration on plasma corticosterone level.

Age	Week 4		Week 6			p Values		
Treatment	CTRL	CORT	CTRL	CORT	SEM	Age	Treatment	Age × Treatment
Plasma CORT (ng/mL)	6.80 [b]	7.65 [a]	6.73 [b]	7.94 [a]	0.37	0.685	0.0004	0.830

[a,b] Means within a row subgroup with no common superscripts are significantly different at $p < 0.05$. SEM = standard error of the mean for main effects (*n* = 20). CTRL = control group; CORT = corticosterone-treated group; Week 4 = 4 weeks of age (2 weeks of CORT administration); Week 6 = 6 weeks of age (4 weeks of CORT administration).

3.5. Absolute Telomere Length

Significant ($p < 0.05$) interaction effect between CORT treatment and age was noted on telomere length for liver and heart but not for whole blood and muscle (Table 7). CORT treatment significantly ($p < 0.05$) shortened liver and heart telomere length at week 4 but not at week 6 of age. As the CORT feeding advanced, telomere length for liver and the heart increased significantly ($p < 0.05$) in the CORT-fed chicken, making the telomere length of the two groups statistically the same at week 6. Telomere length for muscle was significantly ($p < 0.05$) affected by treatment but was not affected by age. Both treatment and age significantly ($p < 0.05$) affected whole blood telomere length. In addition, telomere length in whole blood decreased, both in the CORT-treated and untreated chicken from week 4 to 6.

Table 7. Effect of corticosterone feeding and age on absolute telomere length in whole blood, muscle, liver and heart of broiler chicken.

Age	Treatment	Whole Blood	Muscle	Liver	Heart
Week 4	CTRL	526.40 [a]	446.68 [a]	564.96 [a]	576.89 [a]
	CORT	334.50 [bc]	313.84 [b]	346.84 [b]	264.43 [b]
Week 6	CTRL	417.96 [ab]	463.41 [a]	481.07 [ab]	355.72 [ab]
	CORT	260.09 [c]	282.39 [b]	467.59 [ab]	364.98 [ab]
	SEM	37.27	54.90	58.33	56.03
p **values**	Age	0.016	0.779	0.807	0.316
	Treatment	<0.0001	0.005	0.028	0.009
	Age × Treatment	0.921	0.817	0.047	0.018

[a,b] Means within a column subgroup with no common superscripts are significantly different at $p < 0.05$. SEM = standard error of the mean for main effects ($n = 20$). CTRL = control group; CORT = corticosterone-treated group; Week 4 = 4 weeks of age (2 weeks of CORT administration); Week 6 = 6 weeks of age (4 weeks of CORT administration).

3.6. Telomere Regulatory Genes

3.6.1. Telomeric Repeat Transcriptional Factor 1 (*TRF1*)

The gene expressions of *TRF1* in this study revealed that effects of CORT and duration were significant (Figure 1). It was observed that *TRF1* was significantly downregulated ($p < 0.05$) in the muscle at both week 4 and 6 and in the liver at week 6. However, it was significantly upregulated in the heart at week 4 of CORT treatment. *TRF1* was significantly upregulated in the liver and heart at week 6.

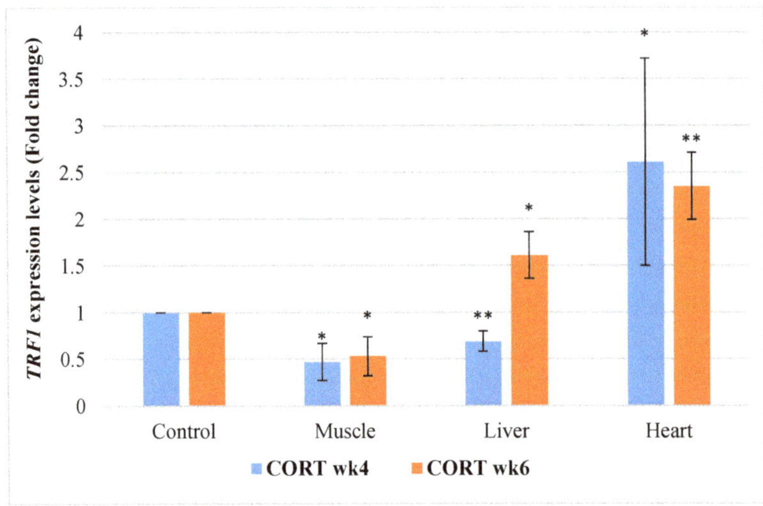

Figure 1. Expression profile of telomeric repeat transcription factor 1 (*TRF1*) in the muscle, liver, and heart at week 4 and 6 of corticosterone-fed chickens compared to the control (in fold change). CORT wk4 = 4 weeks of age (2 weeks of CORT administration); CORT wk6 = 6 weeks of age (4 weeks of CORT administration); n = 20. Probability, * = $p < 0.05$; ** = $p < 0.01$.

3.6.2. Chicken Telomerase Reverse Transcription Factors (*chTERT*)

The gene expression profile of chicken telomerase (*chTERT*) in this study revealed that CORT and duration had a significant effect (Figure 2). *chTERT* was downregulated in the muscle at week 4 and 6 of age (2 and 4 weeks of CORT administration). However, *chTERT* was upregulated at week 4 and 6 in the liver, and at week 6 in the heart of CORT-fed chickens. It was downregulated in the heart at week 4 of the treatment.

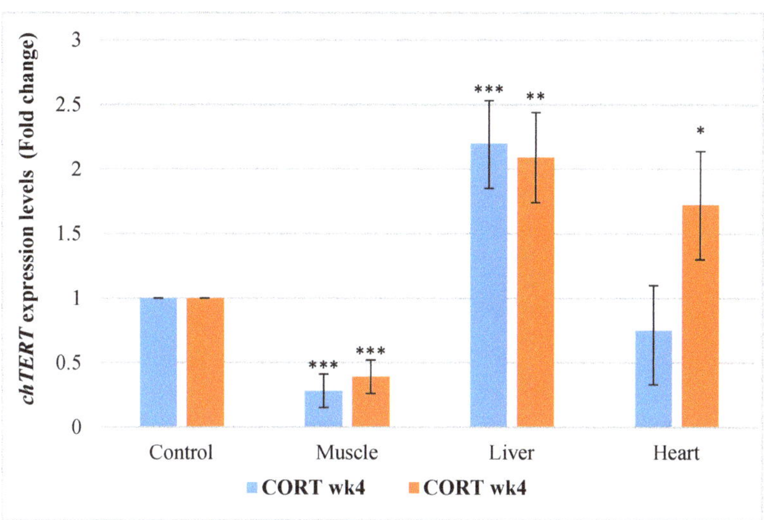

Figure 2. Expression profile of chicken telomerase (*chTERT*) in the muscle, liver, and heart at week 4 and 6 of age in CORT-fed chickens compared to the control. CORT wk4 = 4 weeks of age (2 weeks of CORT administration); CORT wk6 = 6 weeks of age (4 weeks of CORT administration); n = 20. Probability, * = $p < 0.05$; ** = $p < 0.01$; *** = $p < 0.001$.

3.6.3. Telomere Maintenance Gene 2 (*TELO2*)

The gene expressions of *TELO2* are presented in Figure 3. *TELO2* was significantly downregulated in the muscle of the CORT-fed chickens at weeks 4 and 6. However, *TELO2* was significantly upregulated at weeks 4 and 6 in the liver and heart of the CORT-fed chickens.

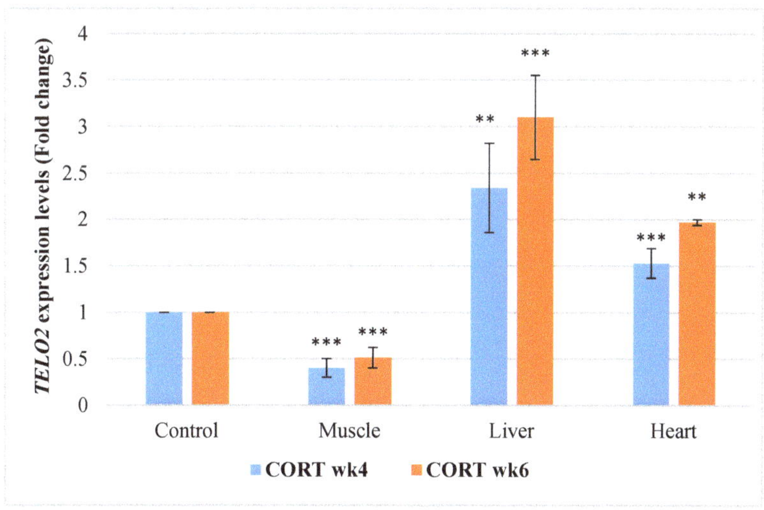

Figure 3. Expression profile of telomere maintenance gene 2 (*TELO2*) in the muscle, liver, and heart at week 4 and 6 of corticosterone-fed chickens compared to the control (in fold change). CORT wk4 = 4 weeks of age (2 weeks of CORT administration); CORT wk6 = 6 weeks of age (4 weeks of CORT administration); n = 20. Probability, * = $p < 0.05$; ** = $p < 0.01$; *** = $p < 0.001$.

3.6.4. Telomeric Repeat-Containing RNA (*TERRA*)

The gene expressions of *TERRA* in this study revealed that impacts of CORT and duration were significant (Figure 4). *TERRA* was downregulated significantly ($p < 0.05$) in the muscle and heart tissue of the CORT-fed chickens at weeks 4 and 6 of CORT administration. Despite this, *TERRA* was upregulated in the liver at week 4 and 6.

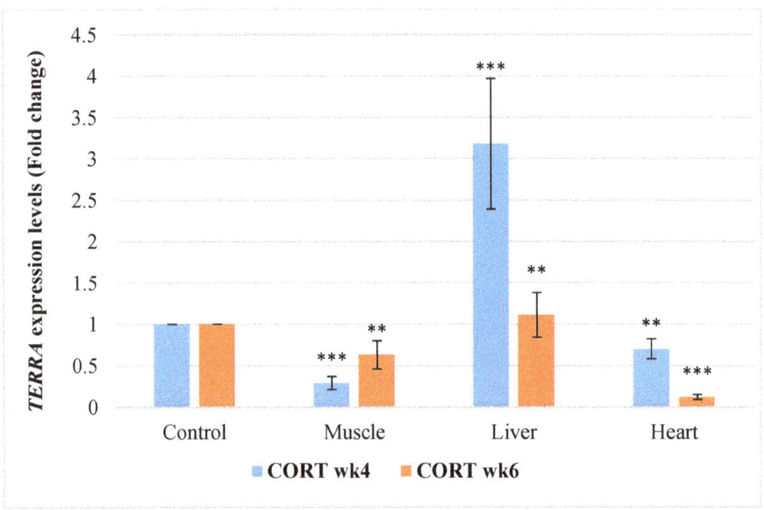

Figure 4. Expression profile of telomeric repeat-containing RNA (*TERRA*) in the muscle, liver, and heart at week 4 and 6 of corticosterone-fed chickens compared to the control (fold change). CORT wk4 = 4 weeks of age (2 weeks of CORT administration); CORT wk6 = 6 weeks of age (4 weeks of CORT administration); n = 20. Probability, ** = $p < 0.01$; *** = $p < 0.001$.

3.6.5. Heat Shock Transcriptional Factor 1 (*HSF1*)

The gene expressions of *HSF1* in this study revealed that impacts of CORT and duration were significant (Figure 5). It was significantly downregulated in the muscle of the CORT-fed chickens at weeks 4 and 6 but was upregulated in the liver of the CORT-fed chickens at weeks 4 and 6. It was, however, downregulated in the heart at week 4 but upregulated at week 6 of the CORT duration.

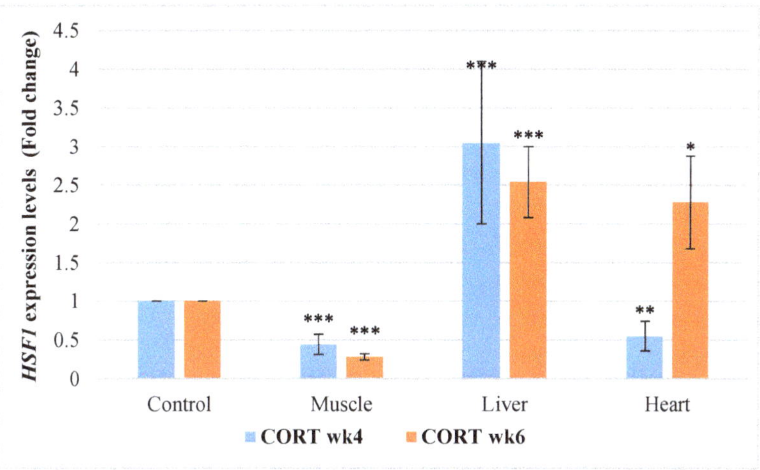

Figure 5. Expression profile of heat shock transcriptional factor 1 (*HSF1*) in the muscle, liver, and heart at week 4 and 6 of corticosterone-fed chickens compared to the control (fold change). CORT wk4 = 4 weeks of age (2 weeks of CORT administration); CORT wk6 = 6 weeks of age (4 weeks of CORT administration); n = 20. Probability, * = $p < 0.05$; ** = $p < 0.01$; *** = $p < 0.001$.

4. Discussion

4.1. Growth Performance

Alterations in the performances of animals due to stress could affect telomere lengths. The results from this study indicated that the administration of corticosterone suppressed body weight and body weight gain, and this agreed with some literature findings [35,41,42]. The weight loss in the CORT-fed chicken could be attributed to activated gluconeogenesis and protein breakdown [43,44] and lower feed consumption. FCR was reduced by CORT administration in the CORT-fed chicken and this could be as a result of poor feed assimilation in the body. The findings revealed that the CORT-fed chicken diverted nutrients meant for growth performance to lipid accumulation and fatty livers [45]. In addition, the poor growth and high mortality rate in the CORT-fed chicken could be attributed to the accumulation of reactive oxygen species (ROS) [18]. The higher mortality in the CORT-fed chicken is expected and this emphasized the association between chronic stress and the increase in mortality rate in broiler chicken.

4.2. Absolute and Relative Weights of Organ

The health status of animals can be diagnosed using their organs [35]. In the present study, we observed a very interesting interaction effect between CORT treatment and age on small intestinal absolute weight. This interaction was evident when CORT-fed chicken exhibited substantial reduction in small intestinal weight at week 6 compared to the control group. In contrast, CORT did not significantly influence small intestinal weight at week 4. CORT had been reported to reduce wet weight of the small intestine in a finding [35]. The small intestine aids in growth and animal performance [46]. The reduction in the absolute weight of the small intestine could lead to a low proportion of nutrient uptake and corresponding loss of protein [35]. According to the present results, there was significant suppression of the absolute weight of the liver, abdominal fat and gizzard due to CORT treatment. Significant improvement in the weights of heart, liver, small intestine, and abdominal fat was noted with advancement in age. Reduction in the sizes of these organs could be attributed to insufficient nutrient uptake in the CORT-treated chicken. Our study revealed that CORT significantly increased the relative weights of heart, small intestine, abdominal fat and gizzard. Relative weights of liver and gizzard significantly reduced with advancement in age but increased in abdominal fat. Increase in the relative weights of the small intestine, liver, and liver fat due to CORT administration in chicken had been reported in most studies [35,41]. This improvement in the relative sizes of organs could be attributed to inflammation acquired during CORT administration. The relationship between the organ's weight and the telomere length has not been uncovered. However, these organs are described as metabolically active with a positive correlation with body weights [20]. Increased metabolic activities could amount to high ROS which is noted to trigger chronic oxidative damage [6] and hence telomere length attrition [19].

4.3. Plasma Corticosterone Levels

Administration of CORT through feeding or implantation had been reported to increase the plasma CORT levels [42]. Meanwhile, influence of CORT treatment and administration period (short versus long) on glucocorticoids had been reported [1]. In the current study, we revealed that CORT administration significantly elevated plasma CORT level and this change was independent of age. Similar levels of CORT plasma had been noticed between CORT-fed sparrows and the control after a week, but the CORT level returned to baseline level two months after implantation [47]. The researchers [47], however, revealed that CORT implantation induced a short-lived negative effect on body weight, blood, and feather weight. In contrast, our data in the current study revealed that influence of CORT administration persisted on the plasma CORT levels, growth performance and other physiological components. In view of removing controversies surrounding the use of CORT as a biomarker of animal welfare due to its inconsistency, telomere length as a more reliable and conserved biomarker of stress is proposed as an alternative in this study.

4.4. Telomere Length

Telomere is a nucleoprotein protecting the end of the chromosome from degradation during cell division and it is highly sensitive to oxidative attack due to its high guanine content [48], the characteristic that prolongs its recovery and gives it stable potential to measure stress conditions. Telomere length is usually shortened with stress and advancement in age and could be used as a biomarker of stress. Its shortening rate predicts a species life span [49]. In the present study, we noticed a noteworthy interaction between the CORT treatment and age for liver and heart telomere length. These interactions were evident when the CORT-fed chicken revealed a drastic loss in liver and heart telomere length at week 4. On the contrary, CORT did not influence liver and heart telomere length at week 6. These results suggest that liver and heart telomere length are more susceptible to CORT treatment at week 4 than week 6. The telomeric DNA shortening in the heart and liver at week 4 was a result of interaction between CORT treatment and age. The telomere length attrition at this early stage of induction could be attributed to the REE and high metabolic rates of these organs [20,21] combined with chronic oxidative damage due to CORT administration [18]. In addition, with advancement in age, liver and heart telomere length improved in the CORT-fed chicken. Based on the current findings, CORT treatment significantly caused telomeric attrition in whole blood and muscle. Influence of age of induction was noticed on whole blood telomere length but was not observed in the muscle. Oxidative stress induced by CORT could be responsible for the poor performance obtained in the CORT-fed chickens and this could be attributed to damaged proteins and DNA [50]. In this study, chickens with short telomere lengths revealed suppressed body weights. The production of ROS via CORT administration usually affects the dynamics of the telomeres [18]. Telomere length has been used to measure individual fitness and survivability [51]. Moreover, the effect of age in 178 single-comb White Leghorns from 10 weeks old was reported for lymphocytes [52]. The amount of telomeric DNA was observed to decrease with the advancement in age in this study. When cell ageing sets in, inadequate amount or absence of some restorative genes will be experienced, leading to aggravated telomeric shortening at senescence [26,53].

Telomere length could be implicated in the high mortality observed in CORT-fed chickens. Reports have attributed telomere length to adaptation and survivability [54,55]. High risk of mortality and low life expectancy have been related to short telomere length [56,57]. Telomere length was sometimes related to longevity and reproductive performances in various studies [13]. In the current study, variation of the telomere in liver, heart and muscle were not affected by age. This implies that telomere length of the tissues could be restored due to the activities of telomerase and the shelterin genes which are usually activated in tumor cells [24,58,59]. This implies that increased telomere length could sometimes be an indicator of unhealthy conditions like cancers. The loss in telomere length in the whole blood due to age could be a result of diminished levels of telomerase which are usually low in somatic tissues [19,60]. Our results implied that with increased duration of treatment, CORT administration could lead to telomere length improvement in tissues suffering from tumors such as liver and heart [24,61]. This improvement in telomere length could be attributed to telomerase activities which are usually higher in damaged or cancer cells and hence maintain telomere lengths [28].

Surprisingly, there was a decline in the telomere length for blood, liver, and the heart from week 4 to week 6 of age in the untreated chickens. Muscle did not reveal such a difference. The difference in the characteristics of telomere length could be attributed to a genetic and tissue effect. Telomere length was previously associated with genetic and non-genetic factors [61]. The results in this study imply that chicken blood, liver and heart telomere length are prone to shortening and this could be the reason for major liver and heart-related diseases usually reported in chicken. Modern strains of broiler chicken have been reported to be highly susceptible to heart failure and heart-related mortalities such as ascites and sudden death syndrome [62]. Short telomere length, to make it clear, has been implicated as the main cause of heart disease and other age-associated diseases [63]. In fact, liver fibrosis and cirrhosis have been reported to be caused by critically short telomeres, a phenomenon known as DNA damage response [64]. Decline in the telomere length has been reported in the human heart along the gestation period and heart development while in other tissue, such as kidney and liver, telomere length remains unchanged [65]. We therefore discovered in this study that age, when combined with stress, could lead to drastic telomere length attrition. Telomere length has been reported to reduce with age in fishes, reptiles, and humans, except in water python and Leach's petrel [66]. Telomere length could therefore be considered as an indicator of aging. Relationships between telomere length and stress [67] hint that telomere length might not be used only for an indicator of chronological age [68], but also for a marker of organism lifestyle [69] and a good proxy of organism fitness or biological age [70]. Here, we reported higher abdominal fat in CORT-fed chicken and increased fat in the untreated chicken with an increase in age. Increase in fat deposit could lead to obesity, ROS, oxidative damage and non-alcoholic fatty liver and these are risk factors for short telomere length.

4.5. Telomere Length Regulatory Genes

The results of the telomere regulatory genes revealed that they all influenced telomere length integrity. They were all implicated in the telomere length shortenings in the blood and the tissues. The expressions of these genes are tissue-dependent. The high variabilities observed in the liver and heart samples showed that these genes vary from cell to cell. It happens in the group of stem cells and cancer or tumor cells. These high variabilities are the cumulative results of intrinsic genetic (inherent) factors, extrinsic factors, and stochastic factors. It has been reported that cell-to-cell variations are often observed within cancerous and embryonic cell samples [71].

In muscle, the shortening of telomere length could be a result of the downregulation of *TRF1, chTERT, TELO2, TERRA* and *HSF1* due to CORT administration. A study has revealed that *TERRA* was involved in telomerase nucleation [72]. The results indicated that the expressions of the telomere regulators were affected by CORT and age. The *TRF1* is one of the components of the shelterin complex that shelters the telomere. Telomerase activities are dependent on the nucleotide sequences from this complex for telomeric DNA formation [72]. Findings suggested that *TRF1* mediated telomere length and function [65]. The downregulation of the *TRF1* by chronic stress in the CORT-fed chickens could affect the telomeric DNA. This effect might initiate chromosomal instability [73]. The deficiency of *TRF1, TELO2, TERRA* and *HSF1* in the muscle could be responsible for the deficiency in the *chTERT* and hence the telomere length shortening due to their reduced nucleation to the *chTERT* [32,59,74]. In the current study, TRF1 was downregulated in muscle. It has been reported that *TRF1* was low in mice skeletal muscle exposed to treadmill running bout stress. The reduced *TRF1* brought about a significant increase in the apoptosis mediator, *P38 MAPK* phosphorylation [75]. The upregulation of *TRF1* led to stability in the liver telomere at week 6, as this gene specializes in promoting replication of telomeres [76]. This implied that in the liver and heart, *TRF1* could be a negative regulator and this assertion agreed with the report obtained in study [73]. Telomerase was not increased by *TRF1* in the heart at week 4, probably as a result of downregulation of *TERRA* and *HSF1*. It has been revealed that when *TRF1* was genetically or chemically inhibited in mouse, increased telomeric DNA damage, reduced produced proliferation and stemness were reported [76,77]. The telomere maintenance gene 2 (*TELO2*) might have a part in telomere length regulation and maintenance as it was found in the pathway of telomerase and in the complex responsible for cellular resistance against DNA damage stress, especially due to radiation, ultraviolet and mitomycin [30,78]. It is being examined and associated with organs and other telomere-regulating genes and we discovered that it has good links with these genes and could be a good candidate for telomeric studies. Moreover, *TERRA* is another gene that plays a pivotal role in the mediation of the heterochromatic marks in the remodeling complex [79]. *TERRA* has been reported to be involved in telomerase nucleation [72]. Though its expression in tissues has not been detailed, its function in telomeric DNA upon heat stress in fibroblast has been revealed [32]. Studies revealed that telomerase activity was very low in almost all somatic tissues or cells with high proliferative potential [59,74]. Cells deficient in *HSF1* had been reported to be deficient in *TERRA* and hence had short telomere lengths [32]. Koskas et al. [32] further reported that *HSF1* promoted *TERRA* transcription and telomere length protection upon heat stress. *HSF1* has been classed among the cancer-related genes group. It is mostly expressed in all tissues in the human body. When the telomere maintenance mechanism was deficient, increased telomere transcription was reported to result in telomere shortening due to DNA replication–dependent loss of the telomere pathways [80].

Telomerase (*TERT*) function is usually high in gamete, stem, and tumor cells. It has been reported to be low in adult somatic cells in mice, whereas it is completely absent in human adult somatic cells [81]. Higher activity of *TERT* has been reported in cancer cells and in most cells suffering from tumors [82] while *TRF1* provided a nucleotide sequence for *TERT*. The heart is made up of cardiac stem cells which help in the self-regeneration of the heart due to high telomerase activities [83]. In addition, activation of *TELO2, HSF1* and *TERRA* in the liver could be responsible for the maintenance of the telomere length in the liver at week 6 in the current study. Telomere shortening was noted to induce *TERRA* expression which then activated telomerase nucleation [72] which could then initiate telomere recovery. Moreover, *HSF1* has been reported to promote *TERRA* transcription and telomere length protection upon heat stress [32]. The observation in this study suggests that *TELO2* could therefore be part of the telomerase pathway, which implies that *TELO2* is a typical novel gene which can be used as a stress biomarker and hence can be considered as a mechanism for telomere synthesis. This observation agreed with the information available in the rat genome database which revealed that *TELO2* was shown to maintain the telomere via telomerase [78] and its role in the telomeric-binding activities had been suggested [30]. The upregulation of *TERRA* in the liver was expected as it was mobilized to areas with short telomeres to cause the nucleation of telomerase. Interactions of *TERRA* with TRFs could mobilize *TERRA* to the telomeres, causing *TERRA* R-loops to be synthesized at the severely shortened telomere, thereby preventing the DNA damage response [31,84].

This observation was noticed in the liver but not in the muscle in the current study. The differences observed in the tissues could be attributed to tissue peculiarities which implied that they have diverse characteristics. Furthermore, *HSF1* activated *TERRA* in the liver but not in the heart. Higher levels of reactive oxygen species (ROS), cleaved caspases and fragmented DNA in the gastrocnemius were reported in muscles of heat-exposed mice but not in control mice and these changes were not observed in the livers of heat-exposed mice [85]. The reason for the downregulation of *TERRA* in the heart and the muscle could be as a result of mutations, as earlier elucidated [86,87].

Generally, telomere length is progressively lost in muscle due to inability of DNA polymerase to completely replicate telomere under stress [26]. The loss in telomere could also be attributed to absence or lower telomere restoration factors such as telomerase activities, shelterin protein complex [24,58], *TERRA* and *HSF1* [32] that regulate the telomeres. These restoration factors tend to be activated in the liver and heart.

5. Conclusions

It was observed that CORT elevated plasma CORT level and altered performances and organ sizes in the CORT-treated chicken, suggesting that both poor conditions were caused by chronic oxidative stress. CORT led to telomere length attrition and altered the tissue telomere length regulators in the CORT-fed group. The expressions of *TRF1*, *chTERT*, *TELO2*, *TERRA* and *HSF1* affected telomere length behavior under chronic stress. The gene expression reports revealed that muscle tissue could be more susceptible to chronic stress and telomeric DNA attritions. Telomere loss in blood is age-dependent, suggesting that it is a potential biomarker of aging. In conclusion, telomere length and its regulators are diverse and can be used as novel biomarkers of stress in broiler chickens.

Author Contributions: Conceptualization, Z.I., K.M.-H., G.Y.M. and K.A.B.; methodology, software, analysis and data curation, K.A.B.; validation, Z.I., K.M.-H., G.Y.M. and K.A.B.; investigation, K.A.B.; resources, Z.I., K.M.-H., A.Q.S. and G.Y.M.; writing original draft, K.A.B. and K.M.-H.; review and editing, Z.I., K.M.-H., G.Y.M. and A.Q.S.; visualization, Z.I. and K.M.-H.; supervision, Z.I., K.M.-H. and G.Y.M.; project administration and funding acquisition, Z.I., K.M.-H., A.Q.S., G.Y.M. and K.A.B. All authors have read and agreed to the published version of the manuscript.

Funding: This study was funded by Universiti Putra Malaysia (Putra Grant No.:UPM/800-3/3/1/9629100).

Institutional Review Board Statement: This study was conducted in compliance with the Animal Utilisation Protocol approved by the Institutional Animal Care and Use Committee (IACUC) of Universiti Putra Malaysia (UPM/IACUC/AUP-RO19/2018).

Data Availability Statement: The data presented in this finding are available upon request from the corresponding author.

Acknowledgments: The authors would like to thank Institute of Tropical Agriculture and Food Security, and Universiti Putra Malaysia for providing funds and facilities for this study, as well as their technical staff.

Conflicts of Interest: The authors declare no conflict of interest.

References

1. Rushen, J. Problems associated with the interpretation of physiological data in the assessment of animal welfare. *Appl. Anim. Behav. Sci.* **1991**, *28*, 381–386. [CrossRef]
2. Fairhurst, G.D.; Marchant, T.A.; Soos, C.; Machin, K.L.; Clark, R.G. Experimental relationships between levels of corticosterone in plasma and feathers in a free-living bird. *J. Exp. Biol.* **2013**, 4071–4081. [CrossRef] [PubMed]
3. Dhabhar, F.S.; McEwen, B.S. Acute stress enhances while chronic stress suppresses cell-mediated immunity in vivo: A potential role for leukocyte trafficking. *Brain Behav. Immun.* **1997**, *11*, 286–306. [CrossRef]
4. Dhabhar, F.S. Acute stress enhances while chronic stress suppresses skin immunity. The role of stress hormones and leukocyte trafficking. *Ann. N. Y. Acad. Sci.* **2000**, *917*, 876–893. [CrossRef]
5. Romero, L.M. Physiological stress in ecology: Lessons from biomedical research. *Trends Ecol. Evol.* **2004**, *19*, 249–255. [CrossRef]

6. Spiers, J.G.; Chen, H.C.; Sernia, C.; Lavidis, N.A. Activation of the hypothalamic-pituitary-adrenal stress axis induces cellular oxidative stress. *Front. Neurosci.* **2015**, *8*, 456. [CrossRef] [PubMed]
7. Bennett, M.C.; Mlady, G.W.; Fleshner, M.; Rose, M.G. Synergy between chronic corticosterone and sodium azide treatments in producing a spatial learning deficit and inhibiting cytochrome oxidase activity. *Proc. Natl. Acad. Sci. USA* **1996**, *93*, 1330–1334. [CrossRef] [PubMed]
8. Barna, J.; Csermely, P.; Vellai, T. Roles of heat shock factor 1 beyond the heat shock response. *Cell. Mol. Life Sci.* **2018**, *75*, 2897–2916. [CrossRef]
9. Jones, T.J.; Li, D.; Wolf, I.M.; Wadekar, S.A.; Periyasamy, S.; Sánchez, E.R. Enhancement of glucocorticoid receptor-mediated gene expression by constitutively active heat shock factor 1. *Mol. Endocrinol.* **2004**, *18*, 509–520. [CrossRef]
10. Houbon, J.M.J.; Moonen, H.J.J.; van Schooter, F.J.; Hageman, G.J. Telomere length assessment: Biomarkers of chronic oxidative state? *Free Radic. Biol Med.* **2007**, *144*, 235–246. [CrossRef] [PubMed]
11. Moyzis, R.K.; Buckingham, J.M.; Cram, L.S.; Dani, M.; Deaven, L.L.; Jones, M.D.; Meyne, J.; Ratliff, R.L.; Wu, J.R. A highly conserved repetitive DNA sequence, (TTAGGG), present at the telomeres of human chromosomes. *Proc. Natl. Acad. Sci. USA* **1988**, *85*, 6622–6626. [CrossRef]
12. Kotrschal, A.; Ilmonen, P.; Penn, D.J. Stress impacts telomere dynamics. *Biol. Lett.* **2007**, *3*, 128–130. [CrossRef]
13. Angelier, F.; Vleck, C.M.; Holberton, R.L.; Marra, P.P. Telomere length, non-breeding habitat and return rate in male American redstarts. *Funct. Ecol.* **2013**, *27*, 342–350. [CrossRef]
14. Heidinger, B.J.; Blount, J.D.; Boner, W.; Griffiths, K.; Metcalfe, N.B.; Monaghan, P. Telomere length in early life predicts lifespan. *Proc. Natl. Acad. Sci. USA* **2012**, *109*, 1743–1748. [CrossRef]
15. Blackburn, E.H.; Chan, S.; Chang, J.; Fulton, T.B.; Krauskopf, A.; Mceachern, M.; Prescott, J.; Roy, J.; Smith, C.; Wang, H. Molecular manifestations and molecular determinants of telomere capping. *Cold Spring Harb. Symp. Quant. Biol.* **2000**, *65*, 253–263. [CrossRef]
16. Richter, T.; Proctor, C. The role of intracellular peroxide levels on the development and maintenance of telomere-dependent senescence. *Exp. Gerontol.* **2007**, *42*, 1043–1052. [CrossRef] [PubMed]
17. Sohn, S.H.; Subramani, V.K. Dynamics of telomere length in the chicken. *World Poult. Sci. J.* **2014**, *70*, 721–736. [CrossRef]
18. Costantini, D.; Marasco, V.; Møller, A.P. A meta-analysis of glucocorticoids as modulators of oxidative stress in vertebrates. *J. Comp. Physiol. Biol.* **2011**, *181*, 447–456. [CrossRef] [PubMed]
19. Von Zglinicki, T. Oxidative stress shortens telomeres. *Trends Biochem. Sci.* **2002**, *27*, 339–344. [CrossRef]
20. Elia, M. Organ and tissue contribution to metabolic rate. In *Energy Metabolism. Tissue Determinants and Cellular Corollaries*; Kinney, J.M., Tucker, H.N., Eds.; Raven Press: New York, NY, USA, 1992; pp. 61–77.
21. Holliday, M.A. Body composition and energy needs during growth. In *Human Growth: A Comprehensive Treatise*; Fulkner, F., Tanner, J.M., Eds.; Plenum Press: New York, NY, USA, 1986; pp. 101–117.
22. Hsu, A.; Heshka, S.; Janumala, I.; Song, M.; Horlick, M.; Krasnow, N.; Gallagher, D. Larger mass of high-metabolic-rate organs does not explain higher resting energy expenditure in children. *Am. J. Clin. Nutr.* **2003**, *1–3*, 1506–1511. [CrossRef]
23. Bianchi, A.; Smith, S.; Chong, L.; Elias, P.; de Lange, T. TRF1 is a dimer and bends telomeric DNA. *EMBO J.* **1997**, *16*, 1785–1794. [CrossRef] [PubMed]
24. De Lange, T. Shelterin: The protein complex that shapes and safeguards human telomeres. *Genes Dev.* **2005**, *19*, 2100–2110. [CrossRef]
25. Ye, J.Z.S.; Donigian, J.R.; van Overbeek, M.; Loayza, D.; Luo, Y.; Krutchinsky, A.N.; Chait, B.T.; de Lange, T. TIN2 binds TRF1 and TRF2 simultaneously and stabilizes the TRF2 complex on telomeres. *J. Biol. Chem.* **2004**, *279*, 47264–47271. [CrossRef] [PubMed]
26. Blackburn, E.H. Structure and function of telomeres. *Nature* **1991**, *350*, 569–573. [CrossRef]
27. Greider, C.W.; Blackburn, E.H. Identification of a specific telomere terminal transferase activity in Tetrahymena extracts. *Cell* **1985**, *43*, 405–413. [CrossRef]
28. Lansdorp, P.M. Telomeres, stem cells, and hematology. ASH 50th anniversary review. *Blood* **2008**, *111*, 1759–1766. [CrossRef]
29. Taylor, H.A.; Delany, M.E. Ontogeny of telomerase in chicken: Impact of downregulation on pre- and postnatal telomere length in vivo. *Dev. Growth Diff.* **2000**, *42*, 613–621. [CrossRef] [PubMed]
30. Gaudet, P.; Livestock, M.S.; Lweis, S.E.; Thomas, P.D. Phylogenetic-based propagation of functional annotations, within the Gene ontology consortium. *Brief. Bioinform.* **2011**, *12*, 449–462. [CrossRef]
31. Graf, M.; Bonetti, D.; Lockhart, A.; Serhal, K.; Kellner, V.; Maicher, A.; Joilivet, P.; Teixeira, M.T.; Luke, B. Telomere length determines TERRA and R-loop regulation through the cell cycle. *Cell* **2017**, *170*, 72–85. [CrossRef]
32. Koskas, S.; Decottignies, A.; Dufour, S.; Pezet, M.; Verdel, A.; Vourch, C.; Faure, V. Heat shock factor 1 promotes TERRA transcription and telomere protection upon heat stress. *Nucleic Acids Res.* **2017**, *45*, 6321–6333. [CrossRef]
33. Blasco, M.; Schoeftner, S. Developmentally regulated transcription of mammalian telomeres by DNA-dependent RNA polymerase II. *Nat. Cell Biol.* **2008**, *10*, 228–236.
34. Romano, G.H.; Harari, Y.; Yehuda, T.; Podhorzer, A.; Rubinstein, L.; Shamir, R.; Gottlieb, A.; Silberberg, Y.; Pe'er, D.; Ruppin, E.; et al. Environmental stresses disrupt telomere length homeostasis. *PLoS Genet.* **2013**, *9*, e1003721. [CrossRef] [PubMed]
35. Hu, X.F.; Guo, Y.M.; Huang, B.Y.; Zhang, L.B.; Bun, S.; Liu, D. Effect of corticosterone administration on small intestinal weight and expression of small intestinal nutrient transporter mRNA of broiler chickens. *Asian Aust. J. Anim. Sci.* **2010**, *23*, 175–181. [CrossRef]

36. Cawthon, R.M. Telomere measurement by quantitative PCR. *Nucleic Acids Res.* **2002**, *30*, e47. [CrossRef]
37. Herborn, K.A.; Heidinger, B.J.; Boner, W.; Noguera, J.C.; Adam, A.; Daunt, F.; Monaghan, P. Stress exposure in early post-natal life reduces telomere length: An experimental demonstration in a long-lived seabird. *Proc. R. Soc. Lond.* **2014**, *281*, 20133151. [CrossRef]
38. O'Callaghan, N.J.; Fenech, M. A quantitative PCR method for measuring absolute telomere length. *BMC* **2011**, *13*, 3. [CrossRef]
39. Livak, K.J.; Schmittgen, T.D. Analysis of relative gene expression using real-time quantitative PCR and the $2^{-\Delta\Delta CT}$ method. *Methods* **2001**, *25*, 402–408. [CrossRef] [PubMed]
40. SAS Publishing. *Statistical Analysis System Multiple Incorporation. Users Guide Statistical Version*; SAS Institute Inc.: Cary, NC, USA, 2002.
41. Gross, W.B.; Siegel, P.B. Some effects of feeding deoxycorticosterone to chickens. *Poult. Sci.* **1981**, *60*, 2232–2239. [CrossRef]
42. Zulkifli, I.; Najafi, P.; Nurfarahin, A.J.; Soleimani, A.F.; Kumari, S.; Anna Aryan, A.; O' Reilly, E.L.; Eckersall, P.D. Acute phase proteins, interleukin 6, and heat shock protein 70 in broiler chickens administered with corticosterone. *Poult. Sci.* **2014**, *93*, 3112–3118. [CrossRef]
43. Lin, H.; Sui, J.; Jiao, H.; Buyse, H. Impaired development of broiler chickens by stress mimicked by corticosterone exposure. *Comp. Biochem. Phys. A* **2006**, *143*, 400–405. [CrossRef] [PubMed]
44. Hayashi, K.; Kaneda, S.; Otsuka, A.; Tomita, Y. Effects of ambient temperature and thyroxine on protein turnover and oxygen consumption inchicken squeletal muscle. In Proceedings of the 19th World's Poultry Congress, Amsterdam, The Netherlands, 19–24 September 1992; Volume 2, pp. 93–96.
45. Gross, W.B.; Siegel, P.B.; DuBose, R.T. Some effects of feeding corticosterone to chicken. *Poult. Sci.* **1980**, *59*, 516–522. [CrossRef]
46. Ziegler, T.R.; Evans, M.E.; Fernandez-Estivariz, C.; Jones, D.P. Trophic and cytoprotective nutrition for intestinal adaptation, mucosal repair, and barrier function. *Annu. Rev. Nutr.* **2003**, *23*, 229–261. [CrossRef]
47. Vágási, C.I.; Pătraş, L.; Pap, P.L.; Vincze, O.; Mureşan, C.; Németh, J.; Lendvai, Á.Z. Experimental increase in baseline corticosterone level reduces oxidative damage and enhances innate immune response. *PONE* **2018**, *13*, 1–17. [CrossRef] [PubMed]
48. Kawanishi, S.; Oikawa, S. Mechanism of telomere shortening by oxidative stress. *Ann. N. Y. Acad. Sci.* **2004**, *1019*, 278–284. [CrossRef] [PubMed]
49. Kurt, W.; Vera, E.; Martinez-nevado, E.; Sanpera, C.; Blasco, M.A. Telomere shortening rate predicts species life span. *Proc. Natl. Acad. Sci. USA* **2019**, *116*, 15122–15127. [CrossRef]
50. Klaunig, J.E.; Zemin, W.; Xinzhu, P.; Shaoyu, Z. Oxidative stress and oxidative damage in chemical carcinogenesis and multistage carcinogenesis. *Toxicol. Appl. Pharmacol.* **2011**, *254*, 86–99. [CrossRef]
51. Reichert, S. Determinants of Telomere Length and Implications in Life History Trade-Offs. Ph.D. Thesis, University De Strasbourg, Strasbourg, France, 2013; p. 344.
52. Kim, Y.J.; Subramani, V.K.; Sohn, S.H. Age prediction in the chickens using telomere quantity by quantitative fluorescence in situ hybridisation technique. *Asian Aust. J. Anim. Sci.* **2011**, *24*, 603–609. [CrossRef]
53. Harley, C.B.; Futcher, A.B.; Greider, C.W. Telomeres shorten during ageing of human fibroblasts. *Nature* **1990**, *345*, 458–460. [CrossRef] [PubMed]
54. Martin, T.E.; Riordan, M.M.; Repin, R.; Mouton, J.C.; Blake, W.M.; Fleishman, E. Apparent annual survival estimates of tropical songbirds better reflect life history variation when based on intensive field methods. *Glob. Ecol. Biogeogr.* **2017**, *26*, 1386–1397. [CrossRef]
55. Muñoz, A.P.; Kéry, M.; Martins, P.V.; Ferraz, G. Age effects on survival of Amazon forest birds and the latitudinal gradient in bird survival. *Auk Ornithol. Adv.* **2018**, *135*, 299–313. [CrossRef]
56. Wilbourn, R.V.; Moatt, J.P.; Froy, H.; Wally, C.A.; Nussey, D.H.; Broonekamp, J.J. The relationship between telomere length and mortality Risk in non-model vertebrate system: A Meta Analysis. *Phil. Trans. R. Soc. B Biol. Sci.* **2018**, *373*, 20160447. [CrossRef] [PubMed]
57. Tricola, G.M.; Simons, M.J.P.; Atema, E.; Boughton, R.K.; Brown, J.L.; Dearborn, D.C.; Divoky, G.; Eimes, J.A.; Huntington, C.E.; Kitaysky, A.E.; et al. The rate of telomere loss is related to maximum lifespan in birds. *Phil. Trans. R. Soc. B Biol. Sci.* **2018**, *373*, 1741. [CrossRef] [PubMed]
58. Blackburn, E.H. Switching and signalling at the telomere. *Cell* **2001**, *106*, 661–673. [CrossRef]
59. Dunham, M.A.; Neumann, A.A.; Fasching, C.L.; Reddel, R.R. Telomere maintenance by recombination in human cells. *Nat. Genet.* **2000**, *26*, 447–450. [CrossRef]
60. Greider, C.W. Telomerase activity, cell proliferation, and cancer. *Proc. Natl. Acad. Sci. USA* **1998**, *95*, 90–92. [CrossRef] [PubMed]
61. Martinez, P.; Blasco, M.A. Heart-Breaking telomeres. *Circ. Res.* **2018**, *123*, 787–802. [CrossRef] [PubMed]
62. Olkowski, A.A. Pathophysiology of heart failure in broiler chickens: Structural, biochemical and molecular characteristics. *Poult. Sci.* **2007**, *85*, 999–1005. [CrossRef] [PubMed]
63. Strong, M.A.; Vidal-Cadenas, S.L.; Karim, B.; Yu, H.; Guo, N.; Greider, C.W. Phenotypes in mTERT$^{+/-}$ and mTERT$^{-/-}$ mice are due to short telomeres, not telomere-independent functions of telomerase reverse transcriptase. *Mol. Cell Biol.* **2011**, *31*, 2369–2379. [CrossRef] [PubMed]
64. Armanios, M. Telomeres and age-related disease: How telomere biology informs clinical paradigm. *J. Clin. Investig.* **2013**, *123*, 996–1002. [CrossRef]
65. Booth, S.A.; Charchar, F.J. Cardiac telomere length in heart development, function and disease. *Physiol. Genom.* **2017**, *49*, 368–384. [CrossRef] [PubMed]

66. Yip, B.W.; Mok, H.O.; Peterson, D.R.; Wan, M.T.; Taniguchi, Y.; Ge, W.; Au, D.W. Sex-dependent telomere shortening, telomerase activity and oxidative damage in marine medaka Oryzias melastigma during aging. *Mar. Pollut. Bull.* **2017**, *124*, 701–709. [CrossRef] [PubMed]

67. Epel, E.S.; Blackburn, E.H.; Lin, J.; Dhabhar, F.S.; Adler, N.E.; Morrow, J.D.; Cawthon, R.M. Accelerated telomere shortening in response to life stress. *Proc. Natl. Acad. Sci. USA* **2004**, *101*, 17312–17315. [CrossRef]

68. Monaghan, P.; Haussmann, M.F. Do telomere dynamics link lifestyle and lifespan? *Trends Ecol. Evol.* **2006**, *21*, 47–53. [CrossRef]

69. Monaghan, P. Telomeres and life histories: The long and the short of it. *Ann. N. Y. Acad. Sci.* **2010**, *1206*, 130–142. [CrossRef] [PubMed]

70. Bauch, C.; Becker, P.H.; Verhulst, S. Telomere length reflects phenotypic quality and costs of reproduction in a long-lived seabird. *Proc. R. Soc. Lond.* **2013**, *280*, 20122540. [CrossRef]

71. Yip, S.H.; Sham, P.C.; Wang, J. Evaluation of tools for highly variable gene discovery from single-cell RNA-seq data evaluation of tools for highly variable gene discovery from single-cell RNA-seq data. *Brief. Bioinform.* **2018**, *20*, 1583–1589. [CrossRef] [PubMed]

72. Cusanelli, E.; Angelica, C.; Romero, P.; Chartrand, P. Telomeric noncoding RNA TERRA is induced by telomere shortening to nucleate telomerase molecules at short telomeres. *Mol. Cell* **2013**, *51*, 780–791. [CrossRef]

73. Okamoto, K.; Iwano, T.; Tachibana, M.; Shinkai, Y. Distinct roles of TRF1 in the regulation of telomere structure and lengthening. *J. Biol. Chem.* **2008**, *283*, 23981–23988. [CrossRef]

74. Kolquist, K.A.; Ellisen, L.W.; Counter, C.M.; Meyerson, M.; Tan, L.K.; Weinberg, R.A.; Haber, D.A.; Gerald, W.L. Expression of TERT in early premalignant lesions and a subset cells in normal tissues. *Nat. Genet.* **1998**, *19*, 182–186. [CrossRef]

75. Ludlow, A.T.; Spangenburg, E.E.; Chin, E.R.; Cheng, W.H.; Roth, S.M. Telomeres shorten in response to oxidative stress in mouse skeletal muscle fibers. *J. Gerontol. Biol. Sci. Med. Sci. J.* **2014**, *69*, 821–830. [CrossRef]

76. Schmutz, I.; de Lange, T. Shelterin. *Curr. Biol.* **2016**, *26*, R397–R399. [CrossRef]

77. Bejarano, L.; Schuhmacher, A.J.; Squatrito, M.; Blasco, M.A. Inhibition of TRF1 telomere protein impairs tumor initiation and progression in glioblastoma mouse models and patient-derived xenografts. *Cancer Cell* **2017**, *32*, 590–607. [CrossRef] [PubMed]

78. RGD (Rat Genome Database). (N.D). Telomere Maintenance 2 Gene. 2020. Available online: https://www.rgd.mcw.edu (accessed on 3 June 2019).

79. Scheibe, M.; Arnoult, N.; Kappei, D.; Bucholz, F.; Decottignies, A.; Butter, F.; Mann, M. Quantitative interaction screen of telomeric repeat-containing RNA reveal novel TERRA regulators. *Genome Res.* **2013**, *23*, 2149–2157. [CrossRef] [PubMed]

80. Maicher, A.; Kastner, L.; Dees, M.; Luke, B. Deregulated telomere transcription causes replication—Dependent telomere shortening and promotes cellular senescence. *Nucleic Acid Res.* **2012**, *40*, 6649–6659. [CrossRef] [PubMed]

81. Prowse, K.R.; Greider, C.W. Development and tissue specific regulation of mouse telomerase and telomere length. *Proc. Natl. Acad. Sci. USA* **1995**, *92*, 4818–4822. [CrossRef]

82. Shay, J.W.; Bacchetti, S. A survey of telomerase activity in human cancer. *Eur. J. Cancer* **1997**, *33*, 787–791. [CrossRef]

83. Goichberg, P.; Chang, J.; Liao, R.; Leri, A. Cardiac stem cells: Biology and clinical applications. *Antioxid. Redox Signal.* **2014**, *21*, 2002–2017. [CrossRef] [PubMed]

84. Bettin, N.; Pegorar, C.O.; Cusanelli, E. The emerging roles of TERRA in telomere maintenance and genome stability. *Cells* **2019**, *8*, 246. [CrossRef]

85. Yifan, C.; Tianzheng, Y. Mouse liver is more resistant than skeletal muscle to heat-induced apoptosis. *Cell Stress Chaperones* **2021**, *26*, 275–281. [CrossRef]

86. Lu, A.L.; Li, X.; Gu, Y.; Wright, P.M.; Chang, D.Y. Repair of oxidative DNA damage:mechanisms and functions. *Cell Biochem. Biophys.* **2001**, *35*, 141–170. [CrossRef]

87. Marnett, L.J. Oxyradicals and DNA damage. *Carcinogenesis* **2000**, *21*, 361–370. [CrossRef] [PubMed]

Communication

Behavioral Monitoring Tool for Pig Farmers: Ear Tag Sensors, Machine Intelligence, and Technology Adoption Roadmap

Santosh Pandey [1,*], Upender Kalwa [1], Taejoon Kong [2], Baoqing Guo [3], Phillip C. Gauger [3], David J. Peters [4] and Kyoung-Jin Yoon [3,*]

1 Department of Electrical and Computer Engineering, Iowa State University, Ames, IA 50011, USA; upender_kalwa@outlook.com
2 Center for Defense Acquisition and Requirements Analysis, Korea Institute for Defense Analyses, 37 Hoegi-ro, Dongdaemun-gu, Seoul 02455, Korea; taejoonkong@gmail.com
3 Department of Veterinary Diagnostic and Production Animal Medicine, Iowa State University, Ames, IA 50011, USA; bqguo@iastate.edu (B.G.); pcgauger@iastate.edu (P.C.G.)
4 Rural Sociology, Department of Sociology and Criminal Justice, Iowa State University, Ames, IA 50011, USA; dpeters@iastate.edu
* Correspondence: pandey@iastate.edu (S.P.); kyoon@iastate.edu (K.-J.Y.)

Simple Summary: In a pig farm, it is challenging for the farm caretaker to monitor the health and well-being status of all animals in a continuous manner throughout the day. Automated tools are needed to remotely monitor all the pigs on the farm and provide early alerts to the farm caretaker for situations that need immediate attention. With this goal, we developed a sensor board that can be mounted on the ears of individual pigs to generate data on the animal's activity, vocalization, and temperature. The generated data will be used to develop machine learning models to classify the behavioral traits associated with each animal over a testing period. A number of factors influencing the technology adoption by farm caretakers are also discussed.

Abstract: Precision swine production can benefit from autonomous, noninvasive, and affordable devices that conduct frequent checks on the well-being status of pigs. Here, we present a remote monitoring tool for the objective measurement of some behavioral indicators that may help in assessing the health and welfare status—namely, posture, gait, vocalization, and external temperature. The multiparameter electronic sensor board is characterized by laboratory measurements and by animal tests. Relevant behavioral health indicators are discussed for implementing machine learning algorithms and decision support tools to detect animal lameness, lethargy, pain, injury, and distress. The roadmap for technology adoption is also discussed, along with challenges and the path forward. The presented technology can potentially lead to efficient management of farm animals, targeted focus on sick animals, medical cost savings, and less use of antibiotics.

Keywords: precision swine farming; ear tag pig sensor; behavioral monitoring; machine intelligence; technology adoption

Citation: Pandey, S.; Kalwa, U.; Kong, T.; Guo, B.; Gauger, P.C.; Peters, D.J.; Yoon, K.-J. Behavioral Monitoring Tool for Pig Farmers: Ear Tag Sensors, Machine Intelligence, and Technology Adoption Roadmap. *Animals* **2021**, *11*, 2665. https://doi.org/10.3390/ani11092665

Academic Editors: Melissa Hempstead and Danila Marini

Received: 31 July 2021
Accepted: 7 September 2021
Published: 10 September 2021

Publisher's Note: MDPI stays neutral with regard to jurisdictional claims in published maps and institutional affiliations.

1. Introduction

The economics of a pig farm is dependent on the health and welfare status of pigs [1]. Therefore, all stakeholders in the pig industry want to ensure that pigs display normal behavior and physiological functioning with the absence of lesions, diseases, or malnutrition. This requires frequent assessment of their well-being status and disease symptoms. Good well-being is reached when the animal is in harmony with itself and its environment, whereas poor well-being happens when the animal is exposed to infections and adverse conditions resulting from different management practices. In general, poor well-being is manifested by behavioral changes (e.g., abnormal movement, reduced feeding or drinking, lethargy, or aggressive nature), physiological changes (e.g., increased heart rate or respiration rate), and pathological changes (e.g., lesions, stress-related biomarkers, and other clinical signs). Reduced well-being may negatively influence the pig's health, growth, behavior, and emotional state.

Today, manual surveillance is generally conducted few times during the day in pig farms to visually screen for apparent signs of illness in their pigs, such as lethargy, lameness, and coughing [2]. Farm animal caretakers can know the extent of their pigs' well-being with respect to their mental state (i.e., being calm, satisfied, relaxed, curious, playful, scared, stressed, or grunting) and physical state (i.e., healthy, medicated, or injured). However, human observations are subjective, as it is difficult to delineate the factors associated with the pig's mental and physical state (e.g., in stress or pain) [2]. Manual surveillance may miss out to detect early signs of sickness (e.g., fever), particularly at night when the disease symptoms can be elevated. Moreover, it is difficult to give proper attention to all the animals and to identify subtle traits indicative of poor well-being in a timely manner. Without timely detection and intervention, the health of the pigs in a herd can be compromised, resulting in additional costs of medications [2], diagnostics [3], and therapy.

To complement manual surveillance, an automated behavioral assessment tool can be built that operates on a set of indicators to provide continuous surveillance, recognize stressful management practices, and detect early signs of sickness in individual pigs to promptly take corrective measures for the betterment of the animal and the entire population [1,4]. However, the identification of all relevant indicators is difficult because pig well-being is essentially a multidimensional and complex concept [5]. To accomplish automated behavioral assessment at the single animal resolution, a high-throughput data acquisition system is needed to measure the indicators noninvasively, under varying conditions, and in real time with minimal human intervention [6,7]. Furthermore, data science approaches are needed to identify relative value and relationships amongst the indicators [8,9].

Within the category of wearables, a variety of smart collars are commercially available that incorporate motion sensors, cameras, and/or microphones to record the daily activities of animals (e.g., dogs, cats, cattle, laboratory mice, sheep, goats, monkeys, marine life, and avian) [10]. The smart collars gather information about the physical location, movement, body temperature, grazing, and drinking behavior of these animals (Table 1). Most smart collars have an accompanying smartphone app through which the user can visualize, store, and analyze the gathered data [4]. In addition to smart collars, smart capsules (also called video endoscopy capsules) are used to visualize the gastrointestinal tract for disease diagnosis in a minimally invasive manner (e.g., Olympus EndoCapsule 10 System^{TM} and Medtronic PillCam SB 3 System^{TM}). Moreover, in vitro microchip technologies have been custom designed for several applications in veterinary parasitology, such as for screening the efficacy of anthelmintics and chemical compounds against nematodes and enteroparasites for the pig host often tested by larval migration assays or fecal egg counts [11–14]. Imaging and/or thermal cameras have been used to record the movement and body heat emission of pig populations, along with subsequent machine learning analysis of recorded data [7,9]. Farm automation equipment is routinely deployed to control the indoor ventilation, feeding and watering, housing units, and thermal control systems (with regulated heating and cooling).

Most of the available technologies in human wearables and animal health monitoring are not feasible for direct adoption by the pig industry. For example, the wearables for human healthcare (e.g., smartwatches and fitness trackers) are cost prohibitive and priced over $50 per device. The anatomic conformation and behavior of the pigs make it virtually impossible to design wearables that can be used for extended periods of time (hours/days or more). Furthermore, skin-worn adhesive sensors are not suited for hairy pigs [15]. The commercialized smart collars designed for cows and companion animals are very expensive for monitoring a large number of pigs [10,16]. Moreover, the smart collars for cows and sheep are too large and heavy to be adopted for small piglets, while the electronic pet neck collars can be easily damaged because of the chewing behavior of pigs. Most sensor systems in pig farms (such as imaging cameras, microphones, climate control) cater to herd-level monitoring and do not have the resolution to pick up vital signs of individual pigs.

Table 1. Commercialized products for tracking humans, pets, and livestock.

Products	Description
1. Tractive, FitBark, Whistle 3	Real-time GPS activity tracker for cats and dogs
2. TekVet, FeverTag	Temperature sensor for livestock
3. Abilify, EndoCapsule EC-10, Vital Herd	Ingestible pill to track core, heart rate, respiration rate, pH levels, drug doses
4. Cowlar, CowCollar, Cattle Watch	Electronic collar for cows to track body motion, cud-chewing, location, create invisible fence
5. E-Shepherd	Electronic collar for sheep to generate irritating sound and lights signals to deter predators
6. Apple Watch, Fitbit, Garmin, Samsung Gear	Track fitness, performance, heart rate, respiration, stress, blood oxygen levels, sleep quality

Aside from sensing and control units, there is a dearth of swine behavioral data to generate machine learning models that stand validated through multiple checks. The tools for gathering field data from swine farms, storing the acquired data, and applying predictive analytics are still premature but are constantly improving [9,17,18]. One could look for inspiration within the field of medical diagnostics, where there is a growing trend in low-cost, easily accessible, and field-deployable technology platforms for remote patient monitoring and decision support. In this regard, smartphone-based tools have been demonstrated for the objective assessment of physical attributes for clinical decision-making that could be adapted for swine behavioral monitoring, such as smartphone-based tools for the visual grading of pelvic asymmetry in equines [19] and skin cancer diagnostics in humans [20]. In addition to building a reliable data collection and data communication pipeline, there is a need to investigate new paradigms and models in swine health and welfare assessment that are fundamentally more robust and intelligent than the traditional models designed to estimate the future value of the animal (e.g., body mass, growth rate, genetic or breeding parameters).

Our motivation lies in the objective assessment of swine behavior using a comprehensive set of indicators. The behavioral indicators should be meaningful and realistic, easy to collect on the farm in a noninvasive and unobtrusive manner, reliable to predict wellness levels, scalable to different animal classes (i.e., group size, growth phase), integrative with on-farm decision tools, and easily validated through alternate behavioral assessment tools [5,21]. In this paper, we demonstrate a proof-of-concept data acquisition system, based on a cluster of electronic sensors, to collect data on behavioral attributes on individual pigs (Figure 1). In Section 2, we describe the sensors to monitor external body temperature, physical movement, and vocalization of individual pigs. In Section 3, we show the results obtained from characterizing the sensors in the laboratory and tests from animal studies. In Section 4, we present a discussion on possible machine intelligence methodology to identify normal versus abnormal behavioral attributes in pigs, along with behavioral indicators that provide information about the pigs' well-being. We discuss the challenges, limitations, and path forward for the development and dissemination of this technology, including the social–economic–ethical considerations for technology adoption.

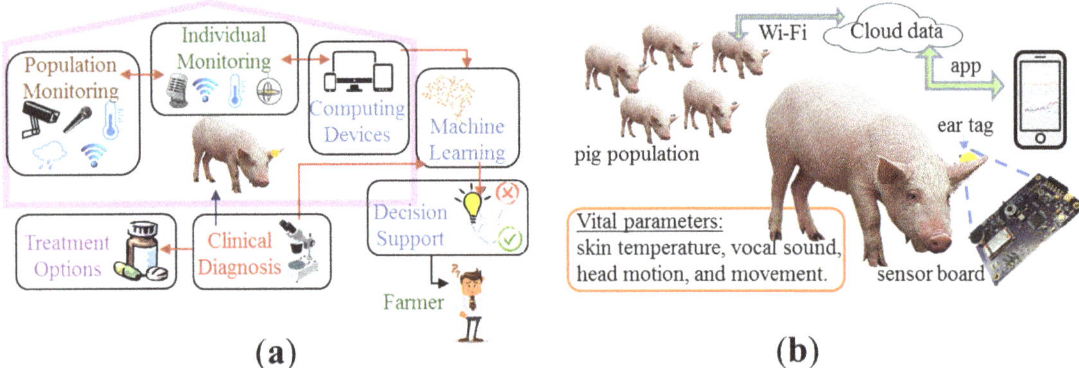

Figure 1. Overall theme for behavioral health monitoring in pig farms: (**a**) the pig farm is equipped with technologies to monitor individual animals and populations with mobile devices, data computing and storage units, machine learning, decision support tools, clinical diagnosis, and treatment options; (**b**) our measurement setup is shown. The ear tag with sensor board is attached to the ear lobe of individual pigs to measure vital parameters and send the data through a smartphone app to the cloud.

2. Materials and Methods

Electronic Sensor Boards

One cluster of sensors (soldered on a custom printed circuit board (PCB)) was attached to an ear tag for pigs to record parameters such as external temperature, head tilt, movement, and vocalization. A Bluetooth-enabled, application-specific integrated circuit (ASIC) chip was soldered on the electronic sensor board for the wireless transmission of raw data to a computing device located in or nearby the pen. A replaceable lithium battery was housed within a battery holder on the board. Figure 2 shows the computer-aided design (CAD) software layout of the electronic sensor board, fabricated PCB, and its attachment to a sample ear tag. The physical dimensions of the electronic sensor board were 3.05 cm × 3.05 cm × 1.27 cm (L × W × H), with a weight of 5 g and a development cost of around USD 20 per sensor board.

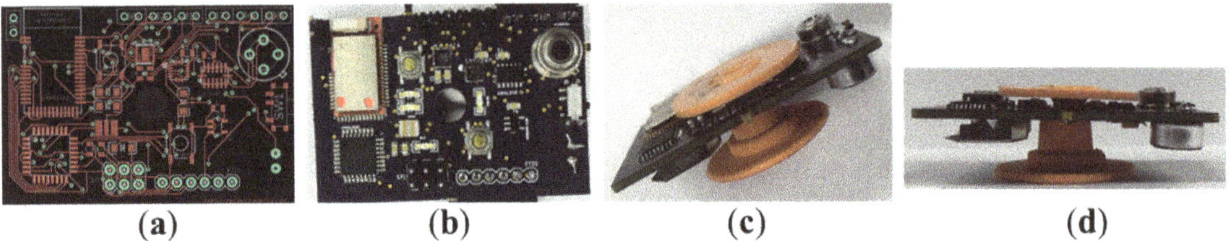

Figure 2. Prototype design and assembly: (**a**) CAD layout of the electronic sensor board; (**b**) the fabricated PCB with soldered electronic chips; (**c**,**d**) inclined view and side view of an assembled sensor board sandwiched between the orange-colored ear tag.

The electronic sensor board had the following sensors:

(i) A contactless, infrared temperature sensor (MLX90614) to measure both the ambient temperature and external body temperature of the pig;

(ii) A 3-axis accelerometer (ADXL345) and gyroscope (ADXRS300) to record data on head tilt and movement, which is correlated to aggression, lethargy, lameness, or neurological disorders;

(iii) A sound sensor (MAX9814) to capture the vocalizations for the identification and classification of call types or aggressive behavior;

(iv) A Bluetooth low energy (BLE) module (NRF51822 SoC) to receive data from the abovementioned sensors at a preselected rate (e.g., 5 samples per second). This BLE module incorporates a chip antenna, 32-bit ARM Cortex M0 CPU, 256 kB Flash Memory/16 kB RAM, and 32 GPIO. The BLE module is suited for ultralow-power wireless communication at 2.4 GHz by supporting third-party I2C, SI, UART, and PWM interfaces.

A 3.3 V, 230 mAh coin cell battery (CR 2032) provided power to the sensors and BLE module on the PCB. The electronic sensor boards were sealed in polymeric coatings (e.g., SU-8, PDMS) before being put within ear tags.

3. Results

3.1. Laboratory Testing

All the sensors and the Bluetooth module were tested in our device characterization laboratory according to the manufacturer's specifications. The software development kit (SDK) from Nordic Semiconductor was downloaded (nRF5 SDK v12.3.0) prior to testing the Bluetooth module. We prepared 10 electronic sensor boards. Within every board, data from each sensor were read by placing the board next to the receiver or at a certain distance (roughly 1 m). We tested different sampling rates (from five samples per second to one sample per 10 min) and sleep times of the sensors (1 s to 10 min) to maximize battery life while still having adequate data resolution. With a sampling rate of five samples per second, the Bluetooth module and gyroscope consumed the maximum current (6 mA), followed by the sound sensor (3.1 mA), temperature sensor (1.5 mA), and accelerometer (0.14 mA). Assuming that the CR2032 had a capacity of 230 mAh, this translates to a battery life of 13.74 h. We found that the battery life as built was increased to over 3 days by prolonging the sampling rate and sleep time. The battery life can be extended by incorporating adaptive sampling and power management schemes.

3.2. Animal Testing

For animal testing, the electronic sensor board was sandwiched between the two pieces of the ear tags and attached to the ear of pigs. Tests were conducted on two growing pigs (body weight: around 18 kg) (Figure 3). We did not observe any unnatural behavior from the pigs during or after mounting the electronic sensor boards on the ear. The data from the growing pigs were recorded for around 4500 s or 75 min (Figure 4). The animal tests confirmed that remote data collection from the sensor boards was near real time and fully automated over a distance of 15.25–18.3 m.

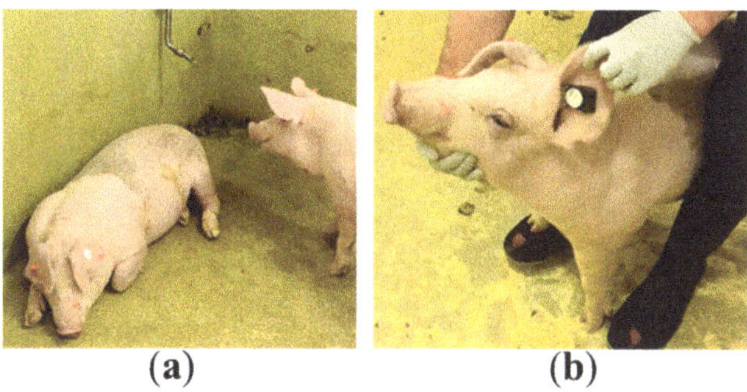

Figure 3. Animal testing: (**a**) two growing pigs were used for testing the electronic sensor boards; (**b**) the ear tag with the sandwiched electronic sensor board was attached to the left ear of the pig. The sensors and ASIC chip were remotely controlled by predefined awake/sleep cycles.

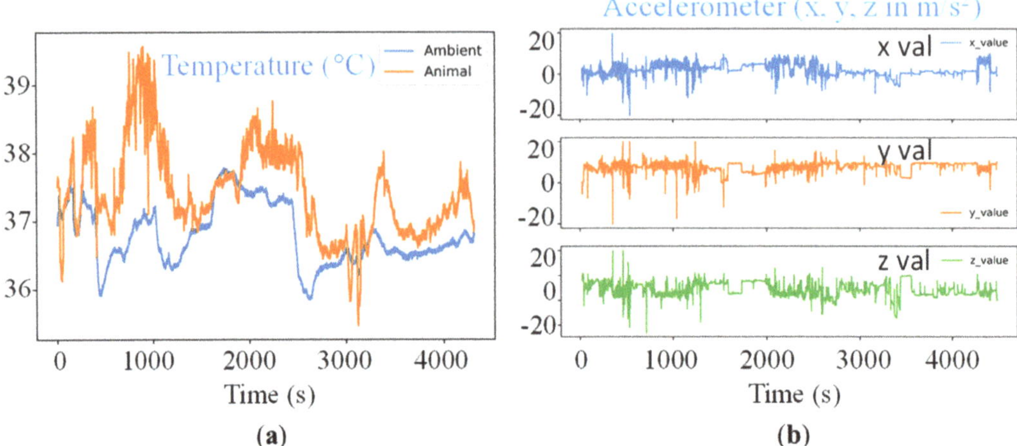

Figure 4. Time-series data recorded from pig testing using the electronic sensor board. Sample data from the animals is shown here as obtained from the on-board (**a**) temperature sensor and (**b**) 3-axis accelerometer. The sound sensors collected the baseline curves, but there were no useful events (e.g., coughing or screaming) during the recorded time period.

4. Discussion

4.1. Machine Intelligence for Swine Behavioral Health Monitoring

A single behavioral health indicator can misjudge the well-being status, and therefore, a realistic set of indicators are required for accurate assessment [22]. In this context, machine intelligence is valuable to assimilate and process data from multiple sensors and human observations for subsequent decision support [8,23]. Currently, our data collection activities are underway with the intent to generate sufficiently large datasets to map the raw sensors' data into predictable behavioral traits, as described in Figure 5. Here, the activity sensors measure the raw data related to posture and gait (e.g., the nature, duration, rate, and ease of events). Thereafter, machine intelligence based on linear regression will be used to analyze the raw data and distinguish different postures and gaits. The sound sensors measure vocalization in terms of the call duration, call rate, main frequency, peak amplitude, and peak frequency. This will help to distinguish the nature and type of call. Abnormal behavioral traits could include prolonged rest, reduced eating and drinking, lameness, pain, aggression, escape events, screams, coughing, and fever. These abnormal traits could be identified by machine learning algorithms such as random forest, decision tree, K-nearest neighbor, support vector machine, artificial neural networks, naïve Bayes, linear regression, logistic regression, and gradient boosting.

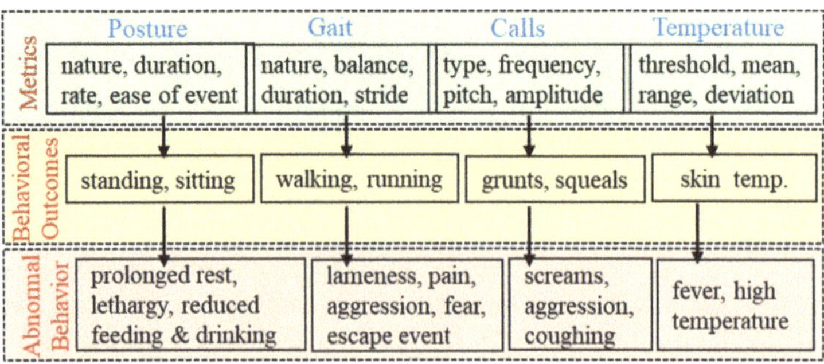

Figure 5. Flowchart for metrics, behavioral outcomes, and indicators of poor well-being.

One suggested scheme for processing the raw data from different electronic sensors is shown in Figure 6. The data processing starts by segmenting the incoming sensors' data into separate time windows, as described in our previous work on deep learning [24]. We compute a set of features that give a good spectral and temporal representation of the data (such as mean, standard deviation, energy, skewness, entropy, and auto-regressive coefficients). Next, we employ classifiers on the computed set of features. When a feature exceeds a certain threshold, the behavior is categorized by a classifier into one of the behavioral classes, and an alert is sent to the farmer. The choice of behavioral traits to train machine learning models should be those that are most relevant to the pig farmer. Some examples of behavioral classes related to pig activity are walking, standing, sitting, lying, head down, grunting, screaming, feeding, and drinking. Choosing an appropriate threshold level is often beneficial, as a low threshold leads to many false positives and false alerts, while a high threshold gives many false negatives and missed true events [17,25]. In most cases, the choice of the optimal threshold for a classifier depends on the datasets, operating conditions, type of model used, receiver operating characteristic (ROC) curve, precision-recall curve, and the intuitive sense of the model developer.

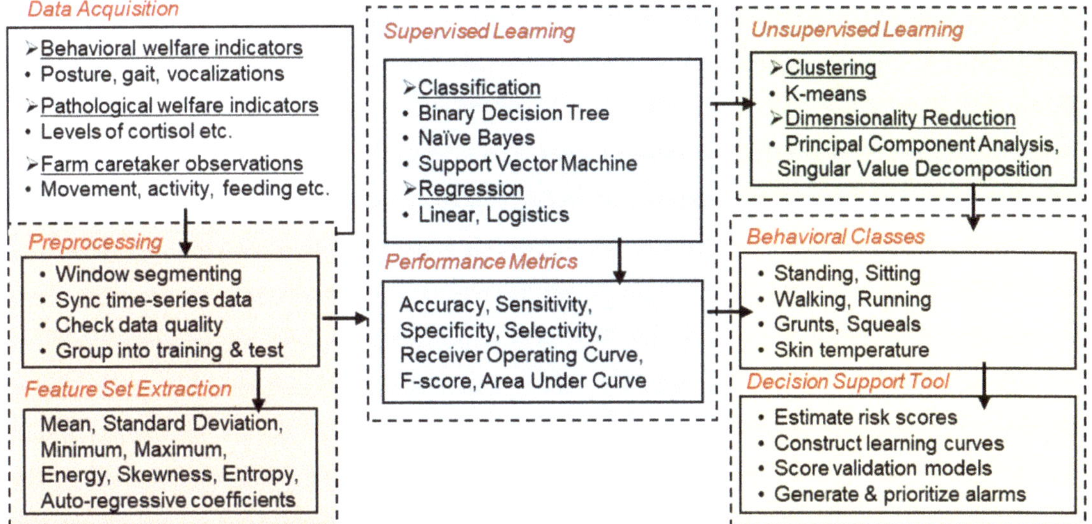

Figure 6. Process flow during machine intelligence involves data acquisition, preprocessing, feature set extraction, supervised or unsupervised learning, class definitions, performance metrics, and decision support.

Table 2 lists the hardware and software features, and potential benefits of our ear tag sensor board for the pig farmer. The technical features are designed to be modular and easily upgradable to support autonomous and remote monitoring of behavioral health indicators. There are near-term challenges during technology development, such as establishing validity with large amounts of real-world data from experimental and field settings. Building reliable machine learning models and decision support tools for large-scale pig farms is also nontrivial [6,26,27]. False alarms increase costs, reduce the trust of operators, and decrease efficiency. Thus, it is advisable to identify farm situations of high prevalence or priority to the farmer where autonomous monitoring and alert notification will complement or supersede manual surveillance [5,27].

Table 2. Attributes and benefits of the ear tag sensor board for the pig farmer.

Attributes	Benefits of the Ear Tag Sensor Board
Hardware features	Light weight, long battery life, low cost, wireless capability, easily upgradable, easy programming, modular, reusable, safe
Software features	Ease of use, multiuser access, autonomous, remote operations, real-time data acquisition and management, decision support
Potential benefits	Real-time awareness of pig vital signs, early disease detection, timely isolation of diseased animal from the herd, cross-platform data compatibility, upgradable and customizable platform, data transparency and privacy, reduced labor, increased profits
Challenges during development	Scope of clinical trials, safety, and health regulations, sustained technical improvements, fluctuations in market trends and surveillance technologies, user acceptability and trust

4.2. Roadmap for Technology Adoption

For technology adoption, bringing the pig farmers on board to test the ear tag wearable technology is critical but challenging. The pig farmer may have misconceptions about digital technologies with respect to their perceived relative value, complexity, adaptability, and compatibility [10,16]. Table 3 lists the inducements and impediments related to technology adoption of the presented ear tag sensor boards.

Table 3. Perceived inducements and impediments in the adoption of ear tag sensor boards by pig farmers.

Adoption Factors	Inducements and Impediments
Technical factors	Learning curve, automation needs, safety, ethics, maintenance support, data accessibility and privacy, trust in decision support
Existing farming practices	Management practices, herd size, labor and service inputs, surveillance techniques, managerial skills, clinical access
Sociocultural factors	Trust groups, local practices, community size, farm size, intent of rearing, labor constraints, digital literacy, communication modes
Economic factors	Cost–benefit clarity, competitiveness, fluctuating markets, financing options, income growth, profit margin, wealth, supply and demand
Health factors	Control over disease outbreaks, early illness detection, reliability of health data, clinical validation, treatment options
External factors	Government and trade policies, farming regulations, transportation, environmental legislation, carbon footprint, bioethics

Technology adoption is often facilitated by communicating its role in promoting the overall well-being of human–animal–environment [1,10]. In this regard, the ethical issues raised by the ear tag sensor boards can be assessed by a two-dimensional bioethics matrix, originally proposed by Dr. Ben Mepham (Special Professor in Applied Bioethics, University of Nottingham) for food biotechnologies [28,29].

In the bioethics matrix, as shown in Table 4, there are three prima facie principles: respect for well-being, respect for autonomy, and respect for justice/fairness. There are five principal stakeholders whose interests are implicated—pigs, pig farmers, consumers, the public at large, and the environment. The matrix framework and components are based on the original work by Dr. Ben Mepham [28,29]. On the one hand, the pork industry is evolving to make the swine production process more rapid and more profitable with minimal human–pig interaction. On the other hand, consumers prefer meat from happy and healthy animals, while the public at large wants transparency in information about animal well-being and handling during production [30]. Respect for the well-being of pigs is achieved by avoiding causes of pain or harm, improving health or welfare, and mitigating risks or costs [31]. Respect for autonomy ensures freedom of choice and behavior for the pigs. Respect for fairness or justice refers to norms regarding fair distribution of costs, risks, and benefits for the animals to protect the intrinsic value of the pigs. The well-being of the pig farmer is reached by satisfactory profits and improved working conditions, autonomy by having the freedom to manage actions, and fairness by access to fair trade deals. Farmers who have more access to data about the health and well-being of their pigs may be expected to benefit and should have increased opportunities to intervene to promote the health and welfare of their animals. However, pig farmers may feel pressurized by the politics and media and skeptical of participating in welfare programs [32]. Modern consumers are interested in the quality and safety of their food, as well as in the humane treatment of animals raised for slaughter and the environmental sustainability of the food they consume [33]. The public at large, and even those who do not consume pork products, may have an interest in the process of pork production and improving the health and safety of pork production facilities. Pork production involves risks to the environment that must be taken into account in the production process, and wherever possible, environmental risks should be minimized [31].

Table 4. Bioethics matrix for developing swine welfare assessment tools.

Respect for	Well-Being	Autonomy	Fairness
Pigs	Avoid pain, improved welfare	Behavioral freedom	Intrinsic value
Farmers	Satisfactory profit, better work condition	Managerial freedom of action	Fair trade laws and practices
Consumers	Better quality of life, food safety	Democratic, informed choices	Availability of affordable pork
The Biota	Conservation	Biodiversity	Sustainability

Source. Motivated and adapted from the pioneering works of Dr. Ben Mepham [28,29].

Social studies can help understand the pig farmers' behavior, existing farm practices, sociocultural perceptions, and decision-making methods while communicating the realistic deliverables [1,16]. Through open discussions and social studies with the potential users, it is possible to understand the factors that influence their decisions on adopting the presented wearable technology. As with other new technologies, the user inputs will help identify effective ways to educate them about the technology's merits and risks from different sociocultural and economic perspectives. By demonstrating the technology at work and rationalizing its cost-versus-benefit analysis in video format and simple language, it may be possible to slowly gain the trust of potential users and ensure that our technology makes its way from the laboratory to the real world.

However, the window for voluntary adoption by hog producers may be fast closing. Animal welfare and food safety is an increasingly important part of purchasing decisions by consumers. For the former, consumers are insisting on minimal welfare standards [34,35], while the latter is centered around consumer concerns over antimicrobial resistance in meat [36,37]. Concentration in food wholesale and retail means consumer pressure can be brought to bear on a few firms that have large market power [38]. As consumers demand more stringent animal welfare and antibiotic use rules, wholesalers and retailers will likely force changes down the supply chain onto livestock producers. In short, the adoption of intelligent sensing systems to improve animal health and reduce antibiotic use may become an economic necessity for hog producers to stay in business.

Additionally, the devastation wrought by coronavirus disease 2019 (COVID-19) highlights how improving animal health is an important part of promoting public health and resilience to pandemics. Large-scale hog production provides opportunities for the generation and transmission of novel viruses from hogs to humans [39]. Although rare, certain variants of swine influenza could emerge to become a regional epidemic, if not a pandemic [40]. Workers in hog production and processing are at risk of contracting these novel viruses and may transmit the disease to the broader community [41]. Thus, early detection of diseases in large-scale swine production facilities is more important than ever in the light of COVID-19.

On the technology development front, there are possibilities to expand the breadth of sensing functionalities tailored to the needs and expectations of the swine farms. The electronics sensors and communication systems are constantly evolving, and there are many options for low-cost and high-performance sensors manufactured by various semiconductor companies [42,43]. Therefore, in conjunction with the presented electronic sensor board, there is scope to equip the farms with other types of farm deployable sensors in swine farms. For instance, ambient and environmental sensors can be installed to monitor different locations in the barn and transmit data to the local server via Bluetooth modules. These sensors can record environmental parameters in the barn such as humidity, room temperature, airflow, and levels of common gases. Some examples of commercial environmental sensors that can be used in pig farms are the capacitive humidity sensor (DHT11 sensor, Adafruit Industries, New York City, NY, USA) to report the dampness estimate in the room, miniaturized airflow transducers (Kanomax^{TM} Model 0965-01, Kanomax USA Inc., Andover, NJ, USA) to measure the air velocity and indoor airflow, wireless gas sensors to monitor the levels of CO_2, CH_4, H_2S, and NH_3 (Monnit^{TM}, Agrologic^{TM}), and infrared cameras (FLIR^{TM}) and webcams (Pig Cam^{TM}) to take images of the pigs.

There are limitations in the ear tag sensor-board technology presented here. Firstly, the current development cost of the electronic sensor boards (approximately USD 20 to USD 30 per board) is a significant investment for swine producers in the short term, even though these reusable boards may have a justifiable return on investment in the long term. Secondly, extending the battery life of wearable devices is a challenge considering the high-power consumption for BLE communication and inaccessibility to recharge or replace the batteries during animal testing. Thirdly, there is a risk of damage to the sensor boards and loss of data communication while the boards are mounted on the pigs. The dimensions of the sensor board are greater than the ear tags used in our tests, which may interfere with the pigs. Using larger ear tags or reducing the size of the sensor boards would be advisable. Fourthly, there are realistic limits on the number of behavioral patterns that can be accurately characterized by machine intelligence. Additionally, one set of machine learning classifiers may not work on different herds and different farms, and some degree of customization may be required. Lastly, winning and retaining the trust of active users is challenging, as there can be considerable variability in user expectations and the performance of decision-making tools for different pig farms.

5. Conclusions

Intelligent sensing systems have the potential to automate and integrate various activities in swine production, such as breeding, feeding, handling, housekeeping, and disease management. However, the pace of technology adoption in pig farming is slow because of three main technological deficiencies—lack of adoptable sensor technologies, limited availability of meaningful data, and unproven means to translate the data science into actionable items. Our work here was primarily motivated to address the first deficiency. We showed a proof-of-concept working of an electronic sensor board that can be attached to the ear tags of individual pigs to record physical activity. Aside from the components used in our board, there are several other options for off-the-shelf sensors and communication modules that can be purchased and tested depending on their functionality, features, compatibility, physical dimensions, and pricing. We recognized a number of inducements and impediments for its user acceptance as a reliable behavioral assessment tool. For technology adoption, one key hurdle lies in collecting and analyzing massive quantities of field data with cross validation by farm animal caretakers. This is a prevalent challenge for the wearables industry at large and requires cooperative efforts from all stakeholders, from swine producers to pork consumers.

Author Contributions: Conceptualization, S.P. and K.-J.Y.; methodology, U.K. and T.K.; software, U.K. and T.K.; validation, U.K., S.P., T.K. and K.-J.Y.; formal analysis, S.P.; investigation, S.P.; resources, D.J.P., B.G., P.C.G., S.P. and K.-J.Y.; data curation, U.K., T.K., D.J.P., P.C.G. and B.G.; writing—original draft preparation, S.P. and K.-J.Y.; writing—review and editing, S.P., D.J.P., B.G. and K.-J.Y.; visualization, U.K. and S.P.; supervision, S.P. and K.-J.Y.; project administration, S.P. and K.-J.Y.; funding acquisition, S.P. and K.-J.Y. All authors have read and agreed to the published version of the manuscript.

Funding: This research was partially supported by National Science Foundation (NSF), Grant Number IDBR-1556370 and by the United States Department of Agriculture, Grant Number 2020-67021-31964.

Institutional Review Board Statement: The animal study was reviewed and approved by the Institutional Committee for Animal Use and Care from Iowa State University (ICAUC Approval #7-17-8561-S).

Acknowledgments: We are grateful to Dalton Pottebaum for providing access to the animals, and to Clark Wolf for our discussions and collaborative efforts.

Conflicts of Interest: The authors declare no conflict of interest.

References

1. Benjamin, M.; Yik, S. Precision livestock farming in swine welfare: A review for swine practitioners. *Animals* **2019**, *9*, 133. [CrossRef]
2. Lekagul, A.; Tangcharoensathien, V.; Liverani, M.; Mills, A.; Rushton, J.; Yeung, S. Understanding antibiotic use for pig farming in Thailand: A qualitative study. *Antimicrob. Resist. Infect. Control* **2021**, *10*, 1–11. [CrossRef] [PubMed]
3. Kong, T.; Backes, N.; Kalwa, U.; Legner, C.; Phillips, G.J.; Pandey, S. Adhesive Tape Microfluidics with an Autofocusing Module That Incorporates CRISPR Interference: Applications to Long-Term Bacterial Antibiotic Studies. *ACS Sens.* **2019**, *4*, 2638–2645. [CrossRef] [PubMed]
4. Bailey, D.W.; Trotter, M.G.; Tobin, C.; Thomas, M.G. Opportunies to Apply Precision Livestock Management on Rangelands. *Front. Sustain. Food Syst.* **2021**, *5*, 93. [CrossRef]
5. Velarde, A.; Fàbrega, E.; Blanco-Penedo, I.; Dalmau, A. Animal welfare towards sustainability in pork meat production. *Meat Sci.* **2015**, *109*, 13–17. [CrossRef]
6. Tzanidakis, C.; Simitzis, P.; Arvanitis, K.; Panagakis, P. An overview of the current trends in precision pig farming technologies. *Livest. Sci.* **2021**, *249*, 104530. [CrossRef]
7. Kapun, A.; Adrion, F.; Gallmann, E. Case study on recording pigs' daily activity patterns with a uhf-rfid system. *Agriculture* **2020**, *10*, 542. [CrossRef]
8. Benos, L.; Tagarakis, A.C.; Dolias, G.; Berruto, R.; Kateris, D.; Bochtis, D. Machine Learning in Agriculture: A Comprehensive Updated Review. *Sensors* **2021**, *21*, 3758. [CrossRef] [PubMed]
9. Zhang, Y.; Cai, J.; Xiao, D.; Li, Z.; Xiong, B. Real-time sow behavior detection based on deep learning. *Comput. Electron. Agric.* **2019**, *163*, 104884. [CrossRef]
10. Groher, T.; Heitkämper, K.; Umstätter, C. Digital technology adoption in livestock production with a special focus on ruminant farming. *Animal* **2020**, *14*, 2404–2413. [CrossRef] [PubMed]
11. Beeman, A.Q.; Njus, Z.L.; Pandey, S.; Tylka, G.L. Chip technologies for screening chemical and biological agents against plant-parasitic nematodes. *Phytopathology* **2016**, *106*, 1563–1571. [CrossRef]
12. Carr, J.A.; Lycke, R.; Parashar, A.; Pandey, S. Unidirectional, electrotactic-response valve for Caenorhabditis elegans in microfluidic devices. *Appl. Phys. Lett.* **2011**, *98*, 143701. [CrossRef]
13. Pandey, S.; Joseph, A.; Lycke, R.; Parashar, A. Decision-making by nematodes in complex microfluidic mazes. *Adv. Biosci. Biotechnol.* **2011**, *2*, 409–415. [CrossRef]
14. Legner, C.M.; Tylka, G.L.; Pandey, S. Robotic agricultural instrument for automated extraction of nematode cysts and eggs from soil to improve integrated pest management. *Sci. Rep.* **2021**, *11*, 3212. [CrossRef]
15. Legner, C.; Kalwa, U.; Patel, V.; Chesmore, A.; Pandey, S. Sweat sensing in the smart wearables era: Towards integrative, multifunctional and body-compliant perspiration analysis. *Sens. Actuators A Phys.* **2019**, *296*, 200–221. [CrossRef]
16. Groher, T.; Heitkämper, K.; Walter, A.; Liebisch, F.; Umstätter, C. Status quo of adoption of precision agriculture enabling technologies in Swiss plant production. *Precis. Agric.* **2020**, *21*, 1327–1350. [CrossRef]
17. Domun, Y.; Pedersen, L.J.; White, D.; Adeyemi, O.; Norton, T. Learning patterns from time-series data to discriminate predictions of tail-biting, fouling and diarrhoea in pigs. *Comput. Electron. Agric.* **2019**, *163*, 104878. [CrossRef]
18. Marsot, M.; Mei, J.; Shan, X.; Ye, L.; Feng, P.; Yan, X.; Li, C.; Zhao, Y. An adaptive pig face recognition approach using Convolutional Neural Networks. *Comput. Electron. Agric.* **2020**, *173*, 105386. [CrossRef]
19. Marunova, E.; Dod, L.; Witte, S.; Pfau, T. Smartphone-Based Pelvic Movement Asymmetry Measures for Clinical Decision Making in Equine Lameness Assessment. *Animals* **2021**, *11*, 1665. [CrossRef] [PubMed]
20. Kalwa, U.; Legner, C.; Kong, T.; Pandey, S. Skin cancer diagnostics with an all-inclusive smartphone application. *Symmetry* **2019**, *11*, 790. [CrossRef]
21. Špinka, M. *Advances in Pig Welfare*; Elsevier Inc.: Amsterdam, The Netherlands, 2017; ISBN 9780081011195.

22. Friedrich, L.; Krieter, J.; Kemper, N.; Czycholl, I. Iceberg indicators for sow and piglet welfare. *Sustainability* **2020**, *12*, 8967. [CrossRef]
23. García, R.; Aguilar, J.; Toro, M.; Pinto, A.; Rodríguez, P. A systematic literature review on the use of machine learning in precision livestock farming. *Comput. Electron. Agric.* **2020**, *179*, 105826. [CrossRef]
24. Kalwa, U.; Legner, C.; Wlezien, E.; Tylka, G.; Pandey, S. New methods of removing debris and highthroughput counting of cyst nematode eggs extracted from field soil. *PLoS ONE* **2019**, *14*, e0223386. [CrossRef]
25. Krugmann, K.; Warnken, F.; Krieter, J.; Czycholl, I. Are behavioral tests capable of measuring positive affective states in growing pigs? *Animals* **2019**, *9*, 274. [CrossRef] [PubMed]
26. Czycholl, I.; Kniese, C.; Büttner, K.; Grosse Beilage, E.; Schrader, L.; Krieter, J. Test-retest reliability of the Welfare Quality® animal welfare assessment protocol for Growing Pigs. *Anim. Welf.* **2016**, *25*, 447–459. [CrossRef]
27. Bracke, M.B.M.; Hopster, H. Assessing the importance of natural behavior for animal welfare. *J. Agric. Environ. Ethics* **2006**, *19*, 77–89. [CrossRef]
28. Mepham, B. A framework for the ethical analysis of novel foods: The ethical matrix. *J. Agric. Environ. Ethics* **2000**, *12*, 165–176. [CrossRef]
29. Mepham, T.B. Ethical Analysis of Food Biotechnologies: An Evaluative Framework. In *Food Ethics*; Routledge: Oxfordshire, UK, 1996; pp. 101–119.
30. Haigh, A.; O'Driscoll, K. Irish pig farmer's perceptions and experiences of tail and ear biting. *Porc. Health Manag.* **2019**, *5*, 1–10. [CrossRef]
31. Ingenbleek, P.T.M.; Immink, V.M.; Spoolder, H.A.M.; Bokma, M.H.; Keeling, L.J. EU animal welfare policy: Developing a comprehensive policy framework. *Food Policy* **2012**, *37*, 690–699. [CrossRef]
32. Schukat, S.; von Plettenberg, L.; Heise, H. Animal welfare programs in germany—an empirical study on the attitudes of pig farmers. *Agriculture* **2020**, *10*, 609. [CrossRef]
33. Tonsor, G.T.; Wolf, C.A. On mandatory labeling of animal welfare attributes. *Food Policy* **2011**, *36*, 430–437. [CrossRef]
34. Deemer, D.R.; Lobao, L.M. Public Concern with Farm-Animal Welfare: Religion, Politics, and Human Disadvantage in the Food Sector. *Rural Sociol.* **2011**, *76*, 167–196. [CrossRef]
35. Ashwood, L.; Diamond, D.; Thu, K. Where's the Farmer? Limiting Liability in Midwestern Industrial Hog Production. *Rural Sociol.* **2014**, *79*, 2–27. [CrossRef]
36. Lu, J.; Sheldenkar, A.; Lwin, M.O. A decade of antimicrobial resistance research in social science fields: A scientometric review. *Antimicrob. Resist. Infect. Control* **2020**, *9*, 1–13. [CrossRef] [PubMed]
37. Waluszewski, A.; Cinti, A.; Perna, A. Antibiotics in pig meat production: Restrictions as the odd case and overuse as normality? Experiences from Sweden and Italy. *Humanit. Soc. Sci. Commun.* **2021**, *8*, 172. [CrossRef]
38. Howard, P.H. *Concentration and Power in the Food System: Who Controls What We Eat? Contemporary Food Studies: Economy, Culture and Politics*; Bloomsbury Publishing: London, UK, 2016.
39. Saenz, R.A.; Hethcote, H.W.; Gray, G.C. Confined animal feeding operations as amplifiers of influenza. *Vector Borne Zoonotic Dis.* **2006**, *6*, 338–346. [CrossRef] [PubMed]
40. Moore, T.C.; Fong, J.; Hernández, A.M.R.; Pogreba-Brown, K. CAFOs, novel influenza, and the need for One Health approaches. *One Health* **2021**, *13*, 100246. [CrossRef]
41. Peters, D.J. Community Susceptibility and Resiliency to COVID-19 Across the Rural-Urban Continuum in the United States. *J. Rural Health* **2020**, *36*, 446–456. [CrossRef]
42. Zhang, M.; Wang, X.; Feng, H.; Huang, Q.; Xiao, X.; Zhang, X. Wearable Internet of Things enabled precision livestock farming in smart farms: A review of technical solutions for precise perception, biocompatibility, and sustainability monitoring. *J. Clean. Prod.* **2021**, *312*, 127712. [CrossRef]
43. Goldhawk, C.; Crowe, T.; González, L.A.; Janzen, E.; Kastelic, J.; Pajor, E.; Schwartzkopf-Genswein, K. Comparison of eight logger layouts for monitoring animal-level temperature and humidity during commercial feeder cattle transport. *J. Anim. Sci.* **2014**, *92*, 4161–4171. [CrossRef]

Article

Use of Remote Camera Traps to Evaluate Animal-Based Welfare Indicators in Individual Free-Roaming Wild Horses

Andrea M. Harvey [1,*], John M. Morton [2], David J. Mellor [3], Vibeke Russell [4], Rosalie S. Chapple [5] and Daniel Ramp [1]

1 Centre for Compassionate Conservation, School of Life Sciences, University of Technology Sydney, Ultimo, NSW 2007, Australia; Daniel.Ramp@uts.edu.au
2 Jemora Pty Ltd., P.O. Box 2277, Geelong, VIC 3220, Australia; johnmorton.jemora@gmail.com
3 Animal Welfare Science and Bioethics Centre, School of Veterinary Science, Massey University, Palmerston North 4442, New Zealand; D.J.Mellor@massey.ac.nz
4 Veterinary Contractor, c/o Animal Emergency Australia, P.O. Box 1854, Springwood, QLD 4217, Australia; vibekeanna@yahoo.com.au
5 Blue Mountains World Heritage Institute, 16 Dunmore Lane, Katoomba, NSW 2780, Australia; r.chapple@bmwhi.org.au
* Correspondence: andreaharvey.cat@gmail.com

Citation: Harvey, A.M.; Morton, J.M.; Mellor, D.J.; Russell, V.; Chapple, R.S.; Ramp, D. Use of Remote Camera Traps to Evaluate Animal-Based Welfare Indicators in Individual Free-Roaming Wild Horses. *Animals* **2021**, *11*, 2101. https://doi.org/10.3390/ani11072101

Academic Editors: Melissa Hempstead and Danila Marini

Received: 29 May 2021
Accepted: 13 July 2021
Published: 15 July 2021

Publisher's Note: MDPI stays neutral with regard to jurisdictional claims in published maps and institutional affiliations.

Simple Summary: Knowledge of the welfare status of wild animals is critical for informing debates about how we interact with them. However, methodology to enable assessment of the welfare of free-roaming wild animals has been lacking. In this study, we assessed the use of remote camera traps to non-invasively identify individual free-roaming wild horses and evaluate an extensive range of welfare indicators. Camera trapping was successful in detecting and identifying horses across a range of habitats including woodlands where horses could not be directly observed. Twelve indicators of welfare were assessed with equal frequency using both still images and video, with an additional five indicators assessed on video. This is the first time such a methodology has been described for assessing a range of welfare indicators in free-roaming wild animals. The methodology described can also be adjusted and applied to other species, enabling significant advances to be made in the field of wild animal welfare.

Abstract: We previously developed a Ten-Stage Protocol for scientifically assessing the welfare of individual free-roaming wild animals using the Five Domains Model. The protocol includes developing methods for measuring or observing welfare indices. In this study, we assessed the use of remote camera traps to evaluate an extensive range of welfare indicators in individual free-roaming wild horses. Still images and videos were collected and analysed to assess whether horses could be detected and identified individually, which welfare indicators could be reliably evaluated, and whether behaviour could be quantitatively assessed. Remote camera trapping was successful in detecting and identifying horses (75% on still images and 72% on video observation events), across a range of habitats including woodlands where horses could not be directly observed. Twelve indicators of welfare across the Five Domains were assessed with equal frequency on both still images and video, with those most frequently assessable being body condition score (73% and 79% of observation events, respectively), body posture (76% for both), coat condition (42% and 52%, respectively), and whether or not the horse was sweating excessively (42% and 45%, respectively). An additional five indicators could only be assessed on video; those most frequently observable being presence or absence of weakness (66%), qualitative behavioural assessment (60%), presence or absence of shivering (51%), and gait at walk (50%). Specific behaviours were identified in 93% of still images and 84% of video events, and proportions of time different behaviours were captured could be calculated. Most social behaviours were rarely observed, but close spatial proximity to other horses, as an indicator of social bonds, was recorded in 36% of still images, and 29% of video observation events. This is the first study that describes detailed methodology for these purposes. The results of this study can also form the basis of application to other species, which could contribute significantly to advancing the field of wild animal welfare.

Keywords: welfare assessment; animal-based welfare indicators; camera traps; wild horses

1. Introduction

Knowledge of the welfare status of wild animals is critical for informing ethical, legal, and political debates about the ways we interact with them. The importance of developing scientific methods for capturing data to enable the assessment of the welfare of free-roaming wild animals has recently been highlighted [1]. Many studies have evaluated wild horse behaviours, time budgets, home ranges, body condition scores and social organisation, e.g., [2–19], but to date, an extensive range of welfare indicators has apparently not been assessed.

We recently published a Ten-Stage Protocol for scientifically assessing the welfare of individual non-captive wild animals, using free-roaming horses as an example [1]. Table 1 lists these ten stages.

Table 1. The Ten-Stage Protocol for assessing the welfare of non-captive wild animals [1].

1.	Acquire an understanding of the principles of Conservation Welfare.
2.	Acquire an understanding of how the Five Domains Model is used to assess welfare status.
3.	Acquire species-specific knowledge relevant to each Domain of the Model.
4.	Develop a comprehensive list of potential measurable/observable indicators in each physical domain, distinguishing between welfare status and welfare alerting indices.
5.	Select a method or methods to reliably identify individual animals.
6.	Select methods for measuring/observing the potential welfare indices and evaluate which indices can be practically measured/observed in the specific context of the study.
7.	Apply the process of scientific validation for those indices that can be measured/observed and insert validated welfare status indices into the Five Domains Model.
8.	Using the adjusted version of the Model that includes only the validated and practically measurable/observable welfare status indices, apply the Five Domains grading system for grading welfare compromise and enhancement within each Domain.
9.	Assign a confidence score to reflect the degree of certainty about the data on which welfare status has been graded.
10.	Including only the practically measurable/observable welfare alerting indices, apply the suggested system for grading future welfare risk within each domain.

The protocol uses the Five Domains Model for assessing animal welfare [20–24]. The Five Domains Model comprises four interacting physical/functional domains of welfare—'Nutrition', 'Physical Environment', 'Health' and 'Behavioural Interactions'—and a fifth domain of 'Mental State'. Following measurement of animal-based indices within each physical/functional domain, the anticipated negative or positive affective consequences are cautiously assigned to Domain 5. Three of the ten stages of the Ten-Stage Protocol relate specifically to the Model: Stage 4, develop a comprehensive list of potential measurable or observable indicators of welfare in each physical/functional domain; Stage 5, select methods to reliably identify individual animals; and Stage 6, select methods for measuring or observing potential welfare indices, then evaluate which indices can be practically measured or observed in the specific context of the study. This last stage requires further investigation to identify which indices can be measured or observed using the selected methods, before detailed studies to assess welfare can be conducted. That is the primary focus of this paper.

Current knowledge of wild horse ecology and behaviour has historically relied on direct observations of horses, e.g., [2–19]. The disadvantages of direct observations for assessing indicators of welfare include practical limitations on the number of observations of individual animals that can be made, and usually the need to observe them from long distances. This may make evaluation of some welfare indicators challenging, and it restricts observations to only those animals that can be seen directly, primarily in open areas. Horses residing in woodland habitats and/or challenging terrain, or those that become separated from their band, for example due to debilitation or injury, are less likely to be observed directly and hence may be underrepresented in resulting datasets.

Camera traps have been widely used to study wildlife in a variety of habitats, e.g., [25–28], focusing on a range of variables such as abundance and density, e.g., [29], and demographic measures, e.g., [30,31]. More recently, they have been used to collect behavioural data from a range of captive and wild animals, e.g., [32–35] and to evaluate leprosy-like skin lesions in wild chimpanzees [36]. They have also been used to collect data on some aspects of behaviour of wild Przewalski horses [37,38]. However, to date, there are apparently no published reports of camera trap use directed specifically at evaluating a comprehensive range of animal-based welfare indicators in any mammalian species, whether captive or free-roaming.

The aims of this study were to evaluate the use of camera traps, both still images and video, across a range of habitats, to assess (a) whether free-roaming wild horses could be detected and individually identified, (b) which animal-based welfare indicators could be practically and reliably evaluated, (c) the reasons for not being able to identify horses or assess welfare indicators, and (d) whether behaviour could be quantitatively assessed. We specifically describe our methodological information in detail in order to enable other researchers to replicate these methods. Furthermore, we sought to identify advantages and limitations of the use of camera traps for these purposes.

2. Materials and Methods

2.1. Study Overview

We made camera trap and direct observations of a small and geographically constrained population of free-roaming wild horses in Australia over a 15 month period.

2.1.1. Selection of Welfare Indicators

In line with Stage 4 of the published Ten-Stage Protocol (Table 1) for assessing welfare [1], we developed an extensive list of potential measurable/observable indicators in each physical/functional domain, based on a literature search (using the terms 'horse welfare', 'equine welfare', 'welfare indicators') for previously reported animal-based welfare indicators in domestic horses [13,39–59]. As already noted, the physical/functional domains of the model are: 1. Nutrition; 2. Physical Environment; 3. Health; and 4. Behavioural Interactions [24]. From this list, indicators were divided into specific measures of current welfare status, and alerting measures that draw attention to potential welfare risks [1]. We then selected measurable or observable welfare indicators that may be able to be captured in free-roaming horses using camera trap still images and/or videos or by direct observation (Table 2).

Table 2. List of selected welfare indicators in each domain.

Domain	Animal-Based Indices	
	Still Images and Video	**Video Only**
1. Nutrition	Body condition score	
2. Physical Environment	Sweating Wet from rain Huddling together with other horses	Shivering
3. Health	Body posture Hoof condition Coat condition Wounds or other injuries Limb pathology Skin lesions Nasal discharge Ocular discharge Blepharospasm Quidding Food pouching Facial grimace	Gait at walk Gait at trot Gait at canter Weakness Respiratory rate and effort
4. Behavioural Interactions	Specific quantifiable behaviours (feeding, resting, maintenance, locomotory and social behaviours)	Qualitative assessment of behaviour (dull, relaxed, alert, apathetic, curious, anxious, playful)

2.1.2. The Study Area and Horse Population

Data were collected between December 2015 and March 2017 in the Kedumba Valley, an isolated section of approximately 130 km^2 within the Warragamba Special Area of the Blue Mountains National Park, being part of the Greater Blue Mountains World Heritage Area in New South Wales, Australia. The horse population there was known to be small and geographically constrained since immigration and emigration are inhibited by natural physical boundaries, with Lake Burragorang to the south, high cliff faces to the north and east, and a river and steep terrain covered with dense bush to the west (Figures 1 and 2). In the past, some population exchange has been known to occur when a low water level in Lake Burragorang allowed horses to cross to and from the other side, but this was not possible during the study period. Two dirt roads provided access to the valley at the northern and western borders, locked gates preventing their use by horses. No management interventions had occurred in the preceding eight years. This study was part of a larger investigation of the population ecology and welfare of free-roaming horses conducted over a 15 month period (National Parks and Wildlife scientific license 101626, WaterNSW access license D2015/128332).

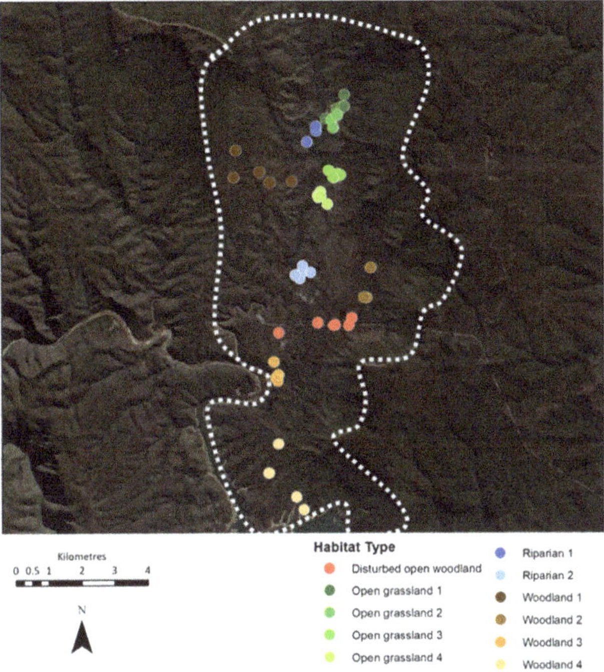

Figure 1. Topographical map of the study area illustrating camera locations and habitat types in the 11 areas. The white dashed line represents the geographic boundaries that inhibit immigration and emigration of this horse population.

Figure 2. View of Kedumba valley facing south. The grassland areas visible near the middle of the picture are the regions named Open grassland 1. and 2. Image A.M. Harvey.

2.1.3. Initial Surveys

Extensive on-ground surveys were performed with a 4WD vehicle and on foot to become familiar with all regions of the valley in order to assess how resources varied spatially and to identify locations of horse tracks and/or faeces. Avenza maps (Avenza Systems Inc., Toronto, ON, Canada) were used on an iPad (Apple Inc., Cupertino, CA, USA) to navigate around the area and record GPS locations of the presence and abundance of horse faeces, horse sightings, water access points and habitat details. There was evidence of horses inhabiting eleven distinct areas dispersed within the valley, comprising four different habitat types; four small areas of open grassland with scattered woodland; two riparian areas; one area of disturbed open woodland situated below powerlines; and four woodland areas where water was accessible within 2 km. Figure 3 illustrates these eleven distinct areas. The remaining habitat comprised eucalyptus woodland and shrubs across undulating and often very steep terrain. Riparian corridors throughout these areas were surrounded by a high cliff, making them inaccessible to horses (Figure 4).

(a)

(b)

(c)

(d)

Figure 3. *Cont.*

Figure 3. *Cont.*

(**k**)

Figure 3. The eleven distinct areas of four main habitat types where horses resided. (**a**) Open grassland 1. (**b**) Open grassland 2. (**c**) Open grassland 3. (**d**) Open grassland 4. (**e**) Disturbed open woodland. (**f**) Riparian 1. (**g**) Riparian 2. (**h**) Woodland 1. (**i**) Woodland 2. (**j**) Woodland 3. (**k**) Woodland 4. Images A.M. Harvey.

(**a**)   (**b**)

Figure 4. The predominant habitat and topography of Kedumba Valley: (**a**) Most of the region consisted of eucalyptus woodland across undulating and often steep terrain. (**b**) Most riparian corridors comprised steep cliffs making those sections of rivers inaccessible to horses. Images A.M. Harvey.

2.1.4. Direct Observations

Direct observations were performed (by A.M.H.) on a total of 29 days, over 14 field trips of 1–3 days each, spread over the 15 month period. During each field trip, all eleven areas were surveyed, with direct observations occurring opportunistically whenever horses were sighted. The duration of direct observations depended mainly on how long horses stayed within sight, and so was not standardized. When horses stayed within sight for more than 1–2 min, they were observed using 10×42 binoculars (Bushnell Powerview FOV 293FT, Bushnell Corporation, Overland Park, KS, USA). A 900 m laser range finder was used to measure distances to horses, thereby enabling estimation of the maximum distances at which each welfare indicator could be assessed. Where possible, magnified photos and video footage (Canon EOS 70D Digital SLR with Canon EF100–400 mm lens) were taken of horses so that welfare indicators could be evaluated more closely at a later stage.

2.1.5. Camera Trap Placement and Settings

Initial on-ground surveys and direct observations were used to identify suitable camera locations to maximise image capture of horses, for example, near drinking locations, pathways to a drinking location or prime grazing area, dirt tracks, horse trails, river crossing points, stallion faecal piles, and open grassland grazing areas where there was an abundance of horse faeces.

Forty-seven cameras (Bushnell Aggressor; Bushnell Corporation, Overland Park, KS, USA) were used throughout the study period, placed in a total of 58 different locations. Thirty-five cameras remained in the same location for the entire study period, five camera locations were changed slightly to improve the viewing angle of horses or to overcome false triggers, and seven cameras were moved to a different area to improve capture of horses in a particular area. Of the 58 different camera locations, 23 were in grassland habitats, 17 in woodland habitats, 13 in riparian habitats, and 5 in a disturbed open woodland habitat (Figure 1). Cameras were placed at a height of 70–110 cm, depending on the terrain. On dirt tracks and creek crossings, cameras were positioned with the aim of capturing the front of the horse approaching the camera, whilst on open grassland areas they were positioned to capture as much of the grassland area as possible. Initially all cameras were set to take still images only. During various periods over the whole study, 18 of the cameras (nine in open grassland, four in riparian habitats, four in woodland habitats and one in the disturbed open woodland habitat) were set to take hybrid video records, which comprised a still image followed by a video clip. Once triggered, videos ran for a specified duration; video clip duration was set at 5, 10, 15, 20, or 30 s. Camera settings were as follows: 14 M pixel image size or 1920 × 1080 video resolution; video sound on; capture number (i.e., number of still images taken in a sequence once the camera was triggered) ranged from one-to-three; camera time delay interval before responding to another trigger varied between 1 and 5 s; light-emitting diode (LED) control high, sensor level auto, night vision shutter speed medium, and camera mode 24 hrs. An identifying number for each camera, date, time and GPS co-ordinates, were entered into each camera. Camera locations were marked on Avenza maps to assist with relocating them. Each camera was checked, and batteries and secure digital (SD) cards changed, during each of the 14 field trips.

2.1.6. Acquisition of Demographic Details and Identification of Individual Horses

Camera trap images were used together with resighted identification marks of horses under direct observation, in order to obtain unique identifying features for each horse, in addition to its sex, age category, and familial relationships. Horses observed early in the study period close together on multiple occasions, in the same geographical area, but distant (>1 km) from other horses were considered to constitute a distinct band.

A list of identifying features was made for each horse, including body coat colour, limb coat colour (if different from the body), mane and tail colour (if different from the body), facial markings (blaze, stripe, star, snip) and their size and shape, and limb markings (including the length the marking extended up the limb). An identification diagram was also completed for each horse to record the presence, size, and shapes of facial and limb markings. A unique identifier was then assigned to each horse.

2.1.7. Assessment of Camera Trap Images and Videos

Images and video footage from all cameras over the 15 month study period were viewed and all 'sightings' where at least one horse could be identified were recorded. Each unique combination of horses captured within a duration of one second on a particular camera on a particular day constituted a separate sighting. From these data, a unique camera-day was defined as each camera-date combination where at least one horse was seen.

A stratified-random subset of camera days from the full dataset of 2071 camera-days was used for detailed evaluation of whether individual horses could be identified and whether selected welfare indicators (Table 2) could be assessed. To do so, the study was divided into five 90-day time periods, broadly aligned with seasons (period 1: summer; period 2: autumn; period 3: winter, period 4: spring; period 5: summer) and for each camera method (still images or video), from each time period, one camera-day was selected from each camera that recorded sightings in the 90-day period, using computer-generated random numbers.

Within each selected camera-day, every observation event was assessed (by A.M.H.). An observation event was defined as a series of consecutive still images or video clips of the same horse on the same camera on the same day, with a maximum of a 3-min interval between consecutive images or video clips where the horse was visible. The observation event ended when the horse left the camera's field of vision and did not return within 3 min. Thus, one horse could have multiple observation events on the same camera on the same day. Other horses could be in view for part or all of another horse's observation event; each horse in view was treated as a separate observation event. Table 3 lists the details recorded for each observation event. For the purposes of this study it was only recorded whether or not it was feasible to assess each indicator, with regard to whether the relevant part(s) of the horse were captured at an appropriate distance to the camera, or for indicators such as gait evaluation and weakness, requiring dynamic records, whether enough of the horse was captured moving within video clips. Precise descriptions that enable the particular indicators to be categorized in terms of potential welfare compromise in each physical/functional domain will be provided in subsequent manuscripts.

Table 3. Details recorded from each observation event.

- Date, method (still image vs. video), camera number
- Identification of the individual horse when possible, if not recorded as unidentified
- Number of images of video clips in the observation event (and video duration setting)
- Start and finish time of the observation event
- Number of images (or for video, time) until the horse identity was determined
- Reason for being unable to identify an individual at all
- Whether the image or video was taken in the daylight or at night
- For each welfare indicator listed in Table 2, whether it was able to be assessed or not, and for those indicators whose status can differ between left- and right-hand sides of the same horse, whether they could be assessed on the left, right, or both sides
- Reason(s) why any of the welfare indicators could not be assessed
- Where there were two or more video clips within an observation event, each video clip was also individually assessed for all of the above

Qualitative behaviour assessment (QBA) [42,46,54,55] was performed on video-observation-events, by an experienced observer (A.M.H.), using the fixed descriptors of 'dull', 'relaxed', 'alert', 'apathetic', 'curious', 'anxious', 'playful'. If such assessment was not possible, the reasons were recorded. In fact, QBA scores have not been reported here because the present purpose was restricted to evaluating whether or not video-observation events captured enough of the 'whole horse' behaviour to enable a QBA to be conducted.

Observations of specific behaviours were also quantified. Behaviours were assigned to five main categories using a previously defined ethogram for free-roaming horses [13]. Categories were: locomotion (subcategorised as walking, trotting, cantering); standing resting; grazing; maintenance behaviours (including grooming, nursing, rolling, lying down, standing alert, drinking, urinating, defaecating); and social behaviours (included allogrooming, play behaviour, nuzzling, sexual interactions, and other affiliative and agonistic social interactions).

A behaviour event was defined as an occurrence or period in which an individual horse performed a particular behaviour without interruption. The duration of these behaviour events was recorded in terms of the number of images and/or times when the behaviour was observed. The end of a behaviour event occurred either when the horse left the camera's field of view or changed to a different behaviour. Table 4 lists additional information recorded from each behavioural event.

These results were used to evaluate the relative proportions of the different behaviours captured by camera traps. For still images, the relative proportions of different behaviours were calculated separately, based on the number of images, and the period of time for which the same behaviour was observed (Table 4).

Table 4. Information recorded from each behavioural event.

- The identifiable behaviour (or recorded as unidentifiable and reasons for that)
- Number of images per behaviour (for still images)
- Start and finish times of the behaviour (for still images this was recorded as the time of the first and last images in a series of consecutive images in which the behaviour was captured)
- The habitat in direct field of view of the camera (i.e., grassland, dirt track, riparian, river crossing, waterhole, horse track through woodland, rest area beneath trees, woodland clearing)

Social bonds were also analysed by reference to 'close spatial proximity' [60]. For the purposes of this study, this was defined for each observation event as occurring if other horses were present at the same time in any image or video frame in addition to the focal horse for that observation event, or in a subsequent image taken within one second of the previous image. Since close spatial proximity could occur concurrently with a range of other behaviours, proximity was analysed separately to other social behaviours. Furthermore, for less frequently performed social behaviours, such as allogrooming, play behaviour, nuzzling, sexual interactions, and other affiliative and agonistic social interactions, an 'all occurrence' method was used where the entire dataset of 2071 camera-days (still images and video clips pooled) was evaluated, and the behaviour recorded each time it was captured on images or videos.

All data were entered into standardized Excel 2011 spreadsheets (Microsoft, Washington, DC, USA) with separate spreadsheets for data from still images, from video, and for quantifying behavioural observations.

2.2. Data Analyses

Statistical analyses were performed using Stata (version 16.1; StataCorp, College Station, TX, USA). The numbers of images required for a horse to be identified via still-image observation events was assessed as time-to-event data using Kaplan–Meier survivor and failure functions [61], where the 'failure' event in this context was identification of the horse. This approach allowed inclusion of observation events where the horse was not identified. For these events, the record was right-censored at the maximum image number for the horse in that observation event. In time-to-event analyses, right-censoring is used when a subject leaves the study before an event occurs (in this case before the horse is identified). The Kaplan–Meier failure function value at a specified number of images is equivalent to the cumulative percentage of observation events where the horse was identified by that number of images if none-were right-censored. Survivor functions were compared using log rank tests, by day/night, habitat, horse coat colour, facial marking and limb marking categories. For horse-level factors, only observation events where the horse was identified were used. The same methods were used to determine the times at which the horse was identified for still-image observation events. Stata's *sts list* and *sts test* commands were used.

Separately for each welfare indicator, proportions of observation events where an indicator could be assessed were compared between habitats and between methods (still images or video) using logistic regression, with both variables fitted simultaneously. Stata's *logistic* command was used. For bilateral welfare indicators (e.g., wounds, limb pathology, and skin lesions), each observation event was classified as one of: could be assessed on both sides/one side/neither side. Then, separately for each welfare indicator, generalised ordered logit models were fitted using the *gologit2* command in Stata. Where the *p*-value for the proportional odds/parallel lines assumption was >0.05, proportional odds were assumed and ordered logit models were fitted using Stata's *ologit* command. Likelihood ratio test *p*-values were used.

Reasons why welfare indicators could not be assessed were compared between habitats and between methods using the same approach as that for evaluating whether non-bilateral welfare indicators could be assessed (i.e., those welfare indicators that could be assessed without seeing both sides of the horse), with one exception. Blurred images were not an impediment in any of the video-observation events, so methods were compared using exact logistic regression with Stata's *exlogistic* command. Habitat was not fitted, and sufficient statistics were used, rather than the other main statistical alternatives of conditional scores tests or conditional probabilities tests. For all analyses, observation events were treated as if they were statistically independent of each other after accounting for the variables fitted in each model.

3. Results

3.1. Demographic Details of the Population

Twenty-nine horses were identified in the population during the study period. These horses were distributed across five bands, with a total of 5 stallions, 16 mares, 3 fillies, 4 colts and one foal of unknown sex (Table 5).

Table 5. Details of the population and unique identifying features.

Unique Identifier [1]	Sex [2]	Age Group	Reproductive Status	Coat Colour	White Facial Markings	White Limb Markings [3]
1A	M	Mature	Band stallion	Black	Star and snip	None
1B	F	Mature	Foaled year 2	Chestnut	Star and stripe	BLH socks
1C	F	Mature	No foal	Chestnut	None	BLH and RF fetlocks
1D	F	Mature	No foal	Black	Small star	BLH pastern spots
1E	F	Mature	No foal	Black	None	BLH and LF pastern
1F	F	Mature	Yearling (1G) at foot	Black	Large star	BLH fetlocks
1G	F	Yearling	Yearling of 1F	Brown	Blaze	BLH socks
1H	F	Mature	Foal (1K) at foot	Black	None	BLH pasterns
1I	F	Mature	Foal (1J) at foot	Dark bay	Star and stripe	BLH fetlocks and LF pastern
1J	M	Foal	Foal of 1I	Black	Large stripe	LH pastern, small spot medial RH pastern
1K	F	Foal	Foal of 1H	Black	Small star and snip	None
1L	Unknown	Foal	Foal of 1B (year 2)	Bay	Star and snip	None
2A	M	Mature	Band stallion	Flaxen chestnut	Blaze	LH stocking
2B	F	Mature	Foal (2C) at foot	Flaxen chestnut	Blaze	None
2C	M	Foal	Foal of 2B	Flaxen chestnut	Blaze	None
2D	M	Yearling	Dam unknown	Flaxen chestnut	Thick blaze	BLF socks, LH stocking
2E	F	Juvenile	No foal	Black	Large star and stripe	LF pastern, BLH fetlocks
3A	M	Mature	Band stallion	Brown	Stripe	BLH socks
3B	F	Mature	Foal (3C) at foot	Black	Star and stripe	BLH and RF pasterns
3C	F	Foal	Foal of 3B	Black	Large stripe	BLH and LF fetlocks
3D	M	Juvenile	Colt with another stallion in Band	Flaxen chestnut	Blaze	LH sock
4A	M	Mature	Band stallion	Black	None	LH fetlock, RF coronet

Table 5. *Cont.*

Unique Identifier [1]	Sex [2]	Age Group	Reproductive Status	Coat Colour	White Facial Markings	White Limb Markings [3]
4B	F	Juvenile	No foal	Black	Very small star	RH fetlock, LF pastern
4C	F	Mature	No foal	Black	Interrupted stripe	LH pastern, RH coronet
4D	F	Mature	No foal	Chestnut	Stripe	RH fetlock
5A	M	Mature	Band stallion	Black	Star	LF pastern, BLH fetlocks
5B	F	Mature	No foal	Chestnut	Very small star	None
5C	F	Mature	No foal	Black	Large star	LH fetlock, RH coronet
5D	F	Mature	No foal	Black	Large star	LH fetlock, RF coronet

[1] Number = band number; Letter = sequential letter within band. [2] M = male; F = female. [3] LH = left hindlimb; RH = right hindlimb; LF = left forelimb; RF = right forelimb; BLH = bilateral hindlimb; BLF = bilateral forelimb.

3.2. Direct Observations

All individuals in Bands 1 (11 horses) and 2 (5 horses), who predominantly frequented open grassland habitats, were identified during the first two of the 14 field trips. The remaining 13 horses in the population were never identified by direct observations during the 15 month study. Figure 5 illustrates the band compositions, their habitats and whether or not they were directly observed.

Horses were observed and identified in open grassland habitats during all 14 field trips and in riparian habitats during 4/14 field trips. Horses were observed in woodland and disturbed open woodland habitats during 4/14 field trips, but only 3 horses were identified (on a single field trip) during observations in woodland habitats, and none were identified in disturbed open woodland. On all other occasions in the woodland and disturbed open woodland habitats, horses moved out of sight quickly, preventing their identification or assessment of welfare indicators.

Welfare indicators could only be assessed directly in the 16 horses (Bands 1 and 2) in the larger open grassland habitats, the primary determinants of success being how long each horse remained in sight and how close it was to the observer. Proximity influenced whether or not individual horses could be identified, and which welfare indicators could be assessed (Figure 6). Using 10× magnification binoculars, the distances where specific attributes were observable were as follows: horses within approximately 150 m could be identified as individuals and their body posture and behaviour assessed; at approximately 100 m, all indicators other than facial indicators (i.e., facial grimace, blepharospasm, ocular discharge, nasal discharge, food pouching and quidding), and hoof condition could be assessed, although closer distances enabled more accurate assessment and detection of more subtle abnormalities; and within 50 m, facial indicators and hoof condition could be assessed. Photographs/videos were acquired with a 400 mm lens, which provided 8× magnification, enabling welfare assessments at similar distances to dynamic live direct observations with binoculars, but with the benefit of less hurried deliberation.

Although welfare indicators may be assessed at longer distances using more powerful magnification devices, the observer needs a clear line of sight, unobscured by trees and undulating terrain. In this study, the maximum such distance was approximately 300 m, which only occurred on one open grassland location occupied by Band 1. Horses were mostly observed from distances of 50–100 m, and so the facial indicators and hoof condition could not always be assessed. Horses in Bands 1 and 2, which were observed regularly, became more habituated to observer presence throughout the study. By the end of the study, all of their welfare indicators could be assessed for longer periods at distances of 20–30 m. In all other habitats, the unobscured line of sight was rarely greater than 20 m, but none of this sub-population of horses would allow such close proximity.

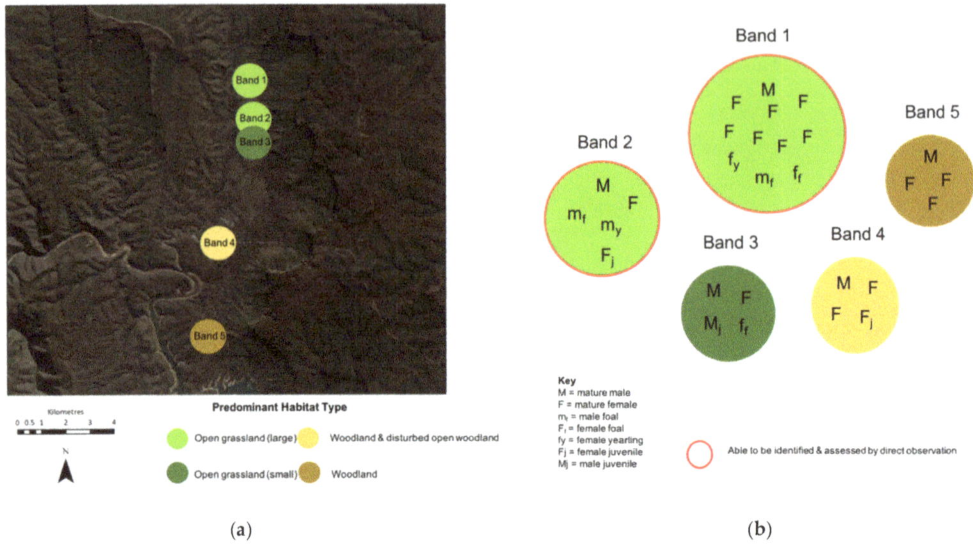

(a)   (b)

Figure 5. Illustration of (**a**) the geographical location and predominant habitat type of the different bands, and (**b**) band compositions. Only horses in Bands 1 and 2 that resided on larger open grassland habitats could be identified and assessed by direct observation. Horses in all other bands and habitats could only be identified and assessed with remote camera traps.

(a)   (b)

Figure 6. *Cont.*

Figure 6. *Cont.*

(i)

(j)

Figure 6. Examples of identification of horses and assessment of welfare indicators with direct observations (with 8–10× magnification), and some reasons for not being able to assess indicators. (**a**) Behaviour can be assessed but other indicators are obscured by vegetation. (**b**) Close spatial proximity with other horses can be assessed, but other indicators are obscured by vegetation. (**c**) At >100 m, behaviour is being observed without disturbing the horses, but the distance is too great to assess other welfare indicators. (**d**) At 100 m from the horse most welfare indicators can be assessed, but in this case the orientation of the horse prevents assessment of facial features and the right side of the horse. (**e**) At approximately 50 m, most welfare indicators can be assessed, but at this distance the observer will be seen by the horse, altering its behaviour. Distal limbs are obscured by vegetation. (**f**) Facial indicators can be assessed from approximately 50 m. (**g**) The detail of facial features is much greater <50 m to the horse but it is rare to be able to approach them so closely. (**h**) Trot is more readily assessed with direct observations. (**i**) Canter is more readily assessed with direct observations. (**j**) A good example of where most welfare indicators can be assessed on the right side of the horse, but at such a close distance behaviour is impacted by the observer's presence. Photographs A.M. Harvey.

3.3. Camera Trapping

3.3.1. General Statistics

Over the 15 month period, a total of 220,836 image/video files (a file constitutes a single image or video clip) were obtained. Of these, 42,925 (19%) image/video files contained horses whilst the remainder contained only other animals (e.g., macropods) or were false triggers due to wind-blown vegetation moving in front of the camera.

One or more horses were seen and identified via at least one camera on 428 days (95%) of the study period. For individual cameras, one or more horses were seen and identified on a mean of 9.5% of the camera's days (range 0–45.9%). By habitat, of the cameras located in open grasslands, the corresponding average was 12.5% of the camera's days (range 0–45.5%), which was similar to cameras located in disturbed open woodland (mean 14.4%; range 0–45.9%). In contrast, horses were seen and identified on lower proportions of days by cameras located in riparian (mean 6.8%; range 1.7–21.6%) and woodland habitats (mean 3.6%; range 0–21.2%).

In all, 199 camera-days were randomly selected from the 2071 camera-days that had detected horses, of which 158 days were still images and 41 were videos. Within the 158 camera-days of still images, there was a total of 538 observation events (range 1–23 observation events per camera day, mean 3.4, SD 3.44, median 2). Within the 41 camera-days of video clips, there was a total of 119 observation events (range 1–11 observation events per camera-day, mean 2.9, SD 3.85, median 1).

For still images, there was a range of 1–36 images per observation event (mean 4.2, SD 4.75, median 3) with the duration of observation events ranging from <1–257 s (mean 47, SD 113, median 4). For the 411 observation events with more than one image, the time between images ranged from <1–163 s (mean 14.5 secs, median 4.3, SD 25.76).

For videos, the duration of observation events ranged from 1 to 252 s (mean 29.3, SD 46.3, median 5). The most common total observation event durations were 5 s in 40/119 (33.6%), 10 s in 12/119 (10.1%), and 30 s in 38/119 (31.9%). Observation events comprised between 1 and 11 separate video clips with 26/119 (20.2%) involving more than one clip (mean 1.63).

Longer video clip durations resulted in the horse being visible for longer (Table 6), such that each additional 5 s added, on average, 3.1 s where the horse was visible. For the 5 s duration, the horse was usually visible for the entire 5 s, whereas, for the 10-, 15- and 30 s durations, the horse was visible for 50–61% of the time (Table 6).

Table 6. Video clip durations and the period during which the horse was visible in the video clip.

Video Clip Duration Setting (Seconds)	Number of Video Clips	Number of Video Clips That a Horse Was Visible for Full Video Duration	Mean Period during Which a Particular Horse Was Visible in Video Clip (Seconds)
5	110	106 (96%)	4.9
10	20	10 (50%)	7.3
15	18	11 (61%)	11.2
20	6	1 (17%)	8.3
30	40	22 (55%)	20.6

3.3.2. Identification of Individual Horses

Of the 29 horses in the study population, 27 (93%) had at least one randomly selected still image observation event, with the number of such events for each of those horses ranging from 2 to 33 (mean 14.9). Twenty horses (69%) had at least one randomly selected video observation event, with the number of observation events per horse ranging from 1 to 11 (mean 3.2).

For still images, the individual horse was identified in 405 of 538 (75%) observation events. In 62% (333/538; 95% CI, 58% to 66%) of observation events, the horse was identified from the first image. The Kaplan–Meier failure function (indicating successful identification) rose to 83% (95% CI, 78% to 87%) by five images. Increases with further images were only small. By 16 images, the Kaplan–Meier failure function was 93% (95% CI 86% to 97%). For the time until identification, the horse was identified immediately in 333/538 (62%) observation events. The Kaplan–Meier failure function was 70% by 10 s (95% CI, 66% to 75%) and 77% (95% CI, 73% to 82%) after 30 s.

Identification of individual horses was more rapid with still images taken during daylight than with images taken at night ($p < 0.001$ for both number of images and time to identification). During the day, 70% of horses were identified with one image, and 88% of horses were identified after 4 images, whereas at night only 42% of horses were identified after one image, and 61% after 4 images (Figure 7). Patterns of time to identification were similar (Figure 7).

There were no substantial differences in the numbers of images or times to identification between horses with different coat colours or markings.

Of the 538 observation events, 300 occurred in open grassland habitats, 99 in riparian areas or at river crossings, 93 in woodland habitats, and 46 in disturbed open woodland habitats. Numbers of images until the horse was identified varied by habitat (overall $p = 0.03$), with the greatest numbers of images required for identification in observation events in open disturbed woodland habitats (Figure 8).

The most common reason for not being able to identify a horse in the first image was being unable to see facial or limb markings in the image (n = 31). Other reasons were only a small part of the horse being in the image, for example only the back, side, neck, shoulder or head side on (n = 24), the horse being too distant +/− at night (n = 5) and horse obscured by another horse or vegetation (n = 2).

The reasons for not being able to identify a horse at all for a whole observation event were only a small part of the horse being in the image (n = 57), the images either being too dark, blurred and/or the horse too distant (n = 31), being unable to see facial or limb markings in the image (n = 25), and the images captured at night (n = 24). Some examples are shown in Figure 9.

For camera trap video, horses were identified in 72% (86/119) of observation events, and when this was the case, it was always possible to identify horses at the beginning of the observation event. Reasons for the horses not being identifiable were only a small part of the horse being in the video (n = 18), the horse being too far from the camera (n = 12), and an inability to see facial markings (n = 1).

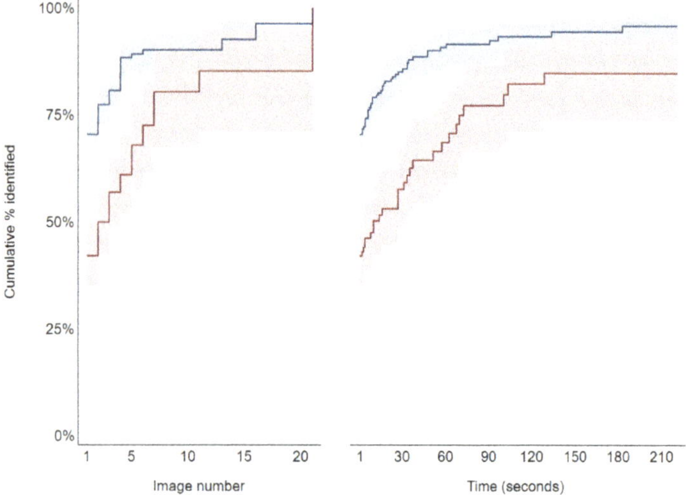

Figure 7. Cumulative percentages of observation events where the horse was identified by number of images (left-hand graph) and time from start of each observation event (right-hand graph) from still image observation events in the daytime (blue) and night-time (red). Kaplan–Meier failure functions are graphed; these are equivalent to cumulative percentages of observation events where each horse had been identified by the specified *x*-axis if no observation events had been right-censored. Shaded areas are point-wise 95% confidence intervals.

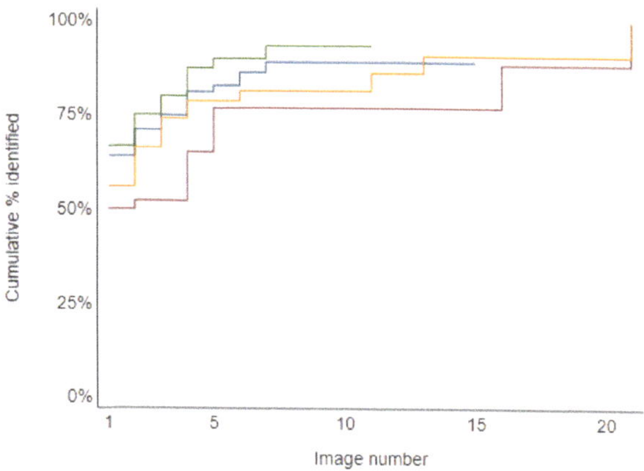

Figure 8. Cumulative percentages of observation events where the horse was identified by number of images from still image observation events in various habitats: riparian (green); open grassland (blue); woodland (orange); open disturbed woodland (red). Kaplan–Meier failure functions are graphed; these are equivalent to cumulative percentages of observation events where the horse was identified by the specified *x*-axis value if no observation events had been right-censored.

Figure 9. *Cont.*

Figure 9. *Cont.*

Figure 9. *Cont.*

Figure 9. Examples of identification of horses, and assessment of welfare indicators with camera trap images, and common reasons for not being able to identify horses or assess indicators. (**a**) This horse was too close to the camera resulting in only the top half of the horse being in the image, therefore it was not possible to identify or assess welfare indicators other than body condition on this particular image. (**b**) Although this image was taken during daylight, shadowing from trees resulted in the image being too dark to identify the horse or welfare indicators. (**c**) A night-time image illustrating the difficulty in identifying horses or indicators in night-time images. (**d**) These horses' distinctive markings make them identifiable, and it can be determined that both horses are cantering, however blurring of the image means no other welfare indicators are able to be assessed. (**e**) This horse can be identified from its colour and hindlimb markings, and it is evident that the horse is grazing, but no other welfare indicators can be assessed due to only the back of the horse being visible. (**f**) This horse can be identified, and it is evident that the horse is walking; body posture, body condition, hoof condition and presence of limb pathology can be assessed, but other welfare indicators are unable to be assessed as the image is too dark. (**g**) This horse cannot be identified on this single image as the shape of the blaze is not evident from the side-on image, and the limb markings are obscured by the water. Body condition and presence of wounds on the right side can be assessed. (**h**) The horse on the left cannot be identified as only the rump and tail are in the image, however it is enough to ascertain that the horse on the right is not alone. The horse on the right can be identified from its limb markings, and is observed to be grazing but the image is too dark and the horse too distant from the camera to be able to assess any other welfare indicators. (**i**) This horse can be identified from his colour, limb markings, sex, and length of tail suggesting he is juvenile. It is evident that he is grazing, but he is too distant from the camera to enable assessment of other welfare indicators. (**j**) A closeup image of a horse just at the perfect distance and orientation to the camera to enable both its identification and assessment of all welfare indicators on the left side. (**k**) This horse is identifiable from her colour and distinctive facial marking, and it is evident that she is grazing. An appreciation of body condition is possible although an accurate assessment of body condition score cannot be made with the limited proportion of the body captured in the image, and she is too distant to the camera to be able to assess facial indicators accurately. No other indicators are able to be assessed from this image. (**l**) This

horse is identifiable from her facial marking, and it is evident that she is cantering. Body posture and body condition can be assessed, but she is too distant from the camera to enable assessment of other welfare indicators. (**m**) This image captures four horses from Band 2 cantering together and illustrates close spatial proximity between them all. Most welfare indicators can be assessed from the left side of the horse in the centre of the image, and a variable number of indicators can be assessed for the other horses due to distance from the camera or only half the horse being in the image. (**n**) This horse is at a good orientation to and distance from the camera to enable most indicators to be assessed on the left side. (**o**) A close up facial image enables assessment of facial indicators on the left side of this horse. (**p**) Although a night-time image, this horse is can be identified, and body condition and body posture assessed. (**q**) Only parts of the horses are visible in this image hindering assessment of welfare indicators, however the distinct facial marking on the foal enables his identification; facial indicators on the right side can be assessed in the foal, whilst body condition can be assessed in the mare. (**r**) This image shows close spatial proximity between two identifiable horses, with the horse on the left grazing and the horse on the right walking in a relaxed posture. Body condition, coat condition, body posture, limb pathology, presence of wounds on the left side, and, in the chestnut horse, some facial indicators on the left side can be assessed. (**s**) The orientation and distance of this horse from the camera enable most indicators to be assessed on the left side, except hoof condition since the front hooves are missing from the image. (**t**) This image shows close spatial proximity between three horses with the two on the right grazing and the chestnut on the left standing alert. Body posture and body condition can be assessed, but distance from the camera precludes assessment of other welfare indicators.

3.4. Assessment of Welfare Indicators

The welfare indicators that can be assessed most frequently on both still camera trap images and video, were body condition score (73% and 79% of observation events, respectively), body posture (76% for both), coat condition (42% and 52%, respectively), and whether or not the horse was sweating excessively (42% and 45%, respectively) (Table 7). Coat condition was the only welfare indicator where there was evidence of differences between still images and videos, where it could be assessed more often via video (Table 7).

Body condition score was less likely to be assessable in open grassland (67% of observation events) than in other habitats (80% to 85%; overall p-value for habitat adjusted for method <0.001). Percentages of observation events where body posture could be assessed were similar for each habitat (74% to 78%) and the p-value for differences between habitats was high (0.860). Hoof condition, coat condition, sweating, and facial grimace were all less able to be assessed in open grassland habitats (for each, $p < 0.001$).

The most common reasons for not being able to assess particular welfare indicators were images/videos being captured at night and the horse being too distant from the camera, such as in open grasslands (Table 8). Assessment of hoof condition was prevented when hooves were obscured by vegetation, mud, or water. Although night-time commonly prevented a range of welfare indicators from being identified, body condition score (68% and 75%, respectively) and body posture (75% for both) could be assessed as frequently in night-time as in daylight observation events, whereas coat condition could only be assessed in 22% of night-time observation events compared with 51% of those in daylight. Some examples are shown in Figure 9.

An additional seven welfare indicators were assessed only in video observation events as we considered a priori that it would never be possible to assess these with still image observation events. Of these, the indicators that were able to be assessed most frequently were presence or absence of weakness (66%), qualitative behavioural assessment (60%), presence or absence of shivering (51%) and gait at walk (50%, Table 7). Gait at trot and canter could not be assessed in any observation events, but respiratory rate and effort could be assessed in 36% of video observation events (Table 7). For assessments of hoof condition and facial grimace in video observation events, numbers were sufficient to assess the usefulness of additional video clips within the observation event. For 15 and 20 observation events, respectively, these indicators could not be assessed in the first video clip nor in any of the further 1 to 10 video clips.

Gait at walk (overall $p = 0.010$), respiratory rate and effort (overall $p = 0.006$), and shivering (overall $p = 0.018$) were less likely to be assessable in open grassland habitats than in woodlands. The most common reasons for not being able to assess video-specific welfare indicators were: only a small part of the horse being in the camera's field of view (n = 28); the horse not moving substantially during the video recording (n = 18); the horse being too distant from the camera (n = 16); and the horse not being in the camera's field of view for long enough (n = 11).

Table 7. Welfare indicators assessed with camera trap still images and video.

Welfare Indicator	Able to Be Assessed with:		*p*-Value for Comparison of Ability to Assess Indicator between Still Image and Video Observation Events [1]
	Still Images: Number of Observation Events (% of 538 Observation Events)	Video: Number of Observation Events (% of 119 Observation Events)	
Body posture	411 (76%)	91 (76%)	0.972
Body condition score	392 (73%)	94 (79%)	0.237
Coat condition	225 (42%)	62 (52%)	0.053
Sweating	226 (42%)	54 (45%)	0.587
Facial grimace	102 (19%)	20 (17%)	0.441
Hoof condition	92 (17%)	25 (21%)	0.382
Wounds:			0.227
Both sides could be assessed	24 (4%)	8 (7%)	
Only one side could be assessed	198 (37%)	49 (41%)	
Limb pathology:			0.611
Both sides could be assessed	144 (27%)	40 (34%)	
Only one side could be assessed	36 (7%)	0 (0%)	
Skin lesions:			0.538
Both sides could be assessed	23 (4%)	7 (6%)	
Only one side could be assessed	180 (33%)	42 (35%)	
Nasal discharge:			0.604
Both sides could be assessed	21 (4%)	15 (13%)	
Only one side could be assessed	89 (17%)	13 (11%)	
Ocular discharge and blepharospasm:			0.550
Both sides could be assessed	17 (3%)	15 (13%)	
Only one side could be assessed	107 (20%)	14 (12%)	
Quidding or food pouching:			0.220
Both sides could be assessed	4 (0.7%)	5 (4%)	
Only one side could be assessed	80 (15%)	8 (7%)	
Gait at walk	NA [2]	60 (50%)	NA
Gait at trot	NA	0 (0%)	NA
Gait at canter	NA	0 (0%)	NA
Weakness	NA	78 (66%)	NA
Respiratory rate and effort	NA	43 (36%)	NA
Shivering	NA	61 (51%)	NA
Qualitative behavioural assessment	NA	71 (60%)	NA

[1] *p*-value for method after adjustment for habitat. [2] NA: not assessed-we considered a priori that it would never be possible to assess these indicators in still image observation events.

Table 8. Reasons for being unable to assess welfare indicators on camera trap still images and video.

Reason for Being Unable to Assess Welfare Indicators	Still Images: Number of Observation Events (% of 538 Observation Events)	Video: Number of Observation Events (% of 119 Observation Events)	*p*-Value for Comparison of Reason for Being Unable to Assess Welfare Indicator between Still Image and Video Observation Events [1]
Night-time images	172 (32%)	31 (26%)	0.345
Only top half of horse in the camera's field of view	96 (18%)	26 (22%)	0.273
Only front of horse in the camera's field of view	26 (5%)	1 (1%)	0.022
Only back of horse in the camera's field of view	59 (11%)	20 (17%)	0.090
Only head in the camera's field of view	35 (7%)	2 (1%)	0.020
Obscured by another horse	13 (2%)	1 (0.8%)	0.213
Hooves obscured by vegetation, mud or water	129 (24%)	18 (15%)	0.037
Horse too distant from the camera	178 (33%)	36 (30%)	0.845
Image too dark	190 (35%)	7 (6%)	<0.001
Image blurred	57 (11%)	0 (0%)	<0.001 [2]

[1] Likelihood ratio test *p*-value for method after adjustment for habitat. [2] Exact *p*-value for method; no adjustment for habitat.

3.5. Quantifying Behavioural Observations

From the random selection of camera days, 601 behaviour events were identified on still images, the specific character of which could be identified in 560/601 (93%) of these (Table 9). On video clips, 213 behaviour events were identified, the specific character of which could be identified in 178/213 (84%) of these. Behaviours were unidentifiable if the horse was still and the head and/or limbs were not in the camera's field of view.

The most commonly observed behaviours were grazing and walking, with grazing occupying a larger proportion of time, being 60% when based on duration of time taken from still images, 43% when derived from number of still images, and 51% when timed from video clips (Table 9). Social behaviours were rarely observed, with only two such events detected, namely, a horse sniffing a stallion faecal pile, and one horse sniffing another.

In addition, close proximity of the focal horse to other horses was recorded in 236/657 (36%) of observation events [202/538 (38%) for still image and 34/119 (29%) for video observation events]. Using the 'all occurrence' approach to further evaluate uncommonly observed behaviours, specific social interactions between two or more horses were only recorded on 39 occasions within the full dataset of 42,925 image/video files. These interactions were mostly affiliative (35 affiliative interactions including allogrooming, being herded or herding other horses, trotting, cantering or galloping with other horses, frolicking, nuzzling, playing, physical contact with another horse whilst walking), with three occasions of reproductive behaviours (mating/being mated, flehmen response, winking), and one agonistic event (chasing another horse).

Table 9. The number and durations of identifiable behaviours calculated from camera trap still images and videos.

Behaviour	Number of Behaviour Events from Still Images (n = 560)	Number of Behaviour Events from Video (n = 178)	Number (%) of Still Images (n = 2159 Images [1]) with This Behaviour	Duration (%) of Time from Still Images (n = 36,711 s [1]) with This Behaviour	Duration (%) of Time from Video (n = 1683 s [1]) with This Behaviour
Locomotion total	341	83	902 (42)	6916 (19)	465 (28)
Walking	325	83	872 (40)	6900 (19)	465 (28)
Trotting	8	0	14 (<1)	8 (<1)	0
Cantering	8	0	16 (<1)	8 (<1)	0
Standing resting	56	31	276 (13)	7669 (20)	299 (18)
Grazing	148	56	938 (43)	22,025 (60)	851 (50)
Maintenance behaviours	13	7	39 (2)	99 (<1)	65 (4)
Social behaviours	2	1	4 (<1)	2 (<1)	3 (<1)

[1] Total numbers are for pooled behaviour events where the behaviour was identified.

4. Discussion

This paper describes, for the first time, camera trapping methodology that can be successfully applied across a range of habitats in order to identify individual horses, and to non-invasively measure or observe a range of animal-based welfare indicators. Furthermore, it demonstrates the advantages and limitations of this methodology.

Methodological information is often lacking in camera trap publications [62]. We have sought to describe methods in sufficient detail to enable other researchers to easily replicate them. Further, as a result of this study we can make recommendations for others wishing to use camera trapping for this purpose. Firstly, extensive ground surveys to facilitate precise and strategic camera placement are key to optimising image quality and detection of horses at appropriate angles and distances from the camera in order to both identify individual horses and assess a range of welfare indicators. Deploying cameras on tracks, grazing areas, and drinking locations within the same region assists in capturing the full range of listed welfare indicators, including a wider range of behaviours. Secondly, the choice of camera settings is important to enhance the data obtained whilst also minimising battery usage and SD card storage. We recommend using an image capture number of one (i.e., the number of still images taken in a sequence once the camera is triggered), and a camera time delay interval of 1 s for cameras placed on tracks where horses may pass quickly, and 3 s for cameras overlooking grazing or drinking locations. For videos, a duration of 10 s is recommended. Other settings recommended are as described in the Materials and Methods. Whether or not to use both day-time and night-time settings is dependent on the precise purpose of the study; although fewer horses and welfare indicators can be identified on night-time images, meaningful information can still be obtained. Use of video will increase the range of welfare indicators that can be assessed, but does significantly increase battery usage, SD card storage, and the time required for processing and analysing the data. Whilst there are strategies for increasing battery power (e.g., solar panels) and storage (higher memory SD cards), these can be costly, and despite security measures, vandalism and theft can be an issue. We therefore recommend prioritising use of video in habitats where horses cannot be reliably observed directly, in regions where repeated capturing of non-target species is less likely, and where cameras are able to be checked within a 1 to 2 month period. In other situations, still images may suffice and complement direct observations.

In woodland habitats, direct viewing of horses is challenging and relying on this would have significantly underestimated the population size in this study. A major advantage of camera trapping is the ability to identify individual horses in any habitat. Furthermore, demographic information such as herd size and reproductive rates varied spatially and across different habitats, so such data obtained from direct observation alone would not have represented the whole population. Understanding population dynamics and the processes that influence them is critical for management of populations. In Australia, there are few demographic studies on wild horses [5,10,16], particularly within woodland habitats [17,18]. In south-east Australia, although free-roaming horses commonly reside in woodland habitats (eucalyptus forests) and undulating and often steep terrain, population and demographic data for horses in these habitats are lacking. Our study suggests that utilising camera trapping methodology could address some of these knowledge gaps.

Practical limitations in the frequency with which direct observations can be performed is another challenge in assessing the welfare of any free-roaming wild species. Both the severity and the duration of welfare impacts are important in assessing welfare [1,20–24], so repeated assessments are advantageous. Beneficially, the continuous collection of data by camera traps permits more frequent assessments to be made. Further advantages are that the horses are not disturbed, and are captured close-up, thereby enabling a range of welfare indicators to be assessed and informative behaviours to be quantified, even in horses that cannot be observed directly.

To date, published studies have predominantly used invasive measures of health that require physically capturing the animal to perform procedures such as physical examination and blood collection (reviewed by [63]). Whilst these measures are informative, if available, it is preferable to use the least invasive methods for welfare assessments. Our study shows that a wide range of indicators of different aspects of welfare can be evaluated non-invasively using camera traps, and for some horses also by distant direct observations.

Twelve indicators of welfare aligned with the first four domains of the Five Domains Model, were able to be assessed with equal frequency on both still images and video, with an additional five indicators able to be assessed on video. The most practically measurable and reliable animal-based welfare indicators able to be detected were: in Domain 1 (Nutrition), body condition score; in Domain 2 (Physical Environment), presence of sweating or shivering; in Domain 3 (Health), body posture, coat condition, gait at walk, and presence/absence of weakness; and in Domain 4 (Behavioural Interactions), qualitative behavioural assessment and assessment of close spatial proximity for evaluation of social bonds. Spatial proximity of horses may have been overestimated on grasslands due to the greater field of vision of cameras in this habitat, compared to woodlands. In future studies, we therefore recommend defining close spatial proximity as animals standing within two body lengths of each other [60], rather than animals simply being in the same image or video frame. Facial indicators and hoof condition could only be assessed infrequently due to the need for clear close-up images/videos of these body parts, the correct angle of view and an absence of obscuring vegetation. Some indicators were less likely to be detectable in open grassland habitats than in other habitats, because the horses were often further away from the camera in grassland habitats. Additionally, for bilateral indicators, most commonly only one side of the horse could be assessed.

Quantification of specific behaviours was also achievable by evaluating the proportions of time that each behaviour was captured, or the proportion of still images demonstrating a particular behaviour. Since camera traps only capture the behaviour being performed at the precise time that the horse is within camera view, these proportions do not necessarily accurately reflect continuously recorded time budget behaviours [3,5,13,14]. Camera location may also bias the behaviour detected, e.g., grazing is more likely to be detected on open grassland and walking more often captured on tracks. However, if cameras remain in the same location over time, useful information may be obtained regarding temporal changes in the proportions of different behaviours being performed.

Despite the many advantages of remote camera traps, there are limitations that need to be considered when interpreting data. Information can only be collected when horses are in the camera's field of view, which is heavily dependent on camera placement and the number of cameras deployed. Problematically on still images, it can be difficult to assess the motivation for a behaviour, and whether it reflects a positive or negative mental state (Domain 5). For example, rolling can be both a maintenance behaviour [13,44] and an indication of abdominal pain [45,59]; lying in lateral recumbency is commonly linked to rest [13,44], but can also be due to abdominal pain [45,59]; and rubbing/scratching is a normal maintenance behaviour [13,44], but when performed excessively can be an indicator of pruritus [64,65]. Similarly, although some features of 'facial grimace' [48,52] could sometimes be identified on still images, the significance of this may be over interpreted when based on a single still image. When these events are observed directly or with video, additional indicators and the context of the behaviour usually assist interpretation.

5. Conclusions

As far as the authors are aware, this is the first study that describes in detail a remote camera trapping methodology that enables identification of individual free-roaming wild horses across a range of habitats and the assessment of an extensive range of animal-based welfare indicators. Camera trap images and video provided valuable information about the horses, particularly those that could not be sighted regularly, sighted for a long enough duration, or approached closely enough to enable direct assessment of welfare indicators, as was the case in woodland habitats.

The next phases of this research include applying the same methodology to larger populations across different geographical areas, in addition to incorporating these methods into a welfare assessment protocol to objectively evaluate how the welfare of wild free-roaming horses varies spatially and temporally. The described methodology can also form the basis of applications to other species.

Author Contributions: Conceptualization, A.M.H., R.S.C., and D.R.; methodology, A.M.H.; data curation, A.M.H. and V.R.; data analysis, A.M.H. and J.M.M.; writing—original draft preparation A.M.H.; writing—review and editing original drafts J.M.M., D.J.M.; writing—review and editing final draft A.M.H., D.J.M., R.S.C., J.M.M., V.R., and D.R.; Supervision, D.R. All authors have read and agreed to the published version of the manuscript.

Funding: This research received no external funding. A.M.H. was supported by an Australian Government Research Training Program scholarship and University of Technology Sydney Chancellor's Research scholarship.

Institutional Review Board Statement: The study was approved by the University of Technology Sydney Animal Ethics Research Authority 2015000490 (date of approval 15 October 2015).

Data Availability Statement: The data presented in this study are available on request from the corresponding author. The data are not publicly available due to additional ongoing analysis.

Acknowledgments: A.M.H. would like to thank WaterNSW and National Parks and Wildlife Service (NPWS) for their support and access of field sites, in particular Chris Banffy of NPWS for orientation and support in the field. Thanks also to all the field volunteers. Finally, thanks to Peter Krockenberger for proof-reading the manuscript.

Conflicts of Interest: The authors declare no conflict of interest.

References

1. Harvey, A.M.; Beausoleil, N.J.; Ramp, D.; Mellor, D.J. A Ten-Stage Protocol for Assessing the Welfare of Individual Non-Captive Wild Animals: Free-Roaming Horses (*Equus Ferus Caballus*) as an Example. *Animals* **2020**, *10*, 148. [CrossRef]
2. Klingel, H. Social organization and reproduction in equids. *J. Reprod. Fertil.* **1975**, *23*, 7–11.
3. Duncan, P. Time budgets of Camargue horses: II. Time budgets of adult horses and weaned sub-adults. *Behaviour* **1979**, *72*, 26–49. [CrossRef]
4. Garrott, R.A.; Taylor, L. Dynamics of a feral horse population in Montana. *J. Wildl. Manag.* **1990**, *54*, 603–612. [CrossRef]
5. Berman, D.M. The Ecology of Feral Horses in Central Australia. Ph.D. Thesis, University of New England, Armidale, Australia, 1991.
6. Linklater, W.L. Social and Spatial Organisation of Horses. Ph.D. Thesis, Massey University, Palmerston North, New Zealand, 1998.
7. Linklater, W.L.; Cameron, E.Z.; Stafford, K.J.; Veltman, C.J. Social and spatial structure and range use by Kaimanawa wild horses (*Equus caballus*: Equidae). *N. Zeal. J. Ecol.* **2000**, *24*, 139–152.
8. Cameron, E.; Linklater, W.; Stafford, K.; Minot, E. Social grouping and maternal behaviour in feral horses (*Equus caballus*): The influence of males on maternal protectiveness. *Behav. Ecol. Sociobiol.* **2003**, *53*, 92–101. [CrossRef]

9. Linklater, W.L.; Cameron, E.Z.; Minot, E.O.; Stafford, K.J. Feral horse demography and population growth in the Kaimanawa Ranges, New Zealand. *Wildl. Res.* **2004**, *31*, 119–1128. [CrossRef]
10. Dawson, M.J. *The Population Ecology of Feral Horses in the Australian Alps–Management Summary*; Australian Alps Liaison Committee: Canberra, Australia, 2005.
11. Cameron, E.Z.; Linklater, W.L.; Stafford, K.J.; Minot, E.O. Maternal investment results in better foal condition through increased play behaviour in horses. *Anim. Behav.* **2008**, *76*, 1511–1518. [CrossRef]
12. Grange, S.; Duncan, P.; Gaillard, J.M. Poor horse traders: Large mammals trade survival for reproduction during the process of feralization. *Proc. Biol. Sci.* **2009**, *276*, 1911–1919. [CrossRef]
13. Ransom, J.I.; Cade, B.S. Quantifying equid behavior: A research ethogram for free-roaming feral horses. In *U.S. Geological Survey Techniques and Methods Report 2-A9*; USGS: Fort Collins, CO, USA, 2009.
14. Ransom, J.I.; Cade, B.S.; Hobbs, N.T. Influences of immunocontraception on time budgets, social behavior, and body condition in feral horses. *Appl. Anim. Behav. Sci.* **2010**, *124*, 51–60. [CrossRef]
15. Scorolli, A.L.; Lopez Cazorla, A.C. Demography of feral horses (*Equus caballus*): A long-term study in Tornquist Park, Argentina. *Wildl. Res.* **2010**, *37*, 207–214. [CrossRef]
16. Dawson, M.J.; Hone, J. Demography and dynamics of three wild horse populations in the Australian Alps. *Austral Ecol.* **2012**, *37*, 97–109. [CrossRef]
17. Zabek, M.A. Understanding Population Dynamics of Feral Horses in the Tuan and Toolara State Forest for Successful Long-Term Population Management. Ph.D. Thesis, University of Queensland, Queensland, Australia, 2015.
18. Zabek, M.A.; Berman, D.M.; Blomberg, S.P.; Collins, C.W.; Wright, J. Population dynamics of feral horses (*Equus caballus*) in an exotic coniferous plantation in Australia. *Wildl. Res.* **2016**, *43*, 358–367. [CrossRef]
19. Scorolli, A.L. Feral horse population model and body condition: Useful management tools in Tornquist Park, Argentina? *J. Wildl. Manag.* **2020**, 1–7. [CrossRef]
20. Mellor, D.J.; Reid, C.S.W. *Concepts of Animal Well-Being and Predicting the Impact of Procedures on Experimental Animals. Improving the Well-Being of Animals in the Research Environment*; ANZCCART: Glen Osmond, Australia, 1994; pp. 3–18.
21. Mellor, D.J.; Patterson-Kane, E.; Stafford, K.J. *The Sciences of Animal Welfare*; Wiley-Blackwell Publishing: Oxford, UK, 2009.
22. Mellor, D.J.; Beausoleil, N.J. Extending the 'Five Domains' model for animal welfare assessment to incorporate positive welfare states. *Anim. Welf.* **2015**, *24*, 241–253. [CrossRef]
23. Mellor, D.J. Operational details of the five domains model and its key applications to the assessment and management of animal welfare. *Animals* **2017**, *7*, 60. [CrossRef] [PubMed]
24. Mellor, D.J.; Beausoleil, N.J.; Littlewood, K.E.; McLean, A.N.; McGreevy, P.D.; Jones, B.; Wilkins, C. The 2020 Five Domains Model: Including Human–Animal Interactions in Assessments of Animal Welfare. *Animals* **2020**, *10*, 1870. [CrossRef] [PubMed]
25. Cutler, L.; Swann, D.E. Using Remote Photography in Wildlife Ecology: A Review. *Wildl. Soc. Bull.* **1999**, *27*, 571–581. Available online: www.jstor.org/stable/3784076 (accessed on 3 May 2021).
26. Silveira, L.; Jácomo, A.T.A.; Diniz-Filho, J.A.F. Camera trap, line transect census and track surveys: A comparative evaluation. *Biol. Conserv.* **2003**, *114*, 351–355. [CrossRef]
27. O'Connell, A.F.; Nichols, J.D.; Karanth, K.U. Evolution of camera trapping. In *Camera Traps in Animal Ecology: Methods and Analyses*; Springer: New York, NY, USA, 2010; pp. 1–8.
28. Meek, P.D.; Ballard, G.A.; Vernes, K.; Flemming, P.J.S. The history of wildlife camera trapping as a survey tool in Australia. *Aust. Mammal.* **2015**, *37*, 1–12. [CrossRef]
29. Silver, S.C.; Ostro, L.E.T.; Ma, L.K. The use of camera traps for estimating jaguar *Panthera onca* abundance and density using capture/recapture analysis. *Oryx* **2004**, *38*, 148–154. [CrossRef]
30. Smit, J.; Pozo, R.; Cusack, J.; Nowak, K.; Jones, T. Using camera traps to study the age-sex structure and behaviour of crop-using elephants in Udzungwa Mountains National Park, Tanzania. *Oryx* **2017**, *53*. [CrossRef]
31. Karanth, K.U.; Nichols, J.D.; Samba Kuma, N. Estimating of Demographic Parameters in a Tiger Population from Long-term Camera Trap Data. In *Camera Traps in Animal Ecology: Methods and Analyses*; O'Connell, A.F., Nichols, J.D., Karanth, K.U., Eds.; Springer: New York, NY, USA, 2010; pp. 145–162.
32. Duggan, G.; Burn, C.C.; Clauss, M. Nocturnal behavior in captive giraffe (*Giraffa camelopardalis*)—A pilot study. *Zoo Biol.* **2016**, *35*, 14–18. [CrossRef] [PubMed]
33. Rose, P.E.; Lloyd, I.; Brereton, J.E.; Croft, D.P. Patterns of nocturnal activity in captive greater flamingos. *Zoo Biol.* **2018**, *37*, 290–299. [CrossRef]
34. Fazio, J.M.; Barthel, T.; Freeman, E.W.; Garlick-Ott, K.; Scholle, A.; Brown, J.L. Utilizing Camera Traps, Closed Circuit Cameras and Behavior Observation Software to Monitor Activity Budgets, Habitat Use, and Social Interactions of Zoo-Housed Asian Elephants (*Elephas maximus*). *Animals* **2020**, *10*, 2026. [CrossRef]
35. Wooster, E.; Wallach, A.D.; Ramp, D. The Wily and Courageous Red Fox: Behavioural Analysis of a Mesopredator at Resource Points Shared by an Apex Predator. *Animals* **2019**, *9*, 907. [CrossRef]
36. Hockings, K.J.; Mubemba, B.; Avanzi, C.; Pleh, K.; Düx, A.; Bersacola, E.; Bessa, J.; Ramon, M. Leprosy in Wild Chimpanzees. *bioRxiv* **2020**. [CrossRef]

37. Zhang, Y.; Cao, Q.S.; Rubenstein, D.I.; Zand, S.; Songer, M.; Leimgruber, P.; Chu, H.; Cao, J.; Li, K.; Hu, D. Water Use Patterns of Sympatric Przewalski's Horse and Khulan: Interspecific Comparison Reveals Niche Differences. *PLoS ONE* **2015**, *10*, e0132094. [CrossRef]

38. Schlichting, P.E.; Dombrovski, V.; Beasley, J.C. Use of abandoned structures by Przewalski's wild horses and other wildlife in the Chernobyl Exclusion Zone. *Mammal Res.* **2020**, *65*, 161–165. [CrossRef]

39. Henneke, D.R.; Potter, G.D.; Kreider, J.L.; Yeates, B.F. Relationship between condition score, physical measurements and body fat percentage in mares. *Equine Vet. J.* **1983**, *15*, 371–372. [CrossRef]

40. Carroll, C.L.; Huntington, P.J. Body condition scoring and weight estimation of horses. *Equine Vet. J.* **1988**, *20*, 41–45. [CrossRef] [PubMed]

41. McDonnell, S.M.; Haviland, J.C.S. Agonistic ethogram of the equid bachelor band. *Appl. Anim. Behav. Sci.* **1995**, *43*, 147–188. [CrossRef]

42. Wemelsfelder, F.; Hunter, E.A.; Lawrence, A.B.; Mendl, M.T. Assessing the 'whole-animal': A Free-Choice-Profiling approach. *Anim. Behav.* **2001**, *62*, 209–220. [CrossRef]

43. McDonnell, S.M.; Poulin, A. Equid play ethogram. *Appl. Anim. Behav. Sci.* **2002**, *78*, 263–290. [CrossRef]

44. McDonnell, S.M. *The Equid Ethogram: A Practical Field Guide to Horse Behaviour*; CAB International: Cambridge, MA, USA, 2003.

45. Ashley, F.H.; Waterman-Pearson, A.E.; Whay, H.R. Behavioural assessment of pain in horses and donkeys: Application to clinical practice and future studies. *Equine Vet. J.* **2005**, *37*, 565–575. [CrossRef]

46. Wemelsfelder, F. How animals communicate quality of life: The qualitative assessment of behaviour. *Anim. Welf.* **2007**, *16*, 25–31.

47. Samuel, E.K.; Whay, H.R.; Mullan, S. A preliminary study investigating the physical welfare and welfare code compliance for tethered and free-ranging horses on common land in South Wales. *Anim. Welf.* **2012**, *21*, 593–598. [CrossRef]

48. Dalla Costa, E.; Minero, M.; Lebelt, D.; Stucke, D.; Canali, E.; Leach, M.C. Development of the Horse Grimace Scale (HGS) as a Pain Assessment Tool in Horses Undergoing Routine Castration. *PLoS ONE* **2014**, *9*, e92281. [CrossRef]

49. Dalla Costa, E.; Murray, L.; Dai, F.; Canali, E.; Minero, M. Equine on-farm welfare assessment: A review of animal-based indicators. *Anim. Welf.* **2014**, *23*, 323–341. [CrossRef]

50. Mullan, S.; Szmaraged, C.; Hotchkiss, I.; Whay, H.R. The welfare of long-line tethered and free-ranging horses kept on public grazing land in South Wales. *Anim. Welf.* **2014**, *23*, 25–37. [CrossRef]

51. Hockenhull, J.; Whay, H.R. A review of approaches to assessing equine welfare. *Equine Vet. Educ.* **2014**, *26*, 159–166. [CrossRef]

52. Gleerup, K.B.; Forkman, B.; Lindegaard, C.; Andersen, P.H. An equine pain face. *Vet. Anaesth. Analg.* **2015**, *42*, 103–114. [CrossRef]

53. Dalla Costa, E.; Dai, F.; Lebelt, D.; Scholz, P.; Barbieri, S.; Canali, E.; Zanella, A.J.; Minero, M. Welfare assessment of horses: The AWIN approach. *Anim. Welf.* **2016**, *2*, 481–488. [CrossRef]

54. Hintze, S.; Murphy, E.; Bachmann, I.; Wemelsfelder, F.; Würbel, H. Qualitative Behaviour Assessment of Horses Exposed To Short-term Emotional Treatments. *Appl. Anim. Behav. Sci.* **2017**, *196*, 44–51. [CrossRef]

55. Minero, M.; Dalla Costa, E.; Dai, F.; Canali, E.; Barbieri, S.; Zanella, A.; Pascuzzo, R.; Wemelsfelder, F. Using qualitative behaviour assessment (QBA) to explore the emotional state of horses and its association with human-animal relationship. *Appl. Anim. Behav. Sci.* **2018**, *204*, 53–59. [CrossRef]

56. Somerville, R.; Brown, A.F.; Upjohn, M. A standardised equine-based assessment tool used for six years in low and middle income countries. *PLoS ONE* **2018**, *13*, e0192354. [CrossRef] [PubMed]

57. Czycholl, I.; Büttner, K.; Klingbeil, P.; Krieter, J. An Indication of Reliability of the Two-Level Approach of the AWIN Welfare Assessment Protocol for Horses. *Animals* **2018**, *8*, 7. [CrossRef] [PubMed]

58. Sigurjónsdóttir, H.; Haraldsson, H. Significance of Group Composition for the Welfare of Pastured Horses. *Animals* **2019**, *9*, 14. [CrossRef] [PubMed]

59. Torcivia, C.; McDonnell, S. Equine Discomfort Ethogram. *Animals* **2021**, *11*, 580. [CrossRef]

60. Wolter, R.; Stefanski, V.; Krueger, K. Parameters for the Analysis of Social Bonds in Horses. *Animals* **2018**, *8*, 191. [CrossRef]

61. Goel, M.K.; Khanna, P.; Kishore, J. Understanding survival analysis: Kaplan-Meier estimate. *Int. J. Ayurveda Res.* **2010**, *1*, 274–278. [CrossRef] [PubMed]

62. Meek, P.; Ballard, G.; Claridge, A.W.; Kays, R.; Moseby, K.; Brien, T.O.; O'Connell, A.; Sanderson, J.B.; Swann, D.E.; Tobler, M.W.; et al. Recommended guiding principles for reporting on camera trapping research. *Biodivers. Conserv.* **2014**, *23*, 2321–2343. [CrossRef]

63. Kophamel, S.; Illing, B.; Ariel, E.; Difalco, M.; Skerratt, L.F.; Hamann, M.; Ward, L.C.; Mendez, D.; Munns, S.L. Importance of health assessments for conservation in noncaptive wildlife. *Conserv. Biol.* **2021**. [CrossRef]

64. Fadok, V.A. Overview of Equine Pruritus. *Vet. Clin. N. Am. Equine Pract.* **1995**, *11*, 1–10. [CrossRef]

65. Metz, M.; Grundmann, S.; Ständer, S. Pruritus: An overview of current concepts. *Vet. Dermatol.* **2011**, *22*, 121–131. [CrossRef] [PubMed]

animals

Review

Welfare Health and Productivity in Commercial Pig Herds

Przemysław Racewicz [1,*], Agnieszka Ludwiczak [2], Ewa Skrzypczak [2], Joanna Składanowska-Baryza [2], Hanna Biesiada [1], Tomasz Nowak [3], Sebastian Nowaczewski [2], Maciej Zaborowicz [4], Marek Stanisz [2] and Piotr Ślósarz [2]

[1] Laboratory of Veterinary Public Health Protection, Department of Animal Breeding and Product Quality Assessment, Poznan University of Life Sciences, Słoneczna 1, 62-002 Suchy Las, Poland; hanna.biesiada@up.poznan.pl

[2] Department of Animal Breeding and Product Quality Assessment, Poznan University of Life Sciences, Słoneczna 1, 62-002 Suchy Las, Poland; agnieszka.ludwiczak@up.poznan.pl (A.L.); ewa.skrzypczak@up.poznan.pl (E.S.); joanna.skladanowska-baryza@up.poznan.pl (J.S.-B.); sebastian.nowaczewski@up.poznan.pl (S.N.); marek.stanisz@up.poznan.pl (M.S.); piotr.slosarz@up.poznan.pl (P.Ś.)

[3] Department of Genetics and Animal Breeding, Animal Reproduction Laboratory, Poznan University of Life Sciences, 60-637 Poznan, Poland; tomasz.nowak@up.poznan.pl

[4] Institute of Biosystems Engineering, Poznan University of Life Sciences, 60-637 Poznan, Poland; maciej.zaborowicz@up.poznan.pl

* Correspondence: przemyslaw.racewicz@up.poznan.pl; Tel.: +48-618-466-687

Citation: Racewicz, P.; Ludwiczak, A.; Skrzypczak, E.; Składanowska-Baryza, J.; Biesiada, H.; Nowak, T.; Nowaczewski, S.; Zaborowicz, M.; Stanisz, M.; Ślósarz, P. Welfare Health and Productivity in Commercial Pig Herds. *Animals* 2021, *11*, 1176. https://doi.org/10.3390/ani11041176

Academic Editor: Melissa Hempstead

Received: 15 March 2021
Accepted: 17 April 2021
Published: 20 April 2021

Publisher's Note: MDPI stays neutral with regard to jurisdictional claims in published maps and institutional affiliations.

Simple Summary: The continuous development of innovative technologies and the large-scale implementation of these solutions on farms are dynamically influencing the so-called precision livestock farming—PLF. Pig producers striving to increase the profitability of production, food safety, and food itself are increasingly willing to invest in rationalisation systems that raise the technological standards of pig breeding. The use of modern systems in livestock management and animal welfare is based on the use of non-invasive monitoring devices such as cameras, microphones, or detectors and information technology-based data archiving and management systems that support farmers/breeders in the daily running of the farm. Precision farming technologies, which are beneficial for animal welfare as well as for the profit of the livestock producer, help to solve the problems of large-scale animal production and satisfy the expectations of food regulators and consumers themselves. The aim of the paper was to gather contemporary knowledge on innovative technologies applied on pig farms. The paper presents and compares methods of controlling herd behavioural parameters with the use of various monitoring systems and their purpose. The paper also includes a review of potential limitations that may occur in the daily use of the above-mentioned devices. The review presents results on the effectiveness of their use.

Abstract: In recent years, there have been very dynamic changes in both pork production and pig breeding technology around the world. The general trend of increasing the efficiency of pig production, with reduced employment, requires optimisation and a comprehensive approach to herd management. One of the most important elements on the way to achieving this goal is to maintain animal welfare and health. The health of the pigs on the farm is also a key aspect in production economics. The need to maintain a high health status of pig herds by eliminating the frequency of different disease units and reducing the need for antimicrobial substances is part of a broadly understood high potential herd management strategy. Thanks to the use of sensors (cameras, microphones, accelerometers, or radio-frequency identification transponders), the images, sounds, movements, and vital signs of animals are combined through algorithms and analysed for non-invasive monitoring of animals, which allows for early detection of diseases, improves their welfare, and increases the productivity of breeding. Automated, innovative early warning systems based on continuous monitoring of specific physiological (e.g., body temperature) and behavioural parameters can provide an alternative to direct diagnosis and visual assessment by the veterinarian or the herd keeper.

Keywords: pigs; welfare; health; herd management; monitoring technologies

1. Introduction

In recent years, the world has seen rapid changes in the dynamics and efficiency of pig production. The general trend of increasing production, with reduced employment, requires optimisation and a comprehensive approach to herd management [1]. Modern pig production should therefore be based not only on a modern infrastructure and a precisely designed feeding program, but also on the use of modern technologies for monitoring health and welfare of the entire herd [2–4].

Herd health programs for swine include biosecurity, routine health control, and other preventive procedures allowing maintenance of a high health status of pig herds [5]. The health of pigs on the farm directly translates into the economics of production. Diseases lead to higher morbidity and mortality in different age groups and higher veterinary costs for the purchase of medicines and vaccines and more frequent veterinary visits [6,7]. Herds with low health status are also characterised by low productivity, reduced growth, and higher feed consumption. As a consequence, the fattening period is extended and the production efficiency is decreased [8].

Productive performance of pigs is a reliable indicator of the efficiency of production under different housing conditions [9,10]. The criteria for assessing animal welfare cover even more characteristics, including indicators of health and ethological parameters [11]. In practice, it is difficult to identify one basic and easy-to-use measure, which demonstrates the imperfections of each indicator and, on the other hand, the complexity of the concept of welfare [12].

In recent years, the market of equipment and systems for continuous, automatic health and behaviour monitoring in pig herds has been enriched by innovative technologies. Modern pig production systems based on intelligent technologies allow for planned, efficient, and thus more cost-effective production [5]. Considering the used methodology and the scope of application, three categories can be distinguished among the available devices. The first category devices are only aimed at detecting specific animal behaviour by means of special sensors. An example of such a solution is the system automatically measuring the frequency of pig visits to the feeder and the time taken to feed by means of radio frequency identification technology—RFID [13,14]. Another example is the use of real-time video visualisation using conventional (2D) monochromatic or colour cameras or 3D cameras to depict activity level, area occupancy, aggression, gait scores, resource use, and posture [15–18]. The second category devices allow for detection and recording of specific behaviours, such as drinking [19], feeding [20], or spatial distribution [21], which are further processed into numeric data and presented, e.g., in the form of a graph on a mobile phone monitor. This type of device allows for identification of changes in animal behaviour but requires farm workers to interpret the data. The last category involves intelligent production systems, automatically analysing the recorded changes in the physiological and behavioural parameters. These systems are based on optimal settings of farm environment, have the ability to extract deviations from theses settings, and automatically make decisions to adapt the production environment to optimal production conditions [22,23]. The goal of this paper was to gather knowledge on novel technologies applied on pig farms in order to promote their health, productivity, and welfare. The article should be in the interest of pig farmers, pork retailers, and researchers working in the field of meat science.

2. Challenges of Pig Farming

There were about 677.6 million pigs worldwide as of January 2020 [24]. Pork is the second-most consumed meat in the world, with the consumption reaching 23.0 kg/capita [25]. Population growth increases demand for meat. The statistics on the projected pork consumption indicate a global increase, by about 17%, predicted for the period from 2021 to 2029 [26].

Modern pig production is characterised by intensification and specialisation of production. These two factors lead to an increase in animal productivity and thus contribute to higher economic efficiency of production. On the other hand, they cause serious ecological problems as well as problems related to animal welfare, herd health, and food safety [27]. A high level of welfare is a guarantee of good health of the animal as well as the elimination of antibiotics or other drugs [28].

A series of programs were launched in European Union in order to evaluate and improve the welfare of farmed animals, with the Welfare Quality® program resulting in the development of welfare protocols for the animal species under large-scale production [29–32]. Welfare Quality® protocols contain major welfare principles targeted at the needs of animals under intensive production and based on the Five Freedoms of animal welfare [33]. The protocols are designed in a manner that allows for their species-specific adaptation. According to the available research results, the protocol dedicated for pigs is a useful and reliable tool for identification of farms keeping pigs at poor level of welfare [34].

In order to prevent economic losses due to diseases occurring in the herd, pig farmers should acquire at least basic skills to diagnose and deal with the appearance of the disease unit in their shed. [35]

Changes in animal behaviour preceding or accompanying subclinical and clinical signs may be of significant diagnostic value. They are often referred to as sickness behaviour, including changes in eating habits, social behaviour, mobility, and posture. By definition, subclinical disease is latent, and thus direct monitoring based on staff observation is ineffective. This is due to the fact that it is time-consuming, inaccurate, and impractical in terms of work organisation [22,36].

Automated early warning systems, based on continuous monitoring of specific physiological (e.g., body temperature) and behavioural parameters, can provide an alternative to direct observation of animals [23,37]. A good example are methods based on artificial intelligence. These methods employ computer tools able to track animal behaviour [38–42] and distinguish individuals from each other [43].

Today, commercial pigs are exposed to a great number of stress factors, including stocking densities, high concentration of animals in a limited area, limited possibilities of movement and motivated behaviour expression, and frequent regrouping of animals. The microclimate in pig buildings also has a huge impact on pig welfare and production results. It affects animal health, reproduction parameters, and feed intake [37]. Harmful consequences of stress depend on the sensitivity of the animals to stressors as well as their severity and duration of action. Among farm animals, pigs are characterised as having the lowest tolerance to high environmental temperatures. This is due, among other things, to low adaptability of the thermoregulatory medium in the brain, low number of sweat glands, presence of the subcutaneous fat layer, and intensive metabolism. Exceeding the body's ability to adapt to high temperatures (hyperthermia) is a threat to health and life [44,45]. Hyperthermia in pigs leads to poor condition of the animals, decreased daily gains, and longer fattening period. Moreover, it negatively affects the quality of pork resulting from the interaction between the muscle pH value and the temperature in the process of post-mortem denaturation of muscle proteins [46].

Among the consequences of stress, one can name reduced appetite, reproductive disorders, and reduced immunity. In the aftermath, the pig farmers observe lower daily gains, poor reproductive results, and the appearance of infectious diseases and high mortality [47].

3. Welfare Monitoring Systems for Pigs

The market of systems designed for permanent, automatic monitoring of farm animal welfare is constantly evolving. Automated innovative early warning systems (PLF—precision livestock farming), based on continuous monitoring of specific behavioural and physiological parameters, are an alternative to direct visual assessment by staff or veterinarian. Fast and accurate acquisition of real-time data on animal movement and feed intake frequency enables early detection of diseases and facilitates further management of the herd. Thanks to sensors (cameras, microphones, accelerometers, RFID sensors, and temperature sensors), behavioural patterns of animals are gathered and combined through algorithms. The data derived from PLF technologies can be used to derive warnings and trigger notifications and alarms [48]. With the development of the Internet of Things (IoT, i.e., the interconnection between computing devices via the Internet), decision making can be better informed by connecting PLF information with other data streams, and components of farm management can be automated or even controlled remotely [15,49,50]. This allows for the ability to detect problems early enough to prevent potential, negative effects on productive performance of animals [51]. There are many benefits of precision livestock production, including increased productivity and profitability, increased safety and quality of animal products, and improved animal welfare, as well as reduced environmental impact and combating climate change. The use of precision livestock production in animal nutrition has been shown to reduce feed costs by up to 25% [52,53]. In 2016, the total turnover in the precision agricultural technology market was estimated at USD 4.8 billion. Current forecasts put the market turnover in this area at USD 12.6 billion by 2025 [54]. The adoption of these technologies varies considerably. RFID and accelerometer technologies are well integrated, but other technologies still have to achieve a viable market share [48].

3.1. Vision-Based Systems

Among the adopted PLF methods, video monitoring seems to be the most commonly implemented. It provides non-invasive and efficient tools to be able to record not only the behaviour of a group of animals but the behaviour of each individual [35]. By means of image analysis, the results are converted to detailed data on animal distribution (location and proximity) [18,55] and activity (position and movement) [56,57]. Imaging is also used in pigs to measure body weight [58–60] and to detect lameness [61], aggressive behaviour [62], and heat [63].

Over recent decades, two-dimensional (2D) monochrome and colour have been widely used in computer vision due to its low cost and high efficiency. Many researchers have proposed different systems to extract livestock characteristics, such as body size or body condition, on the basis of 2D images [64,65]. For example 2D image analysis allows for monitoring and estimation of pig growth rates to an accuracy of 1 kg [58]. In turn, the number of cameras (video sets) intended for animal observation depends on the monitored area and the height at which the camera is placed. The quality of monitoring is also influenced by the number of animals per square meter [22]. As many studies show, simple dome cameras are sufficient to monitor the behaviour of the inmates [41,65–67]. For example, it can be a CCTV camera with IP67 waterproof rating and 2MP (1080) resolution, with f/2.8/F1.6 minimum aperture and with built-in infrared heater [67]. Monochrome cameras typically have greater light sensitivity and are thus more ideal for recording under lower light conditions than colour cameras [15].

In the work of Chen et al. [65], a behaviour identification and monitoring study was conducted on eight pigs in pens of approximately 4 square meters, with the camera positioned at a height of 2.4 m. This study used the neural modelling technique using deep learning algorithms. It should be noted that the authors used a more difficult modelling technique in relation to convolutional networks and searched for their own indexes to describe the image. They obtained 98.5% accuracy in terms of behaviour identification through connection of a relatively simple camera with neural modelling technology. The results of research conducted by Chen et al. [65] allow for high efficiency of algorithms used in systems for monitoring animal behaviour [65]. Reikert et al. [67] applied the deep learning system in combination with 2D cameras in order to detect the position and posture of pigs. The authors obtained slightly lower precision results compared to the previously described study (87.4 and 80.2%), but the area of pens and the stocking of animals were much larger [67].

Despite continuous development, 2D imaging technology still has some limitations. It requires appropriate ambient lighting; provides only a flat projection of the animal [68]; is influenced by distance, wavelength, and applied filters [69]; and also requires a contrasting background, e.g., a bright pig against a dark pen wall [15]. Additionally, data extraction from images taken in various environmental conditions leads to inaccurate operation of computer tools for image processing and analysis [70,71].

Three-dimensional (3D) (RGBD) cameras equipped with high-resolution lenses, infrared sensors, or depth sensors with time of flight (ToF) technology give greater possibilities compared to cheaper two-dimensional (2D) ones [35]. ToF technology sends a pulse of infrared light from LED several times per second and records the delay between the pulse and its return. The 3D cameras can operate regardless of the visual light environment, including in total darkness; are unaffected by changing light conditions including changes in contrast and shadow; and are less prone to errors due to occlusion [15,17]. Three-dimensional technology opens the possibility to reconstruct the geometry of animals' bodies and to link abnormal morphological changes to behavioural changes [7,72]. Cameras equipped with ToF technology (kinect cameras) are extremely useful in precision animal husbandry due to their relatively low cost, ability to handle large databases, low power requirements, and ability to adapt to changing light and background conditions [35,73]. However, these types of equipment have a limited distance range (i.e., up to 4.5 m), and the accuracy of the depth data measured by such devices decreases squarely with increasing distance [74].

Despite these limitations, the ability of kinect cameras to detect the movement of individual animals is satisfactory. This was proven in the study of Kim et al. [74], which shows that one Kinect set installed at a height of 3.8 m is sufficient for accurate (94.47% sensitivity) monitoring of an area measuring 2.4 by 2.7 m.

Although video-based recognition of pig behaviour has made significant progress, there are still some unsolved problems [71]. Vision data may require considerable processing and there have been studies on the trade-off between the video image quality and computational processing requirements [75]. Software challenges include detecting individual pigs on the basis of selected features by means of feature selection algorithms [76]. In addition, cameras are susceptible to dust and damage from ammonia, being part of the pig farms' environment, although this can potentially be negated through ingress protection enclosures and maintenance [77].

To ensure accurate and continuous monitoring of individual animals on a modern livestock farm, farmers today need reliable and inexpensive technology [3]. Such systems already exist in cattle breeding and include GEA CowView system, or Lely Qwes. For pigs, the RO-MAIN Smart Cam and eYeNamicTM system is currently the best-known system used to identify and track the pattern of activity in a group of pigs during the growing period. Other newer solutions are still in the realm of research [48].

3.2. Sound-Based Systems

Real-time monitoring can be carried out not only by camera and image analysis but also by microphone and sound analysis [78]. Audio recordings combined with voice analysis and machine learning algorithms are used to detect heat stress and conditions of illness or suffering in animals [5]. Respiratory diseases and/or discomfort associated with poor air quality can cause changes in vocal characteristics and some acoustic signs, such as coughing and sneezing [79]. Monitoring coughing is particularly useful, as it can be easily distinguished from other sounds [80]. Specialised microphones or groups of microphones (microphone arrays) give the ability to distinguish infectious cough from coughs caused by accumulated ammonia or dust, and allow for automated sound source location [81]. Currently available sound analysis systems are so accurate that they can detect and locate respiratory disease outbreaks between individual pens [82]. Several studies have taken up the subject of animal coughing sounds analysis, under laboratory and farm conditions [82–84]. The first study on cough detection in pigs was conducted by Van Hirtum et al. [85] and was followed by additional research on refining algorithms for pigs' cough detection [82,84]. In the study of Ferrari et al. [86], the authors used cough-sound analysis to identify respiratory tract infections in pigs. They found significant differences in several major acoustic parameters, including peak frequency, duration, and time occurring between consecutive coughs in healthy and infected pigs. In turn Exadaktylos et al. [84] proposed a method to identify sick pigs in real time by analysing the sound of coughing, with a recognition accuracy of 85%. Research results on the application of cough algorithms allowed for the development of a commercial tool, the respiratory distress monitor, able to detect infected pigs 2-12 days before the farmer or veterinarian [87]. Van Hirtum and Berckmans [88] suggested that cough sound recognition could be used as a biomarker of air pollution. This thesis was confirmed in a recent study of Wang et al. [79]. Unfortunately sound detection and analysis on pig farms is impeded by the noisy environment of the farm [80].

3.3. Temperature-Based Systems

Temperature meters typically use thermometers embedded in a data logger or a sensor installed in the ear tag or subcutaneous transponder [89–91]. However, as Hartinger et al. [91] reported, this method is characterised by a high degree of variability, which makes it moderately reliable. It has been reported [92] that subcutaneous implanted transponders show a temperature around 1 °C lower compared to rectal measures. Explanation of this discrepancy lies in the position of the transponder, the amount of adipose tissue in the region of measurement, behavioural factors, environmental changes, heat radiation, or blood perfusion in the connective tissue in the implant area. Lohse et al. [92] have shown in their studies that transponders introduced into skeletal muscles show a better correlation with rectal temperature than transponders introduced subcutaneously. An alternative to invasive body temperature measurement with a transponder is to measure the temperature distribution on the body surface by using thermal imaging. Thermography, also known as thermovision, is a method of remote and non-contact assessment of body surface temperature distribution. This technique allows for visualisation of infrared radiation, and thus can obtain information about physiological and pathological processes taking place in the body of humans and animals [93]. The application of thermovision is absolutely non-invasive and has no risk of spreading infections [94,95]. In the diagnosis of farm animals, thermovision is used to investigate injuries and inflammation of the locomotor system, detect infectious diseases, diagnose heat and pregnancy, and monitor welfare and stress levels [93]. Modern thermovision methods make it possible to determine temperature changes both in terms of values and spatial distribution, both in static and dynamic terms. Thermal imaging cameras can produce high resolution images with a temperature accuracy of up to 0.08 °C [96]. Temperature readings depend on the animal's temperature, environmental conditions, and thermoregulation of the peripheral circulatory system. At higher ambient temperatures, thermoregulation results in increased blood flow to the skin tissue, causing an increase of surface temperature [97]. In adult pigs, the temperature measured on the body surface is lower compared to younger animals due to the insulating effect of subcutaneous fat [97]. Skin surfaces behind the ears or near the sternum are hairless and lacking in fat insulation, and therefore better reflect adult body temperature [94].

3.4. Activity-Based Systems

Accelerometers are among the most promising technologies for monitoring livestock behaviour [35]. These instruments are primarily used to measure linear or angular acceleration, and allow for very accurate monitoring and analysis of animal activity: posture and walking patterns, the length of time it spends standing up, delays in lifting, or even antepartum activity in pens, making it possible to detect the onset of labour in sows [98]. Triaxial accelerometers allow for the possibility of collecting three-dimensional information and measure the earth's force by determining the angle of a device (e.g., wireless acceleration sensor nodes placed on the back to record the three-axis movement of pigs) and by measuring the acceleration forces [35]. Several studies have described automatic detection by accelerometers of standing and walking behaviour in pigs [99–101]. Studies of accelerometer readings installed on ear tags have shown that although the ear is virtually independent of the animal's locomotor system, the range of data provided by the device is sufficient to reliably detect early lameness in pigs [102]. Other research indicate that a combination of data from accelerators with data from body temperature sensors allows for automatic detection of infections 1–3 days before using specific diagnostic methods [23]. Thus far, high accuracies have been found for movement and resting behaviours in cows and pigs, while the development of algorithms for analysing feeding and drinking behaviours in pigs is far behind these developed for cattle [103].

Disease, welfare, and productivity problems can have an impact on the feeding patterns of pigs, and may lead to a reduced feeding time or longer intervals between feed intakes [104,105]. RFID at feeding and drinking areas has been used to measure occurrence and duration feeding and drinking behaviour of individual pigs' [20,106,107]. An RFID system requires an RFID transponder (ear tag) and an RFID antenna or receiver (located at the feeder or drinker) [22]. The device is implanted primarily in the ear tags and stores information such as the animal's unique identification number and farm identification number. These data can be used immediately to identify individuals or can be stored and analysed later [35]. Low-frequency RFID is used, for example, in electronic feeders, and makes it possible to dose individually adjusted feed rations [108]. At the same time, data from RFID readers are also used to analyse the frequency of visits to the feeders and the time taken to feed, which allows for early detection of behavioural signs of health problems [107]. Nevertheless, low frequency RFID has two main disadvantages: low reading range (<1 m) and the impossibility to identify more than one animal at a time within the reader's range [104,109]. There is research on the application of high-frequency UHF readers to track multiple animals simultaneously and at longer ranges (3–10 m) [110,111]. Such systems often include anti-collision algorithms to avoid data loss when multiple tags are within the reading range [112,113]. Thanks to its high sensitivity (88.58%) and specificity (98.34%), the HF RFID system performs well in recording feeding visits of pigs [105]. An example of a commercial system used in pig farming, based on UHF-RFID technology, is the SLIDE® system (Simplum Gliwice, Poland). One of the key elements of this system, which distinguishes it from other similar solutions, is the long reading range of approximately 4–5 m, allowing for the automation of the data acquisition process. The system allows not only for monitoring of animals, but also offers the possibility of a very detailed analysis of the individual indicators of each tagged pig, taking into account the factors influencing them and the relationships between individuals [114].

RFID solutions are eagerly used in various sectors, e.g., in factories and warehouses. However, it is important to be aware that the conditions in pig stables, which are mostly based on concrete and reinforced structures, may disrupt the transmission of waves and data. Other disadvantages of RFID-based technology include frequent loss or failure of tags, pain and stress for the animal during tagging, and the need to remove the tag prior to slaughter [35].

An idea worthy of attention is the use of beacons—microcontrollers equipped with BLE (Bluetooth Low Energy) transmitters in pig farming as devices used to identify behaviour and physiological condition. Studies on these systems have been successfully carried on cattle, although the difference in the daily behaviour of cows and pigs is important [115]. Table 1 summarises the advantages and disadvantages of equipment used in precision pig farming.

Table 1. Advantages and disadvantages of equipment used in precision pig farming.

Equipment	Application	Advantages	Disadvantages
2D (RGB) cameras	• Pig identification based on detection of colours in the image [67]. • Automatically detecting pig locomotion [56]. • Automatically detecting pig position and posture [21]. • Monitoring the environment in a pig pen [21]. • Analyse the group behaviour of pigs [21].	• Non-invasive method [67]. • Possibility of individual or group analysis [67]. • Helps to analyse how often animals visit the feeder [48]. • Helps determine time of animal feed intake [48].	• Performance depends on lighting conditions [67]. • Very similar appearances of pigs and varying statuses of the background [41]. • Vulnerability to errors due to occlusion [15]. • May require protective shielding against environmental factors [71]. • Requires filtering to obtain useful information [70].
3D (RGBD) cameras	• Estimation of pig body weights [60]. • Identification of standing pigs [74]. • Tail biting detection [67]. • Automatically detecting pig locomotion [61].	• Non-invasive method [67]. • Possibility of individual or group analysis [67]. • Ability to handle large datasets [35]. • Ability to adapt to variable light and background conditions [74].	• May require protective shielding [71]. • Limited depth measurement range [67]. • Vulnerability to errors due to occlusion [15].
Microphones	• Detection of sickness and heat stress [2]. • Cough detection [2]. • Group behaviour monitoring [2].	• Non-invasive method [2]. • Monitoring of large groups of animals with a single sensor [2]. • Indirect detection of air pollution [85]. • Can be used indoor and outdoor [2].	• Susceptibility to interference from environmental sounds [82]. • Environmental factors may interfere with the functioning of the microphone [82].
Thermometers (implantable device)	• Measurement of the body temperature [93]. • Monitoring of physiological reactions [94].	• Useful for detecting temperature variations [94].	• Invasive method (transponders) [94]. • Moderately reliable method [92].
Infrared thermal imaging (IR)	• Remote temperature measurement [93]. • Temperature monitoring of the whole herd and individual animals [93]. • Examination of musculoskeletal injuries, detection of infectious diseases, diagnosis of oestrus and pregnancy, and monitoring of welfare and stress levels [93].	• Non-invasive method [97]. • Low light imaging capability [97]. • Useful for the analysis of physiological processes [97].	• Environmental factors may interfere with the measurement results [94]. • High equipment cost [94].

Table 1. Cont.

Equipment	Application	Advantages	Disadvantages
Accelerometer	• Pig movement detection and analysis [35]. • Monitoring of pig activity: posture and walking patterns, the length of time it spends standing up [103].	• Non-invasive method (collars) [102]. • Useful for the analysis of movement [35]. • Provides real-time readings [105].	• Invasive method (ear tags) [102]. • Requires external data analysis [103]. • Sensors are fragile and prone to mechanical failure [103]. • Requirement to remove the identifier before slaughter [35].
RFID transponders R	• Pig identification [105]. • Nutrition management [105].	• Helps to analyse how often animals visit the feeder [104]. • Helps determine time of animal feed intake [104]. • Provides real-time readings [105].	• Low range low frequency RFID reading [106]. • Inability to identify more than one animal at a time within the range of the reader [106]. • Requirement to remove the identifier before slaughter [35].

4. Automatic Health and Welfare Monitoring Systems for Pigs—Farmer and Consumer Perspective

Automatic systems for health problem detection in pigs are practical from the scientific point of view and are undoubtedly a common topic in research on detection of health problems on commercial pig farms. The automatic health and behaviour measurement seem to perfectly fit management of big stables; it allows for recording and storing of valuable data, and thus carrying out continuous observations on large numbers of pigs. However, the major problem remains unsolved: How do we convince pig farmers to adopt novel solutions? Are these solutions economically profitable? The economic aspect of health monitoring systems for pigs is still undefined, as no research has been made to compare the outputs and the costs of all inputs used. Moreover, the popularity and possibility of implementation of automatic systems for pig health monitoring is affected by a group of additional factors: effectiveness and reliability of measures, the awareness and technical knowledge of pig farmers, and the housing system and herd size used [116].

Recent years have seen a series of major leaps forward in the technologies and methods of automated animal observation and monitoring, notably under the generic terminology of PLF whose general aim is to increase the efficiency of livestock farming systems while reducing the workload [2].

Commercial pig farms are an aggregation of technical solutions that allow for increased production while limiting the labour. In times of sustainable agriculture and high welfare farming, the commercial pig farms have limited opportunities to follow these trends and to compete with the good reputation of organic pig farms. The modern pig production should meet the public requirements of animal welfare, and the use of automatic systems for welfare monitoring might be a chance of fulfilling these requirements. PLF technologies have the potential to monitor animal health and behaviour in ways that go beyond those of conventional welfare monitoring and observation. PLF allows for the establishment of welfare indicators that are not dependent solely upon periodic human observation and measurement [2]. In addition, PLF provides the opportunity to observe animal behaviour without interference.

Problems that need to be resolved in the near future include inter alia, technical, and scientific issues related to the definition of welfare. In addition, it is necessary to establish effective welfare indicators, improve the reliability of data collected using new observation technologies, develop welfare automation technology dedicated to extensive farming, regulate matters related to the ownership of data generated by PLF technologies in order to effectively manage these data, and further the possibility of exchanging such information between participants in the food chain. To date, however, this potential is both underdeveloped and under-studied [48].

Balzani and Hanlon [117] underline that animal welfare links a variety of perspectives: animal science, veterinary science, public opinion, and the perspective of the farmer, which is often disregarded. Therefore, implementation of automatic welfare monitoring systems should be preceded by research that relate science with farm practice and social reaction. The practical aspects such as economic profitability and consumer feedback should be carefully analysed as this kind of research is lacking. With growing concern for animal welfare, the pressure on the implementation of PLF technology in pig farms will also increase, both from food chain operators and consumers [48,118]. The question is how will consumers feel about the use of novel technologies to assist the farmer with the monitoring of the welfare of pigs. Are they willing to pay higher price for pork produced with this extra supervision? How much educational input is required to create social attitudes that favour novel technologies? The automatic health and welfare monitoring systems in pig stables should be promoted as one of the paths in sustainable animal production, and one of systems that allow for production at high welfare standards. A pork production chain that employs use of systems supporting health and welfare of pigs should be traceable for consumers and promoted by the authorities of each country.

The farmer's attitude to pig welfare monitoring systems is even more important than the social opinion. Because the EU directive (Council Directive 98/58/EC) defining the rules of pig farming charges farmers and stock-people with the responsibility to inspect animals at regular intervals (usually, at least once a day) to verify their wellbeing, farmers' knowledge of the biological and behavioural needs of pigs is key to bringing about changes that promote welfare automation in the pig industry [119,120]. If the farmer is aware that the stable requires changes to produce pigs at high welfare standards, the decision on implementation of novel technological solutions will depend on their economic profitability.

5. Conclusions

This review concluded that automatic health monitoring systems should be widely implemented into the pig industry in order to increase the effectiveness of healthy pig production. The monitoring systems are developing together with the knowledge on effective animal production, requirements considering the level of welfare, and the developments in available technologies. Implementation of novel technologies for health monitoring may be an answer to the demands of society and animal welfare organisations. The research in the field of swine industry deliver a number of practical solutions. The solutions that are already commonly used are automated weight measurement, electronic identification of pigs, automated measurement of feed and water intake, accelerometers (in ear tags) measuring activity, and systems to monitor and manage the microclimate (humidity, temperature, ventilation). The automatic welfare monitoring systems give much more data on the pig herd and are much more developed that these commonly used technologies. However, without proving their economic profitability and defining the reliable possibilities of application, automatic health/welfare monitoring systems will not gain popularity. Research is lacking in this field. Though the social pressure may be the "drive motor" of changes in animal production, the economic aspects of pig heath monitoring systems will decide on the scale of their implementation. Acquisition of data defining the health on pigs in real-time is the key to early disease recognition and disease prevention. The expectations considering the monitoring systems gradually increase, and with time we can observe new technologies that allow us to trace individuals and monitor pigs without stressing the animals.

Summing up the issues discussed in this review, one last thing is still lacking—a system that will allow us to link the health and welfare measures of an individual pig with the data on the quality attributes of obtained pork. Only the careful analysis of this relation would allow for a reliable assessment of the role of pigs' health and welfare in the economical effectiveness of the swine industry, aimed at the production of high-quality meat.

Author Contributions: Conceptualisation, P.R. and M.S.; methodology, P.R., A.L., M.S., E.S. and J.S.-B.; writing—original draft preparation P.R., T.N., M.Z., P.Ś., H.B. and S.N.; writing—review and editing, P.R., A.L. and H.B. All authors have read and agreed to the published version of the manuscript.

Funding: The article was written as part of research conducted by the statutory funding. 506.569.05.00 of the Faculty of Veterinary Medicine and Animal Science University of Life Sciences, Poland.

Institutional Review Board Statement: Not applicable.

Data Availability Statement: No new data were created or analysed in this study. Data sharing is not applicable to this article.

Conflicts of Interest: The authors declare no conflict of interest.

References

1. Ferguson, N.S. Optimization: A paradigm change in nutrition and economic solutions. *Adv. Pork Prod.* **2014**, *25*, 121–127.
2. Berckmans, D. Precision livestock farming technologies for welfare management in intensive livestock systems. *Rev. Sci. Tech. Off. Int. Epizoot.* **2014**, *33*, 189–196. [CrossRef]
3. Vranken, E.; Berckmans, D. Precision livestock farming for pigs. *Anim. Front.* **2017**, *7*, 32–37. [CrossRef]
4. Wolfert, S.; Ge, L.; Verdouw, C.; Bogaardt, M.-J. Big Data in Smart Farming—A review. *Agric. Syst.* **2017**, *153*, 69–80. [CrossRef]
5. Neethirajan, S. The role of sensors, big data and machine learning in modern animal farming. *Sens. Bio-Sens. Res.* **2020**, *29*, 100367. [CrossRef]
6. Dehove, A.; Commault, J.; Petitclerc, M.; Teissier, M.; Mace, J. Economic analysis and costing of animal health: A literature review of methods and importance. *Rev. Sci. Tech. Off. Int. Epizoot.* **2012**, *31*, 605–617. [CrossRef] [PubMed]
7. Markusfeld, O.N. What are Production Diseases, and How do we Manage them? *Acta Vet. Scand.* **2003**, *44*, 1–12. [CrossRef]
8. Tarasiuk, K. Healthy pigs—A satisfied breeder and healthy consumer: Can this be reconciled? In Proceedings of the Healthy Animals—Healthy Food: 1st National Scientific Symposium, Poznan, Poland, 9–10 March 2016; pp. 16–20.
9. Lyons, C.; Bruce, J.; Fowler, V.; English, P. A comparison of productivity and welfare of growing pigs in four intensive systems. *Livest. Prod. Sci.* **1995**, *43*, 265–274. [CrossRef]
10. Klocek, C.; Madej, T. Behaviour of fatteners housing on straw bedding. *Acta Sci. Pol.* **2008**, *7*, 35–44.
11. Botreau, R.; Veissier, I.; Butterworth, A.; Bracke, M.B.M.; Keeling, L.J. Definition of criteria for overall assessment of animal welfare. *Anim. Welf.* **2007**, *16*, 225–228.
12. Bombik, T.; Bombik, E.; Biesiada-Drzazga, B. Animal welfare in terms of evaluation criteria and methods. *Przegląd Hod.* **2013**, *6*, 25–27.
13. Adrion, F.; Kapun, A.; Eckert, F.; Holland, E.-M.; Staiger, M.; Götz, S.; Gallmann, E. Monitoring trough visits of growing-finishing pigs with UHF-RFID. *Comput. Electron. Agric.* **2018**, *144*, 144–153. [CrossRef]
14. Meiszberg, A.M.; Johnson, A.K.; Sadler, L.J.; Carroll, J.A.; Dailey, J.W.; Krebs, N. Drinking behavior in nursery pigs: Determining the accuracy between an automatic water meter versus human observers12. *J. Anim. Sci.* **2009**, *87*, 4173–4180. [CrossRef] [PubMed]
15. Wurtz, K.; Camerlink, I.; D'Eath, R.B.; Fernández, A.P.; Norton, T.; Steibel, J.; Siegford, J. Recording behaviour of indoor-housed farm animals automatically using machine vision technology: A systematic review. *PLoS ONE* **2019**, *14*, e0226669. [CrossRef] [PubMed]
16. Lind, N.M.; Vinther, M.; Hemmingsen, R.P.; Hansen, A.K. Validation of a digital video tracking system for recording pig locomotor behaviour. *J. Neurosci. Methods* **2005**, *143*, 123–132. [CrossRef] [PubMed]
17. Lee, J.; Jin, L.; Park, D.; Chung, Y. Automatic Recognition of Aggressive Behavior in Pigs Using a Kinect Depth Sensor. *Sensors* **2016**, *16*, 631. [CrossRef] [PubMed]
18. Shao, B.; Xin, H. A real-time computer vision assessment and control of thermal comfort for group-housed pigs. *Comput. Electron. Agric.* **2008**, *62*, 15–21. [CrossRef]
19. Madsen, T.N.; Kristensen, A.R. A model for monitoring the condition of young pigs by their drinking behaviour. *Comput. Electron. Agric.* **2005**, *48*, 138–154. [CrossRef]
20. Fernández, J.; Fabrega, E.; Soler, J.; Tibau, J.; Ruiz, J.L.; Puigvert, X.; Manteca, X. Feeding strategy in group-housed growing pigs of four different breeds. *Appl. Anim. Behav. Sci.* **2011**, *134*, 109–120. [CrossRef]
21. Nasirahmadi, A.; Richter, U.; Hensel, O.; Edwards, S.; Sturm, B. Using machine vision for investigation of changes in pig group lying patterns. *Comput. Electron. Agric.* **2015**, *119*, 184–190. [CrossRef]

22. Matthews, S.G.; Miller, A.L.; Clapp, J.; Plötz, T.; Kyriazakis, I. Early detection of health and welfare compromises through automated detection of behavioural changes in pigs. *Vet. J.* **2016**, *217*, 43–51. [CrossRef] [PubMed]

23. Avilés, M.M.; Carrión, E.F.; García-Baones, J.M.L.; Sánchez-Vizcaíno, J.M. Early Detection of Infection in Pigs through an Online Monitoring System. *Transbound. Emerg. Dis.* **2015**, *64*, 364–373. [CrossRef]

24. Shahbandeh, M. Global Number of Pigs 2012–2020. Available online: https://www.statista.com/statistics/263964/number-of-pigs-in-selected-countries/ (accessed on 24 February 2020).

25. OECD. Meat Consumption. Available online: https://data.oecd.org/agroutput/meat-consumption.htm (accessed on 11 March 2021).

26. Statista. Projected Pig Meat Consumption Worldwide from 2020 to 2029 (in 1000 Metric Kilotons). Available online: https://www.statista.com/statistics/739879/pork-consumption-worldwide/ (accessed on 3 December 2020).

27. Pietrosemoli, S.; Tang, C. Animal Welfare and Production Challenges Associated with Pasture Pig Systems: A Review. *Agric.* **2020**, *10*, 223. [CrossRef]

28. Kolacz, R. Animal welfare control in the food safety supervision system. In Proceedings of the Healthy Animals—Healthy Food: 1st National Scientific Symposium, Poznan, Poland, 9–10 March 2016; pp. 6–11.

29. Blokhuis, H.J.; Miele, M.; Veissier, I.; Jones, B. *Improving Farm Animal Welfare: Science and Society Working Together: The Welfare Quality Approach*; Wageningen Academic Publishers: Wageningen, The Netherlands, 2013; p. 232.

30. Welfare, Q. *Welfare Quality® Assessment Protocol for Poultry (Broilers, Laying Hens)*; Welfare Quality® Consortium: Lelystad, The Netherlands, 2009.

31. Welfare, Q. *Welfare Quality® Assessment Protocol for Pigs (Sows and Piglets, Growing and Finishing Pigs)*; Welfare Quality® Consortium: Lelystad, The Netherlands, 2009.

32. Welfare, Q. *Welfare Quality® Assessment Protocol for Cattle (Fattening Cattle, Dairy Cows, Veal Calves)*; Welfare Quality® Consortium: Lelystad, The Netherlands, 2009.

33. FAWC. *Second Report on Priorities for Research and Development in Farm Animal Welfare*; Department of Environment, Food and Rural Affairs, Farm Animal Welfare Council: London, UK, 1993.

34. Temple, D.; Dalmau, A.; de la Torre, J.L.R.; Manteca, X.; Velarde, A. Application of the Welfare Quality® protocol to assess growing pigs kept under intensive conditions in Spain. *J. Vet. Behav.* **2011**, *6*, 138–149. [CrossRef]

35. Benjamin, M.; Yik, S. Precision Livestock Farming in Swine Welfare: A Review for Swine Practitioners. *Animals* **2019**, *9*, 133. [CrossRef] [PubMed]

36. Hemsworth, P.H.; Coleman, G.J.; Barnett, J.L.; Borg, S. Relationships between human-animal interactions and productivity of commercial dairy cows. *J. Anim. Sci.* **2000**, *78*, 2821–2831. [CrossRef]

37. Jääskeläinen, T.; Kauppinen, T.; Vesala, K.M.; Valros, A. Relationships between pig welfare, productivity and farmer disposition. *Anim. Welf.* **2014**, *23*, 435–443. [CrossRef]

38. Cowton, J.; Kyriazakis, I.; Bacardit, J. Automated Individual Pig Localisation, Tracking and Behaviour Metric Extraction Using Deep Learning. *IEEE Access* **2019**, *7*, 108049–108060. [CrossRef]

39. Psota, E.T.; Mittek, M.; Pérez, L.C.; Schmidt, T.; Mote, B. Multi-Pig Part Detection and Association with a Fully-Convolutional Network. *Sensors* **2019**, *19*, 852. [CrossRef]

40. Nasirahmadi, A.; Sturm, B.; Edwards, S.; Jeppsson, K.-H.; Olsson, A.-C.; Müller, S.; Hensel, O. Deep Learning and Machine Vision Approaches for Posture Detection of Individual Pigs. *Sensors* **2019**, *19*, 3738. [CrossRef] [PubMed]

41. Zhang, L.; Gray, H.; Ye, X.; Collins, L.; Allinson, N. Automatic Individual Pig Detection and Tracking in Pig Farms. *Sensors* **2019**, *19*, 1188. [CrossRef] [PubMed]

42. Bonneau, M.; Vayssade, J.-A.; Troupe, W.; Arquet, R. Outdoor animal tracking combining neural network and time-lapse cameras. *Comput. Electron. Agric.* **2020**, *168*, 105150. [CrossRef]

43. Marsot, M.; Mei, J.; Shan, X.; Ye, L.; Feng, P.; Yan, X.; Li, C.; Zhao, Y. An adaptive pig face recognition approach using Convolutional Neural Networks. *Comput. Electron. Agric.* **2020**, *173*, 105386. [CrossRef]

44. Le Bellego, L.; Noblet, J. Performance and utilization of dietary energy and amino acids in piglets fed low protein diets. *Livest. Prod. Sci.* **2002**, *76*, 45–58. [CrossRef]

45. Quiniou, N.; Renaudeau, D.; Collin, A.; Noblet, J. Influence of high ambient temperatures and physiological stage on feeding behaviour of pigs. *Prod. Anim.* **2000**, *13*, 233–245.

46. Warriss, P.D.; Pope, S.J.; Brown, S.N.; Wilkins, L.J.; Knowles, T.G. Estimating the body temperature of groups of pigs by thermal imaging. *Vet. Rec.* **2006**, *158*, 331–334. [CrossRef] [PubMed]

47. Kyriazakis, I.; Tolkamp, B.; Hutchings, M. Towards a functional explanation for the occurrence of anorexia during parasitic infections. *Anim. Behav.* **1998**, *56*, 265–274. [CrossRef]

48. Buller, H.; Blokhuis, H.; Lokhorst, K.; Silberberg, M.; Veissier, I. Animal Welfare Management in a Digital World. *Animals* **2020**, *10*, 1779. [CrossRef] [PubMed]

49. Nóbrega, L.; Gonçalves, P.; Pedreiras, P.; Pereira, J. An IoT-Based Solution for Intelligent Farming. *Sensors* **2019**, *19*, 603. [CrossRef] [PubMed]

50. Shi, X.; An, X.; Zhao, Q.; Liu, H.; Xia, L.; Sun, X.; Guo, Y. State-of-the-Art Internet of Things in Protected Agriculture. *Sensors* **2019**, *19*, 1833. [CrossRef]

51. Nasirahmadi, A.; Hensel, O.; Edwards, S.A.; Sturm, B. A new approach for categorizing pig lying behaviour based on a Delaunay triangulation method. *Animals* **2017**, *11*, 131–139. [CrossRef]

52. Alexandratos, N.; Bruinsma, J. *World Agriculture towards 2030/2050: The 2012 Revision CEMA, 2015. Towards a New Strategic Agenda for the Common Agricultural Policy (CAP) after 2020*; Agricultural Development Economics Division (FAO): Roma, Italy, 2006.
53. Andretta, I.; Pomar, C.; Rivest, J.; Pomar, J.; Lovatto, P.A.; Neto, J.R. The impact of feeding growing–finishing pigs with daily tailored diets using precision feeding techniques on animal performance, nutrient utilization, and body and carcass composition1. *J. Anim. Sci.* **2014**, *92*, 3925–3936. [CrossRef] [PubMed]
54. Report, M.R. *Precision Farming/Agriculture Market Size, Share & Trends Analysis Report by Offering (Hardware, Software, Services), by Application (Yield Monitoring, Irrigation Management) and Segment Forecasts 2019–2025*; MarketsandMarkets: Northbrook, IL, USA, 2019.
55. da Costa, A.C.P.L.; Ismayilova, G.; Borgonovo, F.; Viazzi, S.; Berckmans, D.; Guarino, M. Image-processing technique to measure pig activity in response to climatic variation in a pig barn. *Anim. Prod. Sci.* **2014**, *54*, 1075–1083. [CrossRef]
56. Kashiha, M.A.; Bahr, C.; Ott, S.; Moons, C.P.; Niewold, T.A.; Tuyttens, F.; Berckmans, D. Automatic monitoring of pig locomotion using image analysis. *Livest. Sci.* **2014**, *159*, 141–148. [CrossRef]
57. Gronskyte, R.; Clemmensen, L.H.; Hviid, M.S.; Kulahci, M. Monitoring pig movement at the slaughterhouse using optical flow and modified angular histograms. *Biosyst. Eng.* **2016**, *141*, 19–30. [CrossRef]
58. Kashiha, M.; Bahr, C.; Ott, S.; Moons, C.P.; Niewold, T.A.; Ödberg, F.O.; Berckmans, D. Automatic weight estimation of individual pigs using image analysis. *Comput. Electron. Agric.* **2014**, *107*, 38–44. [CrossRef]
59. Shi, C.; Teng, G.; Li, Z. An approach of pig weight estimation using binocular stereo system based on LabVIEW. *Comput. Electron. Agric.* **2016**, *129*, 37–43. [CrossRef]
60. Pezzuolo, A.; Milani, V.; Zhu, D.; Guo, H.; Guercini, S.; Marinello, F. On-Barn Pig Weight Estimation Based on Body Measurements by Structure-from-Motion (SfM). *Sensors* **2018**, *18*, 3603. [CrossRef] [PubMed]
61. Stavrakakis, S.; Guy, J.; Syranidis, I.; Johnson, G.; Edwards, S. Pre-clinical and clinical walking kinematics in female breeding pigs with lameness: A nested case-control cohort study. *Vet. J.* **2015**, *205*, 38–43. [CrossRef] [PubMed]
62. Guzhva, O.; Ardö, H.; Herlin, A.; Nilsson, M.; Åström, K.; Bergsten, C. Feasibility study for the implementation of an automatic system for the detection of social interactions in the waiting area of automatic milking stations by using a video surveillance system. *Comput. Electron. Agric.* **2016**, *127*, 506–509. [CrossRef]
63. Nasirahmadi, A.; Hensel, O.; Edwards, S.A.; Sturm, B. Automatic detection of mounting behaviours among pigs using image analysis. *Comput. Electron. Agric.* **2016**, *124*, 295–302. [CrossRef]
64. Bewley, J.; Peacock, A.; Lewis, O.; Boyce, R.; Roberts, D.; Coffey, M.; Kenyon, S.; Schutz, M. Potential for Estimation of Body Condition Scores in Dairy Cattle from Digital Images. *J. Dairy Sci.* **2008**, *91*, 3439–3453. [CrossRef] [PubMed]
65. Chen, C.; Zhu, W.; Steibel, J.; Siegford, J.; Han, J.; Norton, T. Recognition of feeding behaviour of pigs and determination of feeding time of each pig by a video-based deep learning method. *Comput. Electron. Agric.* **2020**, *176*, 105642. [CrossRef]
66. Nasirahmadi, A.; Edwards, S.A.; Sturm, B. Implementation of machine vision for detecting behaviour of cattle and pigs. *Livest. Sci.* **2017**, *202*, 25–38. [CrossRef]
67. Riekert, M.; Klein, A.; Adrion, F.; Hoffmann, C.; Gallmann, E. Automatically detecting pig position and posture by 2D camera imaging and deep learning. *Comput. Electron. Agric.* **2020**, *174*, 105391. [CrossRef]
68. Stajnko, D.; Brus, M.; Hočevar, M. Estimation of bull live weight through thermographically measured body dimensions. *Comput. Electron. Agric.* **2008**, *61*, 233–240. [CrossRef]
69. Pezzuolo, A.; Guarino, M.; Sartori, L.; Marinello, F. A Feasibility Study on the Use of a Structured Light Depth-Camera for Three-Dimensional Body Measurements of Dairy Cows in Free-Stall Barns. *Sensors* **2018**, *18*, 673. [CrossRef] [PubMed]
70. Viazzi, S.; Bahr, C.; Van Hertem, T.; Schlageter-Tello, A.; Romanini, C.; Halachmi, I.; Lokhorst, C.; Berckmans, D. Comparison of a three-dimensional and two-dimensional camera system for automated measurement of back posture in dairy cows. *Comput. Electron. Agric.* **2014**, *100*, 139–147. [CrossRef]
71. Yang, Q.; Xiao, D. A review of video-based pig behavior recognition. *Appl. Anim. Behav. Sci.* **2020**, *233*, 105146. [CrossRef]
72. Le Cozler, Y.; Allain, C.; Caillot, A.; Delouard, J.; Delattre, L.; Luginbuhl, T.; Faverdin, P. High-precision scanning system for complete 3D cow body shape imaging and analysis of morphological traits. *Comput. Electron. Agric.* **2019**, *157*, 447–453. [CrossRef]
73. Spoliansky, R.; Edan, Y.; Parmet, Y.; Halachmi, I. Development of automatic body condition scoring using a low-cost 3-dimensional Kinect camera. *J. Dairy Sci.* **2016**, *99*, 7714–7725. [CrossRef]
74. Kim, J.; Chung, Y.; Choi, Y.; Sa, J.; Kim, H.; Chung, Y.; Park, D.; Kim, H. Depth-Based Detection of Standing-Pigs in Moving Noise Environments. *Sensors* **2017**, *17*, 2757. [CrossRef]
75. Chung, Y.; Kim, H.; Lee, H.; Park, D.; Jeon, T.; Chang, H.H. A Cost-Effective Pigsty Monitoring System Based on a Video Sensor. *KSII Trans. Internet Inf. Syst.* **2014**, *8*, 1481–1498. [CrossRef]
76. Sa, J.; Han, S.; Lee, S.; Kim, H.; Lee, S.; Chung, Y.; Park, D. Image Segmentation of Adjoining Pigs Using Spatio-Temporal Information. *KIPS Trans. Softw. Data Eng.* **2015**, *4*, 473–478. [CrossRef]
77. Ahrendt, P.; Gregersen, T.; Karstoft, H. Development of a real-time computer vision system for tracking loose-housed pigs. *Comput. Electron. Agric.* **2011**, *76*, 169–174. [CrossRef]
78. Berckmans, D. General introduction to precision livestock farming. *Anim. Front.* **2017**, *7*, 6–11. [CrossRef]
79. Wang, X.; Zhao, X.; He, Y.; Wang, K. Cough sound analysis to assess air quality in commercial weaner barns. *Comput. Electron. Agric.* **2019**, *160*, 8–13. [CrossRef]
80. Chung, K.F.; Pavord, I.D. Prevalence, pathogenesis, and causes of chronic cough. *Lancet* **2008**, *371*, 1364–1374. [CrossRef]

81. Hennecke, M.; Plötz, T.; Fink, G.A.; Schmalenstroer, J.; Häb-Umbach, R. A hierarchical approach to unsupervised shape calibration of microphone array networks. In Proceedings of the 2009 IEEE/SP 15th Workshop on Statistical Signal Processing, Cardiff, UK, 31 August–3 September 2009; Volumes 1 and 2, pp. 257–260. [CrossRef]

82. Chung, Y.; Oh, S.; Lee, J.; Park, D.; Chang, H.-H.; Kim, S. Automatic Detection and Recognition of Pig Wasting Diseases Using Sound Data in Audio Surveillance Systems. *Sensors* **2013**, *13*, 12929–12942. [CrossRef]

83. Manteuffel, G.; Puppe, B.; Schön, P.C. Vocalization of farm animals as a measure of welfare. *Appl. Anim. Behav. Sci.* **2004**, *88*, 163–182. [CrossRef]

84. Exadaktylos, V.; Silva, M.; Aerts, J.-M.; Taylor, C.J.; Berckmans, D. Real-time recognition of sick pig cough sounds. *Comput. Electron. Agric.* **2008**, *63*, 207–214. [CrossRef]

85. van Hirtum, A.; Aerts, J.-M.; Berckmans, D.; Moreaux, B.; Gustin, P. On-line cough recognizer system. *J. Acoust. Soc. Am.* **1999**, *106*, 2191. [CrossRef]

86. Ferrari, S.; Silva, M.; Guarino, M.; Aerts, J.M.; Berckmans, D. Cough sound analysis to identify respiratory infection in pigs. *Comput. Electron. Agric.* **2008**, *64*, 318–325. [CrossRef]

87. Berckmans, D.; Hemeryck, M.; Berckmans, D.; Vranken, E.; van Waterschoot, T. Animal sound … talks! Real-time sound analysis for health monitoring in livestock. In Proceedings of the International Symposium on Animal Environment and Welfare, Chongqing, China, 23–26 October 2015; pp. 215–222.

88. van Hirtum, A.; Berckmans, D. Objective cough-sound recognition as a biomarker for aerial factors. *Trans. ASAE* **2004**, *47*, 351–356. [CrossRef]

89. Dunney, S.; Shuster, D.; Heaney, K.; Payton, A. Use of an implantable device to monitor body temperature in research swine. *Contemp. Top. Lab. Anim. Sci.* **1997**, *36*, 62.

90. Goodwin, S. Comparison of Body Temperatures of Goats, Horses, and Sheep Measured with a Tympanic Infrared Thermometer, an Implantable Microchip Transponder, and a Rectal Thermometer. *Contemp. Top. Lab. Anim. Sci.* **1998**, *37*, 51–55. [PubMed]

91. Hartinger, J.; Külbs, D.; Volkers, P.; Cussler, K. Suitability of temperature-sensitive transponders to measure body temperature during animal experiments required for regulatory tests. *Altex-Altern. Tierexp.* **2003**, *20*, 65–70.

92. Lohse, L.; Uttenthal, A.; Enoe, C.; Nielsen, J. A study on the applicability of implantable microchip transponders for body temperature measurements in pigs. *Acta Vet. Scand.* **2010**, *52*, 29. [CrossRef] [PubMed]

93. Racewicz, P.; Sobek, J.; Majewski, M.; Różańska-Zawieja, J. The use of thermal imaging measurements in dairy cow herds. *Rocz. Nauk. Pol. Towar. Zootech.* **2018**, *14*, 55–69. [CrossRef]

94. Soerensen, D.D.; Pedersen, L.J. Infrared skin temperature measurements for monitoring health in pigs: A review. *Acta Vet. Scand.* **2015**, *57*, 5. [CrossRef]

95. Soerensen, D.D.; Clausen, S.; Mercer, J.B.; Pedersen, L.J. Determining the emissivity of pig skin for accurate infrared thermography. *Comput. Electron. Agric.* **2014**, *109*, 52–58. [CrossRef]

96. Costa, L.N.; Stelletta, C.; Cannizzo, C.; Gianesella, M.; Fiego, P.L.; Morgante, M. The use of thermography on the slaughter-line for the assessment of pork and raw ham quality. *Ital. J. Anim. Sci.* **2007**, *6*, 704–706. [CrossRef]

97. Sellier, N.; Guettier, E.; Staub, C. A Review of Methods to Measure Animal Body Temperature in Precision Farming. *Am. J. Agric. Sci. Technol.* **2014**. [CrossRef]

98. Bewley, J.; Schutz, M. Assessing the potential economic value of an automated temperature monitoring system using stochastic simulation. *J. Dairy Sci.* **2010**, *93*, 190.

99. Ringgenberg, N.; Bergeron, R.; Devillers, N. Validation of accelerometers to automatically record sow postures and stepping behaviour. *Appl. Anim. Behav. Sci.* **2010**, *128*, 37–44. [CrossRef]

100. Cornou, C.; Lundbye-Christensen, S. Classification of sows' activity types from acceleration patterns using univariate and multivariate models. *Comput. Electron. Agric.* **2010**, *72*, 53–60. [CrossRef]

101. Escalante, H.J.; Rodriguez, S.V.; Cordero, J.; Kristensen, A.R.; Cornou, C. Sow-activity classification from acceleration patterns: A machine learning approach. *Comput. Electron. Agric.* **2013**, *93*, 17–26. [CrossRef]

102. Traulsen, I.; Scheel, C.; Auer, W.; Burfeind, O.; Krieter, J. Using Acceleration Data to Automatically Detect the Onset of Farrowing in Sows. *Sensors* **2018**, *18*, 170. [CrossRef]

103. Chapa, J.M.; Maschat, K.; Iwersen, M.; Baumgartner, J.; Drillich, M. Accelerometer systems as tools for health and welfare assessment in cattle and pigs—A review. *Behav. Process.* **2020**, *181*, 104262. [CrossRef]

104. Brown-Brandl, T.; Rohrer, G.; Eigenberg, R. Analysis of feeding behavior of group housed growing–finishing pigs. *Comput. Electron. Agric.* **2013**, *96*, 246–252. [CrossRef]

105. Maselyne, J.; Saeys, W.; de Ketelaere, B.; Mertens, K.; Vangeyte, J.; Hessel, E.F.; Millet, S.; van Nuffel, A. Validation of a High Frequency Radio Frequency Identification (HF RFID) system for registering feeding patterns of growing-finishing pigs. *Comput. Electron. Agric.* **2014**, *102*, 10–18. [CrossRef]

106. Andersen, H.-L.; Dybkjær, L.; Herskin, M. Growing pigs' drinking behaviour: Number of visits, duration, water intake and diurnal variation. *Animals* **2014**, *8*, 1881–1888. [CrossRef]

107. Maselyne, J.; Saeys, W.; van Nuffel, A. Review: Quantifying animal feeding behaviour with a focus on pigs. *Physiol. Behav.* **2015**, *138*, 37–51. [CrossRef] [PubMed]

108. Cornou, C.; Vinther, J.; Kristensen, A.R. Automatic detection of oestrus and health disorders using data from electronic sow feeders. *Livest. Sci.* **2008**, *118*, 262–271. [CrossRef]

109. Adrion, F.; Kapun, A.; Holland, E.-M.; Staiger, M.; Löb, P.; Gallmann, E. Novel approach to determine the influence of pig and cattle ears on the performance of passive UHF-RFID ear tags. *Comput. Electron. Agric.* **2017**, *140*, 168–179. [CrossRef]
110. Foerschner, A.; Adrion, F.; Gallmann, E. Practical test and evaluation of optimized UHF ear tags for behavior monitoring of fattening pigs. In Proceedings of the 10th International Livestock Environment Symposium, Omaha, NE, USA, 25–27 September 2018; pp. 1–8.
111. Sales, G.; Green, A.; Gates, R.; Brown-Brandl, T.; Eigenberg, R. Quantifying detection performance of a passive low-frequency RFID system in an environmental preference chamber for laying hens. *Comput. Electron. Agric.* **2015**, *114*, 261–268. [CrossRef]
112. Chawla, V.; Ha, D.S. An overview of passive RFID. *IEEE Commun. Mag.* **2007**, *45*, 11–17. [CrossRef]
113. Ruiz-Garcia, L.; Lunadei, L. The role of RFID in agriculture: Applications, limitations and challenges. *Comput. Electron. Agric.* **2011**, *79*, 42–50. [CrossRef]
114. Małopolska, M.; Schwarz TTuz, R.; Nowicki, J.; Pawlak, M.; Cylupa, M. The use of an automatic data acquisition system (RFID) to analyse the impact of parental genotypes of pigs on the fattening and slaughter traits of their offspring. *Sci. Ann. Pol. Soc. Anim. Prod.* **2014**, *10*, 51–66.
115. Koszela, K.; Mueller, W.; Otrząsek, J.; Łukomski, M.; Kujawa, S. Beacon in Information System as Way of Supporting Identification of Cattle Behavior. *Appl. Sci.* **2021**, *11*, 1062. [CrossRef]
116. Åkerfeldt, M.P.; Nihlstrand, J.; Neil, M.; Lundeheim, N.; Andersson, H.K.; Wallenbeck, A. Chicory and red clover silage in diets to finishing pigs—influence on performance, time budgets and social interactions. *Org. Agric.* **2019**, *9*, 127–138. [CrossRef]
117. Balzani, A.; Hanlon, A. Factors that Influence Farmers' Views on Farm Animal Welfare: A Semi-Systematic Review and Thematic Analysis. *Animals* **2020**, *10*, 1524. [CrossRef]
118. Veissier, I.; Kling-Eveillard, F.; Richard, M.M.; Silberberg, M.; de Boyer des Roches, A.; Terlouw, C.; Ledoux, D.; Meunier, B.; Hostiou, N. Élevage de précision et bien-être en élevage: La révolution numérique de l'agriculture permettra-t-elle de prendre en compte les besoins des animaux et des éleveurs. *INRA Prod. Anim.* **2019**, *32*, 281–290. [CrossRef]
119. Bock, B.B.; Van Huik, M.M. Animal welfare: The attitudes and behaviour of European pig farmers. *Br. Food J.* **2007**, *109*, 931–944. [CrossRef]
120. European Union. *Council Directive 2008/120/EC of 18 December 2008 Laying Down Minimum Standards for the Protection of Pigs (Codified Version)*; L47 2009, 5–13; European Union: Brussels, Belgium, 2009.

Article

Genetic Parameters of Effort and Recovery in Sport Horses Assessed with Infrared Thermography

Ester Bartolomé *[iD], Davinia Isabel Perdomo-González [iD], María José Sánchez-Guerrero and Mercedes Valera [iD]

Departamento de Agronomía, ETSIA, Universidad de Sevilla, Utrera Rd. Km 1, 41013 Sevilla, Spain; daviniapergon@gmail.com (D.I.P.-G.); v32sagum@gmail.com (M.J.S.-G.); mvalera@us.es (M.V.)
* Correspondence: ebartolome@us.es

Simple Summary: The way a horse activates (effort phase-EP) and recovers (recovery phase-RP) during a sport event can affect its sport performance. The aim of this manuscript was to test horses' adaptation to sport performance and its genetic basis, using eye temperature assessed with infrared thermography. EP and RP were measured in 495 Spanish Sport Horses, during a performance test, considering sex (2) and genetic lines (5) as fixed effects. The ranking position obtained on the official sport competition celebrated the day after the performance test was also collected. Differences in variables due to genetic line and sex effects were found, showing that, regardless of the genetic line, stallions tended to recover better than mares after the sport test developed. High positive correlations were found between EP and RP for both fixed effects, so that, the higher the EP, the higher the RP. However, for the ranking position, a low negative correlation was found, so that the higher the eye temperature increase, the better the position. Heritabilities showed medium–high values with a medium positive genetic correlation between them. Thus, breed origins and sex influence horses' effort and recovery during sport performance, showing a genetic basis adequate for selection.

Abstract: The way a horse activates (effort phase-EP) and recovers (recovery phase-RP) during a sport event can affect its sport performance. The aim of this manuscript was to test horses' adaptation to sport performance and its genetic basis, using eye temperature assessed with infrared thermography. EP and RP were measured in 495 Spanish Sport Horses, during a performance test, considering sex (2) and genetic lines (5) as fixed effects. The ranking position obtained on an official sport competition was also collected. Differences in variables due to genetic line and sex effects were found, showing that, regardless of the genetic line, stallions tended to recover better than mares after the sport test developed. High positive intra-class correlations ($p < 0.001$) were found between EP and RP for both fixed effects, so that the higher the EP, the higher the RP. However, for the ranking position, a low negative correlation ($p < 0.01$) was found, so that the higher the eye temperature increase, the better the position. Heritabilities showed medium–high values with a medium positive genetic correlation between them. Thus, breed origins and sex influence horses' effort and recovery during sport performance, showing a genetic basis adequate for selection.

Keywords: eye caruncle temperature; Spanish Sport Horse; performance test; genetic lines; heritability; infrared thermography

Citation: Bartolomé, E.; Perdomo-González, D.I.; Sánchez-Guerrero, M.J.; Valera, M. Genetic Parameters of Effort and Recovery in Sport Horses Assessed with Infrared Thermography. *Animals* **2021**, *11*, 832. https://doi.org/10.3390/ani11030832

Academic Editors: Melissa Hempstead and Danila Marini

Received: 17 February 2021
Accepted: 12 March 2021
Published: 16 March 2021

Publisher's Note: MDPI stays neutral with regard to jurisdictional claims in published maps and institutional affiliations.

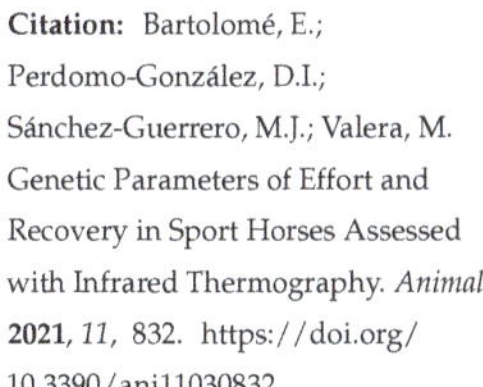

1. Introduction

During sport events, horse metabolism produces a large increase in flux of substrate (as glycogen or glucose) to increase fuel availability, maintain acid–base balance within acceptable limits, and limit body temperature. The supply of these substrates is controlled by the hormonal responses to exercise, which include a reduction in blood insulin concentration and increases in blood catecholamine, cortisol, and glucagon concentrations [1]. The increase in catecholamine (also related with a stress response) enhances an increase in heart rate and a splenic contraction, leading into major blood flow into the central circulation [2].

How effective this system is would be an indicator of a horse's fitness and of a horse's probability to excel at equestrian events, taking into consideration that the stress response developed depends also on the type of exercise performed [2,3].

Monitoring these physiological changes has been assessed with heart rate (HR) response, blood lactate concentrations, oxygen uptake in relation to exercise intensity [4,5], salivary, hair and blood cortisol [3,6–8], or immune cell proliferation [9–11]. However, they are all quite difficult measures to be assessed during regular equestrian competitions as riders and owners are unwilling to allow the experimental procedures to impact on the animal and affect their sport results. Furthermore, these methods either require laboratory conditions (oxygen uptake, salivary, and hair cortisol), a blood sample (lactate, blood cortisol, and immune cells), or to touch and distract the horse during the assessment (heart rate).

Lately, some non-invasive tools for physiological changes measurement, such as eye temperature assessed with infrared thermography (IRT) technology, have shown a great potential to assess physiological changes in horses during equestrian competitions and it is also correlated with traditional invasive physiological measures, such as lactate concentration in blood [12–14]. This tool can detect small changes in temperature by measuring the radiated electromagnetic energy produced by the horse as a physiological response to exercise. Thus, the increase of catecholamine concentrations, in addition to blood flow responses developed during a sport equestrian event, will produce changes in heat production and heat loss in small areas around the medial posterior palpebral border of the lower eyelid and the lacrimal caruncle. Thus, as both of them have rich capillary beds innervated by the sympathetic system, they respond to changes in blood flow [15].

In general, two physiological phases could be described around a competition event: an "effort phase" (EP), which is the physiological difference between the moment when the animal is calm and the moment when the animal develops the sport performance and reaches the peak of the exercise's intensity; and a "recovery phase" (RP), comprising the physiological difference from this sport peak to the moment when body homeostasis is completely restored [13].

Furthermore, as occurs with heart rate or lactate levels [16], eye temperature assessed with IRT is not only dependent on the aerobic capacity of the horse, but on inherited parameters, such as breed or age [13,17,18]. This issue takes a greater importance in crossbreeds, such as the Spanish Sport Horse (CDE) horse. The CDE is a recent composite breed formed from crosses with other horse breeds. It was founded in 2002 and the main goal of its breeding program is "to obtain a horse with a good functional conformation, temperament and health, able to attain a high performance at either national or international sports events in which it participates" [19].

Thus, taking into consideration that previous authors [20,21] have found differences between horse breeds due to the physiological adaptation to sports, the physiological response of these CDE crossbred animals during effort and recovery from exercise would be conditioned by the physiological response of the breeds included on their pedigree.

The aims of this study were, first, to test the influence of different environmental effects on effort and recovery by CDE during a performance test, assessed with IRT, and second, to estimate the genetic parameters and genetic correlations between these variables to test their suitability for genetic selection.

2. Materials and Methods

2.1. Animals

Measurements were obtained from 495 animals (332 stallions and 163 mares) of the Spanish Sport Horse (CDE) breed, aged from 2 to 13 years old. Animals were selected according to their participation in the official show jumping competition celebrated the day after the study. All owners were contacted and informed about the performance test that would be used with their animals. Only those animals whose owners agreed to participate in the study were used for the analyses and were asked to arrive at the competition center one day before the official show jumping competition in order to participate in the performance test. All experiments were performed in accordance with Directive 2010/63/EU guidelines.

According to the composite nature of the CDE, many different breeds were included in the studbook as relatives. In order to consider the influence of the genetic composition on the effort and recovery of the CDE studied, 5 genetic lines were assessed according to their origin. An animal belonged to one of them when the majority of its ancestors was from a breed in this line (genetic contribution $\geq 50\%$) [12]. In total, 14.6% (72 horses) belonged to a 'German' (GE) genetic line (L1), created according to the country of origin of the breeds that conformed it (Holsteiner, Hanoverian, Westphalian, Oldenburger, and Trakehner); 15.2% (75 horses) belonged to a 'Thoroughbred' (TH) genetic line (L2); 28.9% (142 horses) belonged to a 'Trotter' (TR) genetic line (L3); and 24.9% (124 horses) belonged to a 'Pura Raza Española' (PRE) genetic line (L4). In addition, 16.7% (82) of the CDE belonged to a fifth genetic line, referred to as 'Other Breeds' (OT; L5), which included those horses with ancestors from other horse breeds and from those CDE with a minority influence (<50%) of the already classified breeds (Table 1).

Table 1. Description of the Spanish Sport Horse (CDE) genetic lines.

Genetic Line	Name	Description	N (%)
L1	German	More than 50% of the CDE ancestors belonged to German horse breeds: Holsteiner, Hanoverian, Westphalian, Oldenburger, or Trakehner.	72 (14.6%)
L2	Thoroughbred	More than 50% of the CDE ancestors belonged to Thoroughbred breed.	75 (15.2%)
L3	Trotter	More than 50% of the CDE ancestors belonged to trotter horse breeds	142 (28.9%)
L4	Pura Raza Española	More than 50% of the CDE ancestors belonged to Pura Raza Española breed	124 (24.9%)
L5	Other Breeds	Included CDE horses with more than 50% of their ancestors from other sport horse breeds (KWPN, Zangersheide, etc.) and CDE horses with less than 50% of their ancestors from the already classified breeds (TR, PRE, TH or GE).	82 (16.7%)

2.2. Study Design

A retrospective cohort study was used with a performance test specifically developed for this study, in order to obtain performance information under the same conditions. The test was held one day before an official show jumping competition that was performed at the same equestrian center, on the competition arena (30 m^2) with audience in the stands, in order to simulate regular competition environmental conditions. During this specific performance test, the horse developed a 2-min (approximately) routine, beginning with walk, then trot, and followed by gallop movements all around the training arena. Then, the horse finished the performance test by jumping over three simple obstacles 1.00 m, 1.10 m, and 1.25 m high, respectively, all a crossed fence, with an average speed of 250 m/min. All horses analyzed in this study performed the exercise with professional riders (all with Galope 6 or more [22]).

Performance test exercises were held at the same equestrian center, during different show jumping competitions held between 2014 and 2018 in September, hence sharing similar environmental and housing conditions. Temperature ranged from 14 to 24 °C and relative humidity between 40% and 50%. However, for the horses to adapt to the new environmental conditions of the center before the performance test, owners were asked to arrive at the equestrian center at least one day before the performance test (two days before the official show jumping competitions). During their stay, the animals were housed in

3×3 m^2 stall boxes with dry straw as bedding material and were fed with hay, concentrate, and water ad libitum, thus providing standardized environmental and housing conditions.

2.3. Physiological Data

The physiological changes of effort and recovery developed by the participating animals during the performance test were assessed with eye temperature (ET) measurements. For this, samples were collected three times per horse during the test day (one day before the official competition): one hour before the performance test (BT) inside the stall box of the animal, just after the test (JAT) within five minutes after the end of the test at the finish line, and one hour after it (AT), when the animal was resting on the stall box. For the infrared photographs collection, all horses were handled by the owner or their regular horse keeper on the place of the collection (stall boxes for BT and AT measurements and the entrance to the performance arena for JAT measurements). The camera operator was placed 1 m away from the horse, perpendicular to its left eye where the images were taken without touching the horse. In order for the horses to get used to the operators and to the camera itself, a short period of habituation was carried out before the test day with all the horses. During this habituation period, the horse could freely sniff the camera and the camera operator.

To obtain effort and recovery eye temperature measurements, differences between phases were calculated. Thus, the eye temperature of the 'effort phase' (EP) was computed as the difference between JAT and BT eye temperature measurements, whereas the eye temperature of the 'recovery phase' (RP) was computed as the difference between JAT and AT measurements. Hence, for the purposes of this study, statistical analyses were developed just for EP and RP measurements.

Eye temperature images were taken always by the same person, with a portable infrared thermography (IRT) camera (FLIR E60. FLIR Systems AB, Danderyd, Sweden). In order to calibrate the camera results, environmental temperature and relative humidity were recorded with a digital thermohygrometer (Extech 44550, Extech Instruments, Nashua, New Hampshire) every time an eye temperature sample was taken. To determine eye temperature, an image analysis software FLIR Tools 6.0.17046.1002 (FLIR Systems AB, Danderyd, Sweden) was used, measuring the temperature (°C) within an oval area traced around the caruncle of the eye, including approximately 1 cm around it. The program provided the maximum, minimum, and mean temperature of the oval area previously traced (Figure 1), but for the study purposes, only the maximum temperature was used. Three-four images were taken per animal and collection period. Later, the best image was selected to obtain the data for the study.

2.4. Statistical Analyses

A previous Shapiro–Wilk test (results not shown) presented a normal distribution of the variables studied. Hence, parametric statistical analyses and a Bayesian approach for the genetic parameters were used.

In order to test the influence of horse characteristics on the physiological adaptation of the studied horses, a general linear model (GLM) analysis was developed for the EP and RP variables measured, considering the age as a covariate and sex and genetic line as fixed effects, resulting in the two of them being statistically significant for both parameters ($p < 0.05$; results not shown). According to these results, a least square means analysis was used due to the sex and genetic line for EP and RP variables. Moreover, to determine statistically significant ($p < 0.05$) differences between fixed effects' levels, a post-hoc Duncan's test was developed.

To obtain the phenotypic correlations between EP and RP variables, direct and intra-class Pearson's correlations were computed, according to the genetic line and sex effects. Furthermore, in order to analyze the relation of these variables, with the sport performance developed in the official show jumping competitions held the day after the performance test, direct and intra-class Pearson's correlations were computed between the ranking position obtained by the animals on the competition (RANK), and both EP and RP variables obtained on the performance test, according to genetic line and sex effects.

For the statistical analyses, the Statistica software v. 8.0 (Stat Soft. Inc., Tulsa, OK, USA) was used.

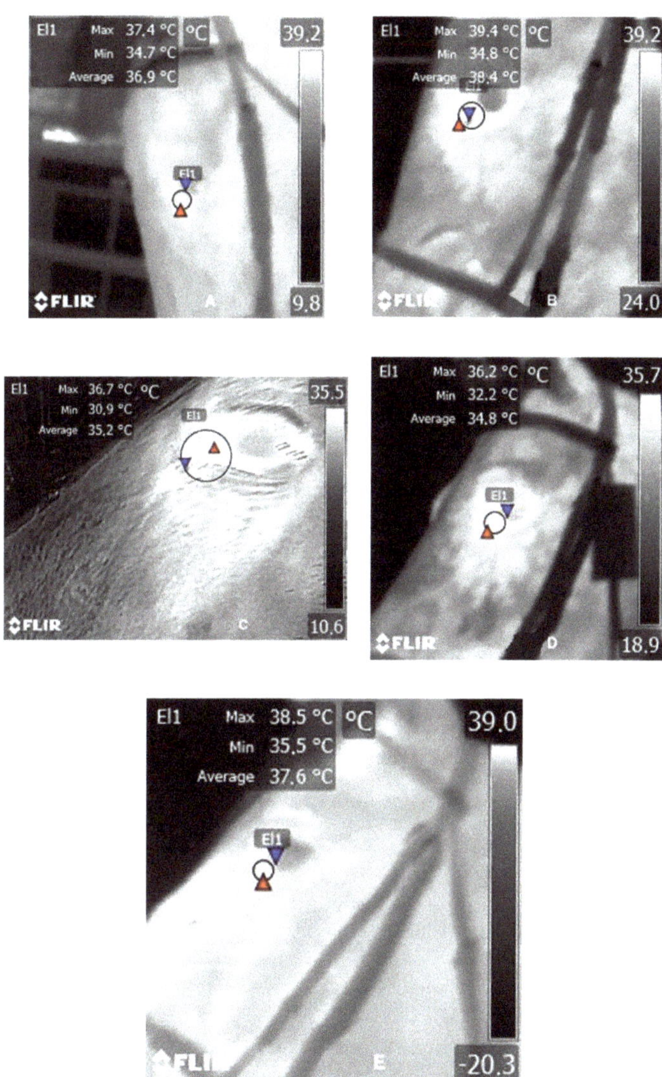

Figure 1. Analyzed eye temperature images of CDE horses according to their genetic line, obtained just after the performance test. (**A**) German genetic line L1; (**B**) Thoroughbred genetic line L2; (**C**) Trotter genetic line L3; (**D**) Pura Raza Española genetic line L4; (**E**) Other Breeds genetic line L5. Where, **El1** is a selected area from the thermographic image (indicated with a circle); ▲ indicates the maximum temperature point; and ▼ indicates the minimum temperature point.

2.5. Genetic Model and Genetic Parameters' Estimation

The pedigree information was gathered from the studbook provided by the National Association of Spanish Sport Horse Breeders (ANCADES). The pedigree matrix was built 4 generations up with the known pedigree of the analyzed horses, to a total of 7907 animals (3000 males and 4907 females). A BLUP genetic evaluation was computed based on a bivariate animal model using a Bayesian approach. The equation in matrix notation to solve the mixed model was:

$$y = Xb + Zu + e, \text{with} \begin{pmatrix} u \\ e \end{pmatrix} \sim N\left(\begin{bmatrix} 0 \\ 0 \end{bmatrix}, \begin{bmatrix} A\sigma_u^2 & 0 \\ 0 & I\sigma_e^2 \end{bmatrix} \right),$$

where y is the vector of observations, X is the incidence matrix of systematic effects, Z is the incidence matrix of animal genetic effects, b is the vector of systematic effects, u is the vector of direct animal genetic effects, e is the vector of residuals, σ_u^2 is the direct genetic variance, σ_e^2 is the residual variance, I is an identity matrix, and A is the numerator relationship matrix.

The fitted model included EP and RP as continuous variables, age as a covariate, two fixed effects: sex (2 levels) and genetic group (5 levels), and two random effects: rider (182 levels) and the rider–horse interaction (275 levels).

As regards to the genetic evaluation computed, phenotypic variance, heritability of the animal, the rider, the rider–horse interaction, and the residual effect for both EP and RP variables and genetic correlations between them were estimated with TM software [23].

2.6. Ethic Statement

No specific ethical approval was required for this study due to the performance test being just warm-up exercises that the animals performed regularly before official competitions, for which we did not have to develop any specific protocol or to use specific animals. As regards to temperature samples obtained from the animals, since the eye temperature assessed with infrared thermography was a non-invasive physiological measure collected from a minimum distance of 0.5–1 m away from the animal, it was not necessary to obtain a specific permit for animal experimentation since no pain nor physical stress was afflicted on the animal. Likewise, all the owners of the animals were previously informed of the entire procedure and the type of samples to be taken, obtaining the approval of all the owners of the animals participating in the study.

3. Results

3.1. Influence of Sex and Genetic Line of Effort and Recovery

A least square means analysis was computed for EP and RP variables (Figure 2), indicating statistically significant ($p < 0.05$) differences between the means for the genetic line and sex effects.

First of all, it has to be noticed that no mares were registered for L4 CDE horses (PRE genetic line), thus Figure 2B shows results just for the L1, L2, L3, and L5 CDE genetic lines. As regards to the EP variable, L1 and L2 stallions (PRE and TH genetic lines, respectively) showed the highest ET increases (1.59 and 1.57, respectively) and differed significantly to L1 and L2 mares, which showed lower effort increases (0.70 and 0.71, respectively), indicating lower physiological differences from the rest to performance moment. As regards the RP variable, L1, L2, and L5 mares (GE, TH, and OT genetic lines, respectively) showed values below 0 (−0.39, −0.18, and −0.04, respectively), indicating a bad physiological recovery of the mares from those genetic lines, from the exercise and the animal's need for longer recovery periods. On the other hand, L3 stallions and mares (TR genetic line) showed the greatest statistically significant recovery from the exercise, with 1.28 and 1.60, respectively.

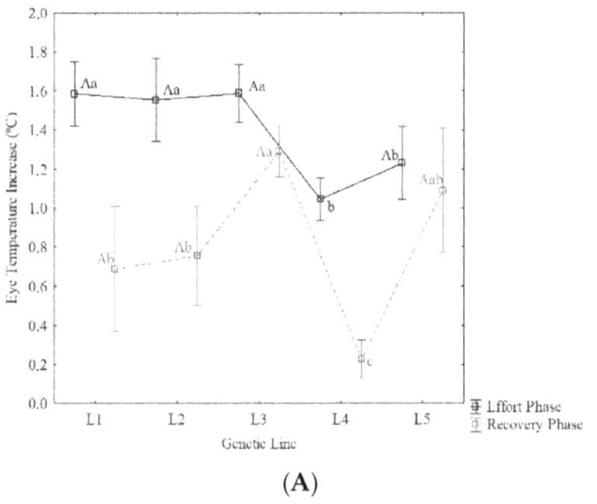

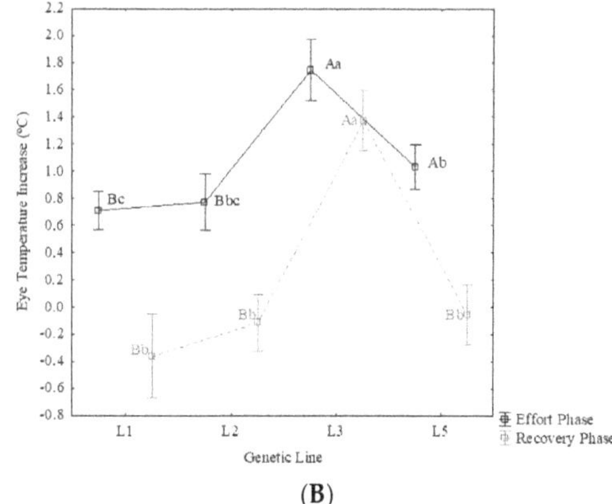

(A)   **(B)**

Figure 2. Least square means analysis (means ± standard deviation) according to sex and genetic line and post-hoc Duncan's test between means. (**A**) Stallions and (**B**) mares. Where L1 is the German genetic line; L2 is the Thoroughbred genetic line; L3 is the Trotter genetic line; L4 is the Pura Raza Española genetic line; and L5 is the Other Breeds genetic line. Different capital letters indicate statistically significant differences ($p < 0.05$) between sexes and within variables, whereas different lowercase letters indicate statistically significant differences ($p < 0.05$) between genetic lines and within variables.

3.2. Phenotypic Correlations between Effort and Recovery

Phenotypic Pearson's correlations (Table 2) between RP and EP revealed a medium positive and statistically significant ($p < 0.001$) correlation (+0.53).

Table 2. Direct and intra-class Pearson's correlations (±standard error) according to sex and genetic line effects, between effort phase (EP) and recovery phase (RP) and the ranking position (RANK), where S refers to the stallions, M to the mares, L1 is the German genetic line, L2 is the Thoroughbred genetic line, L3 is the Trotter genetic line, L4 is the Pura Raza Española genetic line, and L5 is the Other Breeds genetic line. * $p < 0.05$; n.s. not statistically significant.

	EP			RP		
	Direct	**Intra-Class**		**Direct**	**Intra-Class**	
		Sex	**Genetic Line**		**Sex**	**Genetic Line**
RANK	−0.16 (±0.044) *	S: −0.10 (±0.045) n.s. M: −0.34 (±0.042) *	L1: −0.46 (±0.040) * L2: −0.45 (±0.040) * L3: −0.11 (±0.045) n.s. L4: −0.02 (±0.045) n.s. L5: −0.13 (±0.045) n.s.	−0.14 (±0.045) *	S: 0.06 (±0.045) n.s. M: −0.32 (±0.043) *	L1: −0.29 (±0.043) n.s. L2: −0.05 (±0.045) n.s. L3: −0.01 (±0.045) n.s. L4: −0.08 (±0.045) n.s. L5: −0.05 (±0.045) n.s.
RP	0.53 (±0.38) *	S: 0.46 (±0.040) * M: 0.65 (±0.034) *	L1: 0.82 (±0.026) * L2: 0.63 (±0.035) * L3: 0.53 (±0.038) * L4: 0.41 (±0.041) * L5: 0.28 (±0.043) n.s.			

According to genetic line and sex intra-class correlations, positive and statistically significant ($p < 0.001$) correlations were found for L1, L2, L3, and L4 genetic lines and both sexes, ranging from +0.41 for L4 (PRE genetic line) to +0.82 for L1 (GE genetic line) and from +0.46 for stallions to +0.65 for mares. On the other hand, the intra-class Pearson's correlations between EP and RP variables with RANK position showed negative and statistically significant correlations ($p < 0.01$) between RANK and both EP and RP variables (−0.16 and −0.14, respectively). Finally, intra-class negative and statistically significant correlations ($p < 0.01$) were found with mares (−0.34) and L1 and L2 genetic lines (−0.46 and −0.45, respectively) for EP and just for mares (−0.32) for RP variables. These negative correlations indicated that the higher the temperature increase obtained between exercise and rest (hence, the higher the effort response), the lower the RANK position obtained by the animal on the show jumping competition developed the day after the performance test, thus the best classification.

3.3. Genetic Parameters

Finally, the genetic parameters for both physiological variables (EP and RP) are shown in Table 3. The heritabilities obtained for EP (0.26) and RP (0.52) showed a medium to high genetic basis, respectively, for these physiological variables. On the other hand, the ratios for the rider–horse interaction effect (0.37 and 0.33 for EP and RP, respectively) were higher than the ratios for the rider effect (0.15 for both parameters), indicating a greater genetic basis for the interaction than for the rider effect alone. Lastly, the genetic correlation computed between EP and RP (+0.23) indicated a moderate and positive genetic relation between them.

Table 3. Phenotypic variance, mean, and standard deviation of the marginal posterior distributions for the heritabilities of horse, rider, rider–horse interaction, residual effects, and genetic correlation (r_g) between both eye temperature variables analyzed, where EP is effort phase, RP is recovery phase, Vp is phenotypic variance, h^2 is animal heritability, Vr is variance of rider effect/phenotypic variance, Vrh is variance of rider–horse interaction effect/phenotypic variance, Vres is variance of residual effect/phenotypic variance, and s.d. is standard deviation.

	Vp	h^2 ($\pm$s.d.)	Vr ($\pm$s.d.)	Vrh ($\pm$s.d.)	Vres ($\pm$s.d.)	r_g
EP	1.010	0.26 $\pm$ 0.158	0.15 $\pm$ 0.096	0.37 $\pm$ 0.203	0.21 $\pm$ 0.096	0.232
RP	0.918	0.52 $\pm$ 0.073	0.15 $\pm$ 0.057	0.33 $\pm$ 0.080	0.01 $\pm$ 0.003	

4. Discussion

Previous studies have reported that the physiological response of a horse during a sport event is influenced by several factors (sex, breed, environment, rider, fitness of the horse) that would determine its magnitude [1,24,25]. Some of these factors are genetically determined, such as sex or breed. Thus, the knowledge of the magnitude of this response (already settled in the horse from its birth) could be an appropriate tool for the CDE breeding program, as it could help in planning the horses to be mated to obtain animals that could adapt better to the sport competitions they are trained for. Previous studies have also reported differences in the physiological response during competitions due to the horse breed [21,26] that could comprise differences in adapting to new stimuli. These breed differences become key to the selection plans for a composite breed like the CDE, with different breeding recommendations regarding the predominant breed on the CDE analyzed.

Firstly, our results showed that CDE with more than 50% of Trotter breeding in their pedigree (L3) seem to show a greater physiological response before exercise than CDE with other breeds on their pedigree. However, these animals also showed the highest recovery after the exercise, being the fastest to recover their internal homeostasis. This is in accordance with previous studies that found Trotter horses as being selected for speed sport performance (trotting races) and hence, indirectly adapted to become physiologically activated and recover very fast during the performance [27]. Previous studies in humans supported this hypothesis, showing that well-conditioned athletes have shorter recovery times and take longer to become physically fatigued than less fit individuals [28].

Instead, those CDE with more than 50% of PRE (L4) on their pedigree showed the least effort values during this test with a long recovery. This could be due to the PRE influence, which gives calmer behavioral attributes [18], that would be expected to develop a small physiological response between phases, and hence, a small EP and RP phases.

As regards to CDE differences due to sex, a better physical condition would be expected in stallions, due to the more athletic predisposition in males than in females [29]. However, [30] reported a greater proportion of fatal humeral fracture in TH males than in females during races, thus expecting a higher physical level due to a greater effort for the first. This would be in accordance with our results, which showed lower EP and RP results for mares regardless of their genetic line.

As regards the estimation of the genetic parameters for EP and RP variables, the moderate positive genetic correlation found between variables indicated a good potential to include both in the CDE breeding program, either as a selection criterion or as a fixed effect in the genetic model, as it indicates that, despite there being some relation, they can be selected independently. Thus, this breed could be selected due to physiological parameters that would be an indicator of the welfare state of the animal. On the other hand, these results were slightly higher than those reported by [18] in young PRE horses during dressage competitions, probably due to the higher number of animals studied in this paper, which would lead to better genetic estimation of the data. However, these results were in accordance with the results from [31], with heritability values ranging between 0.2 and 0.5 for different behavioral traits (reactivity reactions and learning abilities), which would certainly condition the magnitude of the physiological response of the animal during exercise.

High and statistically significant phenotypic intraclass correlations were found between most genetic lines (L1 to L4) and both sexes denoted that there was a tendency within breeds and sexes to react and recover from the exercise with a similar magnitude. Hence, CDE with genealogical influence from breeds that tended to develop a big physiological response to exercise tend to develop a fast recovery, like the L3 (TR) and L2 (TH) genetic lines. On the other hand, those with a genetic influence from breeds that tended to show a small physiological response to exercise will be expected to show a long recovery from it, like the L4 (PRE) and L1 (GE) genetic lines. Previous studies about differences in horse breeds due to physiological reactions support our results, with PRE and GE breeds showing more proactive and calm responses compared to TR and TH breeds, which are selected for explosive performances that would demand more reactive and temperamental animals [21,26]. Furthermore, [32] reported differences in digital tendons' predisposition to injury in sport horses due to the breed, with Warmblood and TH horses failing more at a higher strain and load than Friesian horses. This supports our findings, with differences in mechanical sport properties conditioning also resulting in differences in physiological response to exercise, due to horse breed.

In order to ascertain how the physiological response developed during effort and recovery phases could affect the sport performance of these CDE horses, EP and RP values were correlated with the ranking obtained by the animals the day after the test, in a regular show jumping competition. Our results indicated a low to medium relation with both parameters, so that the higher the physiological response in the EP and RP phases, the better the ranking position. This is in accordance with previous studies in sport horses that found that the best show jumping individuals were the most reactive ones [11,21]. As regards to intra-class correlations due to sex and genetic line, it appeared that only mares for both variables and L1 and L2 CDE (GE and TH genetic lines, respectively) for EP were influenced by the magnitude of their physiological response when competing. This could be due to the temperament, mechanical sport properties, and reactivity differences between breeds and sexes reported previously [26,32], which could bias the sport results of these animals. Otherwise, these results were considered exploratory, and it must be considered that type I errors were used, thus determining the results highlighted in this study. However, before these can be considered as new selection criteria, more research is required, including these variables as a regular measurement on the equestrian events. Furthermore, large studies carried out over several years and containing more animals are needed before any precise measures concerning the influence of the genetic and environmental effects can be determined.

5. Conclusions

The results obtained in this study indicated that eye temperature assessed with IRT appears as an adequate tool to evaluate the physiological effort and recovery developed by

CDE during exercise and thus, its fitness. Heritability and genetic correlations obtained for EP and RP variables indicated an adequate potential to be included in the CDE breeding program. However, more research is required to confirm the results found here.

Author Contributions: Conceptualization, M.V., M.J.S.-G. and E.B.; methodology, M.V.; formal analysis, D.I.P.-G., M.J.S.-G.; writing—original draft preparation, E.B.; writing—review and editing, M.V., M.J.S.-G., D.I.P.-G. and E.B.; supervision, M.V. All authors have read and agreed to the published version of the manuscript.

Funding: This research received no external funding.

Institutional Review Board Statement: Ethical review and approval were waived for this study, due to the performance test being just warm-up exercises that the animals performed regularly before official competitions, for which we did not have to develop any specific protocol or to use specific animals (Check Section 2.6 of this manuscript).

Acknowledgments: We wish to thank the Spanish Royal Equestrian Federation for providing the performance information for the horses analyzed and the Spanish Sport Horse Breeders Association for providing the pedigree information.

Conflicts of Interest: The authors declare no conflict of interest.

References

1. Moberg, G.P. Biological response to stress: Implications for animal welfare. In *The Biology of Animal Stress: Basic Principles and Implications for Animal Welfare*; CABI Publishing: Wallingford, UK, 2009; pp. 1–21.
2. Pösö, A.R.; Hyyppä, S.; Geor, R.J. Metabolic responses to exercise and training. In *Equine Exercise Physiology*; Elsevier: Amsterdam, The Netherlands, 2008; pp. 248–273.
3. Witkowska-Piłaszewicz, O.; Grzędzicka, J.; Seń, J.; Czopowicz, M.; Żmigrodzka, M.; Winnicka, A.; Cywińska, A.; Carter, C. Stress response after race and endurance training sessions and competitions in Arabian horses. *Prev. Vet. Med.* **2021**, *188*, 105265. [CrossRef] [PubMed]
4. Evans, D.L. Physiology of equine performance and associated tests of function. *Equine Vet. J.* **2007**, *39*, 373–383. [CrossRef] [PubMed]
5. Bitschnau, C.; Wiestner, T.; Trachsel, D.S.; Auer, J.A.; Weishaupt, M.A. Performance parameters and post exercise heart rate recovery in Warmblood sports horses of different performance levels. *Equine Vet. J.* **2010**, *42*, 17–22. [CrossRef] [PubMed]
6. Gardela, J.; Carbajal, A.; Tallo-Parra, O.; Olvera-Maneu, S.; Álvarez-Rodríguez, M.; Jose-Cunilleras, E.; López-Béjar, M. Temporary Relocation during Rest Periods: Relocation Stress and Other Factors Influence Hair Cortisol Concentrations in Horses. *Animals* **2020**, *10*, 642. [CrossRef] [PubMed]
7. Janczarek, I.; Bereznowski, A.; Strzelec, K. The influence of selected factors and sport results of endurance horses on their saliva cortisol concentration. *Pol. J. Vet. Sci.* **2013**, *16*, 533–541. [CrossRef] [PubMed]
8. Kędzierski, W.; Cywińska, A.; Strzelec, K.; Kowalik, S. Changes in salivary and plasma cortisol levels in Purebred Arabian horses during race training session. *Anim. Sci. J.* **2013**, *85*, 313–317. [CrossRef]
9. Witkowska-Piłaszewicz, O.; Pingwara, R.; Winnicka, A. The Effect of Physical Training on Peripheral Blood Mononuclear Cell Ex Vivo Proliferation, Differentiation, Activity, and Reactive Oxygen Species Production in Racehorses. *Antioxidants* **2020**, *9*, 1155. [CrossRef]
10. Cappelli, K.; Mecocci, S.; Gioiosa, S.; Giontella, A.; Silvestrelli, M.; Cherchi, R.; Valentini, A.; Chillemi, G.; Capomaccio, S. Gallop Racing Shifts Mature mRNA towards Introns: Does Exercise-Induced Stress Enhance Genome Plasticity? *Genes* **2020**, *11*, 410. [CrossRef]
11. Won-Seob, K.; Jalil, G.N.; Sang-Gun, R.; Hong-Gu, L. Heat-shock proteins gene expression in peripheral blood mononuclear cells as an indicator of heat stress in beef calves. *Animals* **2020**, *10*, 895.
12. Cook, N.J.; Schaefer, A.L.; Warren, L.; Burwash, L.; Anderson, M.; Baron, V. Adrenocortical and metabolic responses to ACTH injection in horses: An assessment by salivary cortisol and infrared thermography of the eye. *Can. J. Anim. Sci.* **2001**, *81*, 621.
13. Valera, M.; Bartolomé, E.; Sánchez, M.J.; Molina, A.; Cook, N.; Schaefer, A.L. Changes in Eye Temperature and Stress As-sessment in Horses During Show Jumping Competitions. *J. Equine Vet. Sci.* **2012**, *32*, 827–830. [CrossRef]
14. Witkowska-Pilaszewicz, O.; Masko, M.; Domino, M.; Winnicka, A. Infrared Thermography correlates with lactate concentra-tion in blood during race training in horses. *Animals* **2020**, *10*, 2072. [CrossRef]
15. Stewart, M.; Webster, J.; Verkerk, G.; Schaefer, A.; Colyn, J.; Stafford, K. Non-invasive measurement of stress in dairy cows using infrared thermography. *Physiol. Behav.* **2007**, *92*, 520–525. [CrossRef] [PubMed]
16. Marlin, D.; Nankervis, K. *Equine Exercise Physiology*; Blackwell Publishing: Hoboken, NJ, USA, 2002.
17. Bartolomé, E.; Sánchez, M.J.; Molina, A.; Schaefer, A.L.; Cervantes, I.; Valera, M. Using eye temperature and heart rate for stress assessment in young horses competing in jumping competitions and its possible influence on sport performance. *Animals* **2013**, *7*, 2044–2053. [CrossRef] [PubMed]

18. Sánchez, M.J.; Bartolomé, E.; Valera, M. Genetic study of stress assessed with infrared thermography during dressage competitions in the Pura Raza Español horse. *Appl. Anim. Behav. Sci.* **2016**, *174*, 58–65. [CrossRef]

19. Bartolome, E.; Cervantes, I.; Valera, M.; Gutiérrez, J. Influence of foreign breeds on the genetic structure of the Spanish Sport Horse population. *Livest. Sci.* **2011**, *142*, 70–79. [CrossRef]

20. Clark, J.D.; Rager, D.R.; Calpin, J.P. Animal well-being. IV. Specific assessment criteria. *Lab. Anim. Sci.* **1997**, *47*, 586–597.

21. Lloyd, A.S.; Martin, J.E.; Bornett-Gauci, H.L.I.; Wilkinson, R.G. Horse personality: Variation between breeds. *Appl. Anim. Behav. Sci.* **2008**, *112*, 369–383. [CrossRef]

22. Royal Spanish Equestrian Federation (RFHE). Galopes Program. 2018. Available online: http://www.rfhe.com (accessed on 20 March 2019).

23. Legarra, A.; Varona, L.; López de Maturana, E. TM: Threshold Model. 2008. Available online: http://snp.toulouse.inra.fr/~{}alegarra (accessed on 10 July 2019).

24. Desmecht, D.; Linden, A.; Amory, H.; Art, T.; Lekeux, P. Relationship of plasma lactate production to cortisol release following completion of different types of sporting events in horses. *Vet. Res. Commun.* **1996**, *20*, 371–379. [CrossRef]

25. Negro, S.; Bartolomé, E.; Molina, A.; Solé, M.; Gómez, M.D.; Valera, M. Stress level effects on sport performance during trotting races in Spanish Trotter Horses. *Res. Vet. Sci.* **2018**, *118*, 86–90. [CrossRef]

26. Bartolomé, E.; Cockram, M.S. Potential Effects of Stress on the Performance of Sport Horses. *J. Equine Vet. Sci.* **2016**, *40*, 84–93. [CrossRef]

27. Thiruvenkadan, A.; Kandasamy, N.; Panneerselvam, S. Inheritance of racing performance of trotter horses: An overview. *Livest. Sci.* **2009**, *124*, 163–181. [CrossRef]

28. Halliday, E.; Willmott, B.; Randle, H. Physiological measures of fitness of riders and non-riders. *J. Vet. Behav.* **2011**, *6*, 300–301. [CrossRef]

29. Sarzynski, M.A.; Loos, R.J.F.; Lucia, A.; Pérusse, L.; Roth, S.M.; Wolfarth, B.; Rankinen, T.; Bouchard, C. Advances in Exercise, Fitness, and Performance Genomics in 2015. *Med. Sci. Sports Exerc.* **2016**, *48*, 1906–1916. [CrossRef]

30. Whitton, R.C.; Walmsley, E.A.; Wong, A.S.M.; Sahnnon, S.M.; Frazer, E.J.; Williams, N.J.; Guerow, J.F.; Hitchens, P.L. Asso-ciations between pre-injury racing history and tibial and humeral fractures in Australian Thoroughbred racehorses. *Vet. J.* **2019**, *247*, 44–49. [CrossRef]

31. Hausberger, M.; Bruderer, C.; Le Scolan, N.; Pierre, J.S. Interplay between environmental and genetic factors in temperament/personality traits in horses (*Equus caballus*). *J. Comp. Psych.* **2004**, *118*, 434–446. [CrossRef] [PubMed]

32. Verkade, M.E.; Back, W.; Birch, H.L. Equine digital tendons show breed-specific differences in their mechanical properties that may relate to athletic ability and predisposition to injury. *Equine Vet. J.* **2019**, *52*, 320–325. [CrossRef] [PubMed]

Article

Farm Animals Are Long Away from Natural Behavior: Open Questions and Operative Consequences on Animal Welfare

Alberto Cesarani [1,*] and Giuseppe Pulina [2]

[1] Animal and Dairy Science Department, University of Georgia, Athens, GA 30602, USA
[2] Department of Agraria, University of Sassari, 07100 Sassari, Italy; gpulina@uniss.it
*   Correspondence: alberto.cesarani@uga.edu

Simple Summary: Animal welfare is a very important issue. One of the tasks of researchers is to provide explanations and possible solutions to questions arising from non-experts. This work analyzes part of the extensive literature on relationships between selection and domestic, mainly farm, animals' behavior and deals with some very important themes, such as the role of regulations, domestication, and selection.

Abstract: The concept of welfare applied to farm animals has undergone a remarkable evolution. The growing awareness of citizens pushes farmers to guarantee the highest possible level of welfare to their animals. New perspectives could be opened for animal welfare reasoning around the concept of domestic, especially farm, animals as partial human artifacts. Therefore, it is important to understand how much a particular behavior of a farm animal is far from the natural one of its ancestors. This paper is a contribution to better understand the role of genetics of the farm animals on their behavior. This means that the naïve approach to animal welfare regarding returning animals to their natural state should be challenged and that welfare assessment should be considered.

Keywords: bioethics; domestication; genetic selection; animal behavior; animal welfare

Citation: Cesarani, A.; Pulina, G. Farm Animals Are Long Away from Natural Behavior: Open Questions and Operative Consequences on Animal Welfare. *Animals* **2021**, *11*, 724. https://doi.org/10.3390/ani11030724

Academic Editors: Melissa Hempstead and Danila Marini

Received: 13 February 2021
Accepted: 5 March 2021
Published: 6 March 2021

Publisher's Note: MDPI stays neutral with regard to jurisdictional claims in published maps and institutional affiliations.

1. Introduction

Ethics in animal production is a hot topic and a huge amount of literature has been published [1]. One of the most debated topics in the past few decades, both at scholarly and public opinion levels, is the question of whether animals farming would be considered morally justified. A recent multisectoral contribution on ethics in animal-sourced food has been published in a monographic issue of Animal Frontiers [2]. Vegetarian and vegan movements are deboarding from their original nutritional ideological beliefs to broadly embrace the animalist party point of view, claiming that animal products, mainly meat, must be banned in human diets [3]. At the heart of the recent public debate on human–animal relationships, there is the growing awareness of citizens that the use of animals for human purposes must include the obligation on the part of farmers to guarantee them a high level of welfare [4].

It is worthy to note that the word "welfare" may assume different meanings depending on the context; the main two refer to a "physical and mental health and happiness, especially of a person" and "to help given, especially by the state or an organization, to people who need it, especially because they do not have enough money" [5]. Under an operational point of view, as needed for scientific purposes, the concept of welfare applied to farm animals has undergone a remarkable evolution until it reached a formulation that includes the following fourfold aspect: (a) biological and technical definitions, that emphasize the basic needs of animals and the freedom that should be given and the possibility of coping with environmental challenge; (b) regulation approaches, according which animals are sensitive creatures so they must be reared in the environment compatible to the biological needs of the species. This leads to the translation of the concept into a legal framework; (c) philosophical approach, discussing the role of animals in the humans' societies; (d) interactive approach, that considers communication between farmers and animals and its impact on livestock systems [6].

Among biological approaches, neurocognitive studies have begun to achieve important results since Charles Darwin [7], and his student Georges Romanes [8], proposed that human mental states represent a continuum with animal ones deriving from them by evolution. Later, several scholars criticized this point of view, arguing that the animals' traits that can be found in humans, and not the other way around. However, Darwin himself was conscious that arguing "the existence of human traits in animals rather than animal traits in human beings allowed him to express his point of view on animal and human continuity in a more acceptable way for contemporary society in which, as Elizabeth Knoll [9] observed, feelings anthropomorphic were popular in the upper middle classes" [10]. Over a century of comparative neurocognitive studies in animals and humans, which summarization is quite difficult and overtake the scope of this contribution, have led the scientific communities to converge on the belief that higher animals (and here the controversy moves to the border that demarcates this category) are endowed with mental states such as to generate internal representations of the external world. This entailed the normative evolution that established for animals a state of sentient organisms (e.g., EU Lisbon Treaty, 2007) [11], that can express basic sensations such as fear, pain and pleasure. Thus, several laws were issued to protect and safeguard their well-being. However, neurocognitive scientists have established that animals raised for economic purposes (e.g., milk and meat production) show differences in cognitive abilities and brain lateralization that can affect adaptive behavioral, physiological, and immune responses to environmental stressors [12]. However, evidence of advanced cognition in animals says little about their sentience (i.e., feeling) [13] and the lack of direct proofs, acquired only through verbal abilities, advise to consider alternative hypotheses before attributing conscious states that can account for animal behavior [10]. On the other hands, at the begin of this century a behaviorist line of thinking emerged; this new point of view has based the evaluation of animal welfare on the "emotions" of farmed livestock [14], opening a controversial field of discussion between animal sensations, feelings and emotions [15].

The classic research of Belyaev [16] and his students on silver foxes have shown that the domestication process, including in the specific case the selection of the animals that presented less fear of man, affects not only the morphological characters of the animals, but also (and especially) the behavioral ones. Recently, Kukakova et al. [17] have sequenced and assembled the genomes of the fox of Belyaev experiment and re-sequenced a subset of foxes from the tame, aggressive and conventional farm-bred populations to identify genomic regions associated with the response to selection for behavior: they found 103 regions with either significantly decreased heterozygosity in one of the three populations or increased divergence between the populations, demonstrating the genetic based behavior in domestic animals.

Therefore, news perspectives could be opened for animal welfare reasoning around the concept of domestic, especially farm, animals as partial human artifacts (in this review we assume for human artifact or construct the meaning of what, in this case animal, modified by human for his own purposes). Therefore, it is important to understand how much behaviors of farm animals are far away from the natural ones of their ancestors. The consequences bring directly to assume a mixed natural and anthropocentric point of view when we discuss about farm animal welfare. A pioneering work aimed at founding a theory of the effects of domestication on animal behavior was carried out by Price [18] who, on the basis of the already abundant literature available at that time, analyzed the factors that mainly influenced the changes in their character.

This review analyses the effects of domestication and artificial selection on animal behavioral traits coupled with morphological ones. A wide literature has been considered to contribute to better understand the role of genetics of the farm animals on their behavior, finalized to demonstrate that as they are partially human artifacts, and so that their welfare should be assessed taking into account the artificial environment where farm animals have been selected. Scopus® database was firstly scanned by using "bioethics, animal behavior, behavioral heritability, animal selection and behavior" as keywords; furthermore, general literature on animal ethics and animal welfare was taken into account. Results coming from the analyzed literature were extracted, classified and discussed, to derive our original points of view.

The hypothesis we want to demonstrate in this review is that the naive approach to animal welfare that brings it back to a natural state is not completely right and that welfare assessments must necessarily consider that the behaviors of farm animals have been partially, but consistently, constructed by humans.

2. Domestication Changed Animal Traits

Domestication is a lively area of scientific research in which no consensus has been reached on its meaning yet. For our purpose we adopted the Zeder co-evolutive point of view [19]: she defines domestication as "a sustained multigenerational, mutual relationship in which one organism assumes a significant degree of influence over the reproduction and care of another organism in order to secure a more predictable supply of a resource of interest, and through which the partner organism gains advantage over individuals that remain outside this relationship, thereby benefitting and often increasing the fitness of both the domesticator and the target domesticate". Under this framework, Zeder [19] observed that human domestications are different from non-human ones because we have been able to opportunistically grasp and modify the traits that have increased the benefits obtained from co-evolutionary relationships with the target species. Moreover, we have transmitted these practices capable of achieving our goals not only within the parental circle, but also to all through social learning. Animal domestication is a cultural issue of Homo sapiens because, until now, no trace of similar practices has been found by archaeologists for the most ancient *Hominidae* species, even though they have been in continuous contact, as hunters, with wild animals for millions of years.

Restricting this field to livestock, when humans became sedentary farmers, they started to domesticate wild animals to fit different needs, such as work, and meat or wool production. This can be considered at all the first step of the artificial selection processes. However, during this long time of human–animal partnership, natural selection has continued to shape the domestic animals but, as Darwin [20] firstly observed, these factors have been overwhelmed by the artificial ones. For this reason, in this review we have considered the effects of human selection only. Beside historical and archeological traces, modern genomics tools allowed to study and define a timeline of the domestication process [21–23]. The domestication of livestock species started in Southwest Asia (the Near East) about 10,000 years ago with goats, sheep, humpless cattle, pigs and then water buffalo, horse, chicken and, more recently, rabbit [24–26]. Horses were domesticated for their key role in warfare and transportation during multiple events somewhere in the Eurasian plains [27]. An interesting case is the rabbit (*Oryctolagus cuniculus*) which is a key species (is the only relevant mammal livestock species originating from Europe at the Roman empire age) because of its role as livestock, game, pet, and experimental animal (as well as pest in several countries, mainly Australia) [28,29].

Domestication process changes phenotypic and behavioral characteristics of wild animals depending on the species. Behavioral and morphological traits of modern livestock breeds resulted from artificial selection of natural useful characters: in fact, domestication process increased the phenotypic and genetic variability of breeds and the modern biodiversity is one the most important outcomes of the domestication before and selection later. Under this point of view, the domestication created prototypes of each species that were subsequently shaped by the artificial selection process leading to an increase in the variability, in terms of both genetic and phenotypic richness. Rasali et al. [30] reported that humans all around the world raise about 200 and 400 pure and composite sheep breeds, respectively. Even if the exact knowledge about the sheep descendant is still debated, all these breeds, characterized by different morphological traits and ability to adapt to extremely different environments, seem to be originated from just a couple of wild ancestors. Let us think also to bovine (Bos taurus) and zebuine (Bos taurus indicus) breeds that, even showing important differences between and within them, share the same wild ancestor, the aurochs (Bos primigenius): starting from this, more than 1000 cattle breeds are nowadays raised worldwide [31]. Of particular interest is also the pig of which 566 different breeds can be enumerated, all originating from the wild boar [32]. Domesticated sheep and goats show different features compare to their wild ancestors, such as reduced body size and horn length [23,33]. Domesticated cattle breeds show smaller size compared to the extinct wild aurochs and they developed the capacity to adapt to various environments [34–36]. However, some cattle breeds have longer horns with particular shape that were selected for religious reasons [27]. Another interesting phenotype shaped during domestication process is the coat color, that in some species (such as horse, cattle and sheep) is a useful trait to discriminate among breeds.

Domestication changed, not only external features, but also behavior and aptitude of livestock species [18,37]. For example, sheep and goats were chosen for their meekness and for their multiple productions (e.g., meat, milk, wool and horns). However, the mechanisms behind this process were not still completely clear, and different theories have been proposed during time, until second half of the twentieth century, when the modern concept of ethology raised [38]. The behavioralists of the past believed that changes in animals' behavior were mainly due to the different environments, learning and nurture. Nowadays, this concept has been abandoned in favor of a more realistic one involving a genetic determinism [39–41]. This is particularly true because a trait without heritable mechanisms could not play a role in the evolution process. Other authors proposed a theory, in which neural crest cells play a key role, called "domestication syndrome" to put together all the differences in phenotypic, and therefore genetic, characteristic between wild and domesticated species [42,43]. The term domestication syndrome applied to animals came from the domesticated crop plants [44]. As specified by Sánchez-Villagra et al. [45], the word "syndrome" is not related to a particular pathological condition or disease, but it came from the literature about plants that humans selected for interrelated syndromes of characteristics during the domestication process [46]. Sánchez-Villagra et al. [45] reported that some modifications are common to almost all domesticated animal species: increased docility, increased skillfulness in using human cues, increased fecundity, reduction of tooth, brain and rostrum size, floppiness of the ears, curliness of the tail and depigmentation of skin and fur. Two main hypotheses were suggested for explaining the differences between wild and domesticated species: the first one state that these differences are due to the new environment in which domesticated animals received improved diets and found better live conditions. The second one suggested that domestication syndrome was related to hybridization between different breeds or even species. Anyway, livestock species undergone domestication syndrome show differences in morphological and behavioral traits compared to their presumed wild ancestors. (Table 1).

Regarding the changes in animals' behavior, all features of farm animals that are currently raised by humans show heritable genetic variations [47,48]. Fear and interactions with humans, sexual maturity and social behavior were also affected by domestication. Farm animals show less fear of humans, and they are more inclined to cooperate with humans with a lower stress level. Farmed pigs reach sexual maturity at about 6–7 months, whereas for wild boars, sexual maturity is reached at approximately two years [49]. Moreover, farm animals are forced to live in larger groups, compared to their wild ancestor, under crowded conditions [48]. York [50] reported a list of the behavioral traits, and the tests used to study them, that have been analyzed in livestock species.

Table 1. Morphological and behavioral traits associated with the domestication syndrome that changed from wild ancestor to domestica livestock species (adapted from Wilkins [42]).

Trait	Livestock Species
Curly tails	Dog, pig
Depigmentation	Cattle, dog, goat, horse, pig, rabbit
Docility	Cattle, dog, donkey, goat, horse, pig, rabbit, sheep
Floppy ears	Cattle, donkey, dog, rabbit
More frequent estrous cycles	Dog, goat
Neotenous (juvenile) behavior	Dog
Reduced ears	Dog
Shorter muzzles	Cattle, dog, goat, pig, sheep
Smaller brain or cranial capacity	Cattle, dog, goat, horse, pig, rabbit
Smaller teeth	Dog, pig

Genomic studies allowed to trace the domestication process and to highlight genomic regions where genes involved in the phenotypic differences among breeds are mapped. For example, Cesarani et al. [51] compared two related sheep breeds, one showing ancestral traits and one selected for milk aptitude, and they found significant genomic regions that harbor gene involved in morphological traits such as coat color (MC1R, MITF), horned/polled (RXFP2) and body size (NPY, VIP). Moreover, the authors identified the HOXB1 gene that was already associated with the *anotia* phenotype in mice [52], that is of interest because the breed with ancestral traits is characterized by microtia, external malformation trait that is often present in wild animals [53,54]. Hornedness/polledness is one of the main traits modified by human selection during the domestication from wild ancestor to modern breeds [55]: e.g., ancestral wild sheep showed big horns useful as defense and injury instrument.

Several studies focused their attention to the different genetic background (e.g., different gene expression) between wild and domesticated animals [56,57]. One of the ideas is that these differences are not related to different environment and methods of farming, but they are heritable traits that can be transmitted to the progeny [58–60]. This theory seems to be in contrast with the first one of Charles Darwin [20] in which the theory of heredity was absent. Domestication is one of the most important events in mankind's history, which has now consequences that are intrinsic in the everyday reality. Recently, some studies highlighted that species undergone domestication for thousands of years, showed changes in behavior, color, morphology and physiology which are related to several independent genomic regions [61,62]. However, studies on species with more recent domesticated history (e.g., cat and salmon) revealed fewer signals [63,64] compared to the older domesticated species. Thus, these changes could be associated with the extension of the domestication process. This hypothesis is strengthened by the fact that a long time is needed to fix the heritable traits, especially those with lower heritability; even under strong selection pressure, only small changes in gene frequencies can be observed over several generations [65]. On the contrary, recent investigations demonstrated that sometimes only few [66] or exceptionally just one [67] generations could be enough to show genetic adaptation to captivity. In order to test this idea, Christie et al. [60] analyzed gene expression between the offspring of first-generation hatchery trout and wild trout reared in an identical environment. The authors found 723 genes differently expressed in the two groups of animals, mostly involved in responses in wound healing, immunity and metabolism.

As aforementioned, the genomic studies were able to identify selection signatures, i.e., genome regions shaped by selection, that can differentiate wild and domesticated animals. In these regions, genes encoding for the phenotypic differences were mapped. Due to involvement of genes and the fact that progenies of wild and domesticated animals show feature like their ancestor and not to the opposite group, genomic mechanisms of the domestication, and therefore its heritable pathway, could be confirmed.

3. Questions and Answers about Modern Farm Animals

Domestication, firstly, and genetic selection subsequently, are shown to be effective tools to change the temperament of farm animals orienting it towards the desired behavioral characteristics (e.g., docility, resistance to adversity, patience, maternal attitude …). Current characters of farm animals were demonstrated to be a human construct, and that they can be further quite rapidly oriented by their insertion among the breeding goals of genetic selection. Nowadays, we are faced with several ethical questions, and subsequently approaches to assess and improve animal welfare.

Ethic is a philosophical rich field that explores the unknown answer by dealing originally with known arguments. According to Floridi [68], philosophers usually ask open questions to which they usually do not provide definitive answers. However, they tend to define the problem space and within it they try to give the most probable answer according to the state of knowledge. Under the recent constructivism point of view, the philosopher is a model builder who reduces the problem space at the minimum manageable to further connect these elements in a most complex and complete framework. In this way, the scientist is a special type of philosopher who deals exclusively with real facts instead of abstract entities, for which however he seeks an answer based on experience and continually revisable according to the evidence of new elements [69].

Question one: Does the character change in domesticated animals by human purposes shifted some of the responsibility for their survival and well-being into human actions?

As domestication is a particular case of coevolution between living organisms, the pervasiveness of humans should modify the environment and consequently the factors influencing the behavior in wild animals also. In fact, animals living near humans (e.g., in contexts like captivity and urbanization) rapidly change their antipredator behavior and become tolerant to people's presence [70]. This justifies the growing awareness to take care for wild animals (e.g., care in wildlife centers in case of disease, foraging in case of food scarcity, etc.), even if this can divert their evolutionary trajectory because of the manipulation of the natural fitness factors. While it is generally accepted that humans are responsible for the welfare of wild animals by manipulating part of the natural environmental factors to protect them, the share of human responsibility on the welfare of farm animals is greater due to the fact that some behaviors have changed due to artificial selection. This means that the breeders, when designing their welfare programs, must consider the distance that exists between the behavioral construct of the animals they reared and that of the natural environment, paying greater attention to the needs of the animals, the greater this distance. In other words, high productivity animals kept in more intensive farming systems have a greater need for attention to their welfare than less productive ones reared in extensive systems. Just a few examples are the superior care associated with intensive systems, such as higher veterinary care, control of the production environment in terms of temperature, humidity and light, and better feeding and managements strategies [71]. Another example is represented by the use of piglet lamps in the sow farrowing pens compared to body heat of the mother in extensive pig systems [72]. It follows that animal welfare is a technology that, like the others used for animal breeding, must be studied and applied not only for ethical purposes, but also to obtain the best coupling between available environmental conditions and animal health and productivity.

Question two: Can the behavior be indicated as genetic goal in animal selection? The starting point of this answer must necessarily rest on the principle, initially elaborated by Rollin [73] and subsequently perfected by Shriver [74], of the conservation of welfare in animals subjected to genetic manipulation: "any animals that are genetically modified through the use of genetic technology [for our intent, this means both by classical genetic selection of genetic manipulation], for purposes other than research, should be no worse off, in terms of suffering, than the parent stock was prior to genetic alterations." Several genetic studies demonstrated that behavioral traits are inheritable: estimates mostly range from 10 to 50% [47,50,75–77]. The interest in estimate heritability of behavioral traits of farm livestock lies in the past: Kjaer and Sørensen [78] estimated the heritabilities of cannibalism and feather pecking in laying hens of White Leghorn breed. In fact, restricted and crowded spaces bring the hens to attack other animals or even injury themselves and farmers try to prevent these behaviors through beak trimming, practice under scrutiny by animals' right associations [79]. Due to these heritable mechanisms, behavioral traits can become (and sometimes are already) the goal of breeding scheme.

As aforementioned, also behavior was shaped by the domestication process and therefore genetic studies analyzed also this range of traits. For example, Ding et al. [80] found a single intronic retroelement involved with variations in courtship dance in Drosophila species. York [50] analyzed more than one thousand genomic loci (of mammals, birds, fish, insects and nematodes) associated with a series of behavioral traits (such as: courtship, feeding, aggression, motor, emotion, temperament, learning, social, parental and circadian) and he confirmed the genetic determinism of such traits. This author reported that courtship and feeding behaviors are determined by genomic regions of significantly greater effect than other traits. Boissy et al. [81] analyzed the effect of genotypes (i.e., different breed composition) of both lamb and dams on the emotional reactivity in sheep. These authors concluded that most of the differences were due to direct additive genetic effects and that females were more active and avoided more the human contact compared to male lambs.

Reduce the aggressivity of animals would have a series of positive benefits both for animals and humans. Let us focus on pigs and on the positive benefits in this species. Turner et al. [82] investigated the genetic aspects of individual aggressiveness in pigs by the association of skin lesions and behavioral traits. Animals living in crowded situations tend to be more aggressive, even if aggressivity to get access to resources can be also found in nature (for example regrouping at pasture) to define a new social order [83,84]; in pigs there are both reciprocal and non-reciprocated aggressions. In both cases, animals suffering these assaults are more stressed, reduce their feed intake and therefore their productions. One of the main problems related to behavior in pigs is the aggressive behavior of sows towards their piglets: this problem has also been largely studied since the past. Already in the late twentieth century, Knap and Merks [85] analyzed the aggressiveness of primiparous sows and reported a decrease of litter size at weaning of almost two piglets per litter. These authors reported heritabilities ranging between 0.40 and 0.90. These numbers show the effectiveness of including the mildness among the breeding goals: through genetic (or now genomic) selection, the aggressiveness of livestock animals can be reduced by having positive effects in terms of animal welfare and production.

Another behavioral trait is the fighting aptitude in South America and European cattle breeds [86]. One example is the Spanish fighting bull (known also as Iberian or Lidia bulls) which is raised in Spain, Portugal, France and South American countries (e.g., Mexico) for bull fighting events [87,88]. Another example are the not cruel traditional events of fighting among cows belonging to Aosta Chestnut and Aosta Black Pied [89] or Valdostana [90–92] cattle breeds in North Italy (Alpine regions). Both tourism and farmers take important economic advantages from this kind of contest, named "battle of queens". At the end of October, pregnant cows are divided in classes according to their size and they fight (by pushing forward the other animal using their weights) without injuring the opponents. The cow winning this tournament becomes the "queen of all queens" and it assures huge incomes to its farmer, who can sell her or her offspring. Fighting aptitude is evaluated and breeding values are estimated for Valdostana Pezzata Nera and Castana cattle breeds; this behavioral trait is already included in the breeding schemes of these breeds.

Further important "behavioral" traits shifted by selection are adaptation and resilience [93]. The animal production is now become more efficient and, therefore, each unit of production shows a smaller environmental footprint [94]. Moreover, animals are not only more efficient, but they adapted to produce in very different environments. However, areas of production are likely going to become harsher (e.g., increased heat, disease challenges) [93]. Thus, if we consider the heat-tolerance among the "behavioral" traits, adaptability to difficult environments is (and will be) a strategic point. Resilience is mostly defined as the capacity to recover quickly from difficulties. Unfortunately, it seems at resilience is negatively correlated with genetic selection pointing to increase production traits, which reduced environmental adaptation [93]. Frankham [95] stated that genetic selection for the adaptation to captivity reduced fitness, which is confirmed when animals are reintroduced into the wild. Thus, selection for resilience will largely acts on morbidity, longevity and mortality rather than productive performance. Resilient animals may show lower performances in optimal environments [93].

It is interesting to note that the more recent domestication of rabbits has made it possible to retain many of the ancestral and wild behaviors, like living in small groups of 2–8 individuals with familiar subgroups, spend the time mainly underground and coming outside only for foraging, stand up to check for any predators, locate a specific area for the true feces (scrapes), etc. When these animals are placed in free-range conditions, they make nests and dig burrows [96].

Finally, two other species in which behavior has a large impact are dogs [97,98], and horses [99,100]. It is worthy to mention these species because of their tight relationship with humans: dogs and horses are raised as companion (both), working (mainly dogs) or sport (mainly horses) animals. Thus, it is compulsory that these animals show low aggressivity towards humans. Years of selection of candidates based not only on performances, but also on conduct, led to animals with tame behavior particularly suitable for staying with children or disabled: just think of dogs for blind people or horses for therapeutic riding. Without selection for behavioral traits, these animals improving the life quality of humans there would not have been.

Question three: How can farmers achieve the highest possible welfare for their animals? The sustainable intensification of livestock chains is the narrow way to simultaneously meet the growing demand for food of animal origin and the reduction of their environmental and social impacts [101]. The economic pressure to reduce costs for maintaining the viability of farming enterprises and the scaling order that assigns more animals per operator, have diverted the perspective in the design and construction of farming environments towards savings and removed the animals from the direct control of the farmers. Furthermore, the need for manpower on farms with the increasing insert of employees of non-rural extraction or coming from work paths other than breeding, has led to the drop in tension for animal welfare based on life experience or on work or training courses adequate to the ethical challenge in game. Finally, over a century of selection based on functional traits has placed character traits in the background, with the consequence that technologies have to chase the well-being of current animals as their behavior has not been subject to selection as had happened since the dawn of domestication.

To address sustainable animal welfare—where the word "sustainable" means a reliable and enduring set of knowledge developed on ethical and economic assumptions applied to livestock farming—the environment (physical and physio-phycological) where the animals will spend their lives must be designed and built with primary welfare in mind reachable by the real subjects (i.e., by the concrete individuals that will inhabit the precise temporal-physical space). Precision livestock farming (PLF), one of the information and communication technologies (ICT) often installed at farm level with no consistency and underutilized by farmers, needs to be designed and geared consistently to reduce stressors and to monitor on an individual level each animal, replacing the eyes and expertise once owned by the breeders.

An attempt to adapt breeding environments to the high performances of current animals is those pursued by the Ethical Barn®, a new way to treat and stay dairy cows in balance between social sustainability (farmer's income), environmental sustainability and respect for the right of the cow, aiming to give a decent life to cows [102]. This holistic approach allows to manage dairy cows with very high production performances—as the 20 tons/head of milk per lactation indicated as phenotypic to achieve in 10 years by the top dairy farms—while minimizing physical and mental stress, sources of infection and environment impact.

Unlike ruminants, in which the extra-resting circadian time budget is mainly spent on feeding-rumination-milking/suckling activities, in monogastric the conditions of captive breeding involve large time spaces that must be filled to avoid alienation phenomena and consequent psychological discomforts. In pigs, environmental enrichment is a rapidly developing technique, also supported by regulatory obligations within the EU [103], but which faces many obstacles globally to be applied correctly [104].

The Info-Farming revolution is under our eyes, supported by the evolution of the PLF, which allows the rapid diffusion of sensor devices with greater process data capacity at lower costs and pervasive ITC, which have become a prevalent part of our life today. In this last aspect, the spread of smartphones capable of collecting real-life snapshots and delivering them to the social media circuit in real time represents a challenge to the adoption of correct animal welfare procedures on farms. At the same time, the creation of social communities among breeders makes possible to exchange technical information and to share welfare procedures capable of rapidly improving the living conditions of animals raised on their farms.

One of the most important aspects is to bring the animals closer to the breeder, especially when large numbers of animals and large breeding facilities prevent real control of the animal welfare conditions [105]. As pointed out by Neethirajan [106], artificial intelligence (AI) can represent the next step to reach the better animal outcomes, processing and making strategic decisions to improve the welfare and health of farmed animals.

Summarizing, the profound changes in animal genetic and breeding environments mean that farmers are increasingly committed to livestock welfare. This goal can be achieved both with an adequate genetic improvement, which also has as its goal the behavioral characteristics associated with the well-being of animals in modern breeding conditions (genetics for environmental adaptation), and with the massive use of PLF technologies capable to meet the growing needs of increasingly performing animals (environment for high-producing animals) (Figure 1).

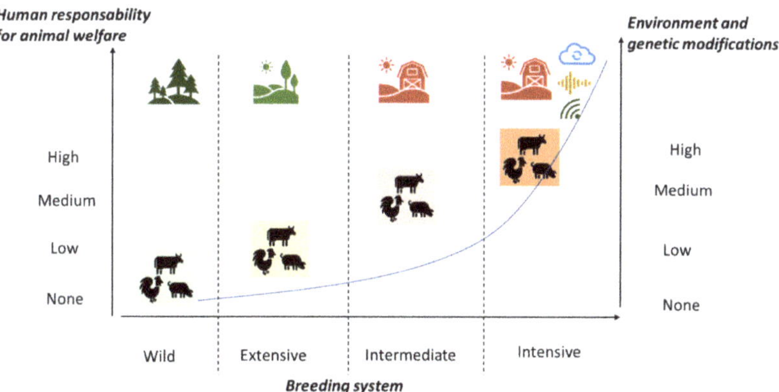

Figure 1. Human responsibility for the welfare of farm animals increases as the genetic modification of livestock and the technological evolution of the farm environment increase.

4. Practical Implications

Current farm animals are the result of millennia of selection which has led to their departure from the physiology and behavior of their ancestors. Cows that produce 12,000 L of milk, steers that grow by 2 kg per day, sows that give birth to 30 piglets per year and hens that lay 280 eggs per year are phenotypes very far from the performance of their wild ancestors (e.g., 1000 L of milk per year per cow, 0.3 kg/d of body weight increase in steers,

8 piglets/year per sow and 20–25 eggs per year per hen) which were commensurate with the turnover rate of the natural population in balance with its environment. The one to two order of magnitude of biological surplus produced by the modern livestock, used for human food interests, places these animals under great metabolic effort which affects their physiological and psychological well-being. With increasing breeding intensity, which is necessary to obtain higher yields, the environment and farm routines are profoundly changing. This means that farmers in intensive systems must pay more attention and make more investments to ensure the welfare of their animals. Farm animals are partially human-made artifacts that bear the responsibilities related to the metabolic fatigue they endure to render the important service of producing high quality food for humanity. Their welfare must be guaranteed in all breeding conditions and in all countries of the world.

In practical terms, current breeders have to deal with an increasing complexity in managing farms and with the containment of production costs. The design of comfortable environments suitable for high production needs is the first requirement for animal welfare, followed by respect for the management of the groups and the circadian needs of the various categories. Highly selected animals have sophisticated dietary needs not only in quantitative terms, but above all in the way they eat their food. Breeders must take into account specific feeding behaviors and schedule and program schedules and rations to meet the specific needs of their animals. High production also implies greater exposure to diseases for which farmers must adopt the best practices suggested by international organizations [107]. Finally, the complexity of managing a modern livestock farm requires continuous training of the staff and the careful vigilance of the entrepreneur so that all legal and company regulations are respected. It is worthy to note that animal welfare certification is recently wide spreading throughout the implementation of the Welfare Quality (WQ) protocols (based on the effects of the farming system on the animal condition-welfare), due to the fact that animal welfare must be guaranteed in all breeding conditions and that people in general, and consumers of animal products in particular, are interested in being sure that products they consume come from animals that have been correctly reared and managed.

5. Conclusions

This review confirms the hypothesis that the behavior of highly selected farm animals is long away from its natural state. There are behavioral changes in some farm animals compared to their wild ancestors due to genetic selection-domestication which should be considered in planning and implementing farm animal welfare standards at the farm level and an upgrade in the regulatory body.

Animal scientists must push governments to ensure that the welfare standards required by the regulations are universal and guide the international trade in products of animal origin; at the same time, they must devote more and more research spaces to make available scientific elements and technologies capable of making this reality operational and at acceptable costs everywhere.

Author Contributions: Conceptualization, G.P.; investigation, A.C. and G.P.; writing—original draft preparation, A.C. and G.P.; writing—review and editing, A.C.; supervision, G.P.; project administration, G.P.; funding acquisition, G.P. All authors have read and agreed to the published version of the manuscript.

Funding: This research was funded by "Fondo di Ateneo 2019–2020".

Institutional Review Board Statement: Ethical review and approval was not needed because no experiments were involved in this study.

Data Availability Statement: This paper did not involve any dataset.

Acknowledgments: The critical reading and comments of Ignacy Misztal from University of Georgia and of Nicolò Macciotta from University of Sassari are greatly acknowledged. The authors thank the anonymous referees for their precious advices that help to improve this review.

Conflicts of Interest: The authors declare no conflict of interest.

References

1. O'Donnell, P. Animals: Ethics, Rights & Law—A Transdisciplinary Bibliography. 2019. Available online: https://www.academia.edu/4843888/Animal_Ethics_Rights_and_Law_bibliography (accessed on 29 December 2020).
2. The Future of Animal-Sourced Foods: Ethical Considerations. *Anim. Front.* **2020**, *10*, 1. [CrossRef]
3. Pulina, G. Ethical meat: Respect for farm animals. *Anim. Front.* **2020**, *10*, 34–38. [CrossRef]
4. European Commission. Attitudes of Europeans towards Animal Welfare. *Spec. Eurobarometer* **2016**, *442*, 22.
5. Cambridge Dictionary. 2020. Available online: https://dictionary.cambridge.org/dictionary/english/welfare (accessed on 30 December 2020).
6. Carenzi, C.; Verga, M. Animal welfare: Review of the scientific concept and definition. *Ital. J. Anim. Sci.* **2009**, *8*, 21–30. [CrossRef]
7. Darwin, C. *The Expression of Emotions in Man and Animals*; Murray, J., Ed.; Oxford University Press: London, UK, 1872.
8. Romanes, G. Animal Intelligence. *Nature* **1884**, *30*, 267. [CrossRef]
9. Knoll, E. Dogs, darwinism and english sensibilities. In *Anthropomorphism, Anecdotes, and Animals*; Mitchell, R.W., Thompson, N.S., Miles, H.L., Eds.; State University of New York Press: Albany, NY, USA, 1997; pp. 12–21.
10. LeDoux, J.E. *The Deep History of Ourselves: The Four Billion Years Story of How We Got Conscious Brains*; Ed. Penguin Books: New York, NY, USA, 2019.
11. EU. Treaty of Lisbon, amending the treaty on European Union and the treaty establishing the European community. *OJEU* **2007**, *17*, 13.
12. Morgante, M.; Vallortigara, G. Animal welfare: Neurocognitive approaches. *Ital. J. Anim. Sci.* **2009**, *8* (Suppl. 1), 255–264. [CrossRef]
13. Vallortigara, G. Sentience does not require "higher" cognition. *Anim. Sentience* **2017**, *30*, 6.
14. Désiré, L.; Boissy, A.; Veissier, I. Emotions in farm animals: A new approach to animal welfare in applied ethology. *Behav. Process* **2002**, *60*, 165–180. [CrossRef]
15. De Waal, F.B.M. What is an animal emotion? *Ann. N. Y. Acad. Sci.* **2011**, *1224*, 191–206. [CrossRef] [PubMed]
16. Belyaev, D.K. Destabilizing selection as a factor in domestication. *J. Hered.* **1979**, *70*, 301–308. [CrossRef]
17. Kukekova, A.V.; Johnson, J.L.; Xiang, X.; Feng, S.; Liu, S.; Rando, H.M.; Kharlamova, A.V.; Herbeck, Y.; Serdyukova, N.A.; Xiong, Z.; et al. Red fox genome assembly identifies genomic regions associated with tame and aggressive behaviours. *Nat. Ecol. Evol.* **2018**, *2*, 1479–1491. [CrossRef]
18. Price, E.O. Behavioral development in animals undergoing domestication. *Appl. Anim. Behav. Sci.* **1999**, *65*, 245–271. [CrossRef]
19. Zeder, M.A. Core questions in domestication research. *Proc. Natl. Acad. Sci. USA* **2015**, *112*, 3191–3198. [CrossRef] [PubMed]
20. Darwin, C. *The Origin of Species*, 6th ed.; John Murray: London, UK, 1859; Volume 570.
21. Vonholdt, B.M.; Pollinger, J.P.; Lohmueller, K.E.; Han, E.; Parker, H.G.; Quignon, P.; Degenhardt, J.D.; Boyko, A.R.; Earl, D.A.; Auton, A.; et al. Genome-wide SNP and haplotype analyses reveal a rich history underlying dog domestication. *Nature* **2010**, *464*, 898–902. [CrossRef] [PubMed]
22. Cagan, A.; Blass, T. Identification of genomic variants putatively targeted by selection during dog domestication. *BMC Evol. Biol.* **2016**, *16*, 10. [CrossRef]
23. MacHugh, D.E.; Larson, G.; Orlando, L. Taming the past: Ancient DNA and the study of animal domestication. *Ann. Rev. Anim. Biosci.* **2017**, *5*, 329–351. [CrossRef]
24. Conolly, J.; Colledge, S.; Dobney, K.; Vigne, J.D.; Peters, J.; Stopp, B.; Manning, K.; Shennan, S. Meta-analysis of zooarchaeological data from SW Asia and SE Europe provides insight into the origins and spread of animal husbandry. *J. Archaeol. Sci.* **2011**, *38*, 538–545. [CrossRef]
25. Asouti, E.; Fuller, D.Q. A contextual approach to the emergence of agriculture in Southwest Asia reconstructing early Neolithic plant-food production. *Curr. Anthr.* **2013**, *54*, 299–345. [CrossRef]
26. Larson, G.; Piperno, D.R.; Allaby, R.G.; Purugganan, M.D.; Andersson, L.; Arroyo-Kalin, M.; Barton, L.; Vigueira, C.C.; Denham, T.; Dobney, K.; et al. Current perspectives and the future of domestication studies. *Proc. Natl. Acad. Sci. USA* **2014**, *111*, 6139–6146. [CrossRef] [PubMed]
27. Gepts, P.; Papa, R. Evolution during domestication. In *Encyclopedia of Life Sciences*; Nature Publishing Group, Ed.; Macmillan Publishers: New York, NY, USA, 2001.
28. Clutton-Brock, J.A. *Natural History of Domesticated Mammals*; Cambridge University Press: Cambridge, UK, 1999.
29. Carneiro, M.; Afonso, S.; Geraldes, A.; Garreau, H.; Bolet, G.; Boucher, S.; Tircazes, A.; Queney, G.; Nachman, M.W.; Ferrand, N. The genetic structure of domestic rabbits. *Mol. Biol. Evol.* **2011**, *28*, 1801–1816. [CrossRef] [PubMed]
30. Rasali, D.P.; Shrestha, J.N.B.; Crow, G.H. Development of composite sheep breeds in the world: A review. *Can. J. Anim. Sci.* **2008**, *86*, 1–24. [CrossRef]
31. Felius, M.; Beerling, M.L.; Buchanan, D.S.; Theunissen, B.; Koolmees, P.A.; Lenstra, J.A. On the history of cattle genetic resources. *Diversity* **2014**, *6*, 705–750. [CrossRef]
32. FAO. Status of Animal Genetic Resources. Available online: www.fao.org/docrep/pdf/010/a1250e/a1250e02.pdf (accessed on 2 December 2020).
33. Ryder, M.L. *Sheep and Man*; Gerald Duckworth & Co. Ltd.: London, UK, 1983.
34. Loftus, R.T.; MacHugh, D.E.; Bradley, D.G.; Sharp, P.M.; Cunningham, P. Evidence for two independent domestications of cattle. *Proc. Natl. Acad. Sci. USA* **1994**, *91*, 2757–2761. [CrossRef] [PubMed]

35. Upadhyay, M.R.; Chen, W.; Lenstra, J.A.; Goderie, C.R.J.; MacHugh, D.E.; Park, S.D.E.; Magee, D.A.; Matassino, D.; Ciani, F.; Megens, H.J.; et al. Genetic origin, admixture and population history of aurochs (Bos primigenius) and primitive European cattle. *Heredity* **2016**, *118*, 169–176. [CrossRef] [PubMed]
36. Zeder, M. Domestication and early agriculture in the Mediterranean basin: Origins, diffusion, and impact. *Proc. Natl. Acad. Sci. USA* **2008**, *105*, 11592–11604. [CrossRef] [PubMed]
37. Harri, M.; Mononen, J.; Ahola, L.; Plyusnina, I.; Rekilä, T. Behavioural and physiological differences between silver foxes selected and not selected for domestic behaviour. *Anim. Welf.* **2003**, *12*, 305–314.
38. Thrope, W.H. *The Origins and Rise of Ethology*; Praeger: London, UK, 1979.
39. Jensen, P. Domestication—From behaviour to genes and back again. *Appl. Anim. Behav. Sci.* **2006**, *97*, 3–15. [CrossRef]
40. Kumar, A.; Faiq, M.A.; Pareek, V.; Kulandhasamy, M. Heritability of Behavior. In *Encyclopedia of Animal Cognition and Behavior*; Vonk, J., Shackelford, T.K., Eds.; Springer International Publishing: Berlin/Heidelberg, Germany, 2017.
41. Dochtermann, N.A.; Schwab, T.; Anderson Berdal, M.; Dalos, J.; Royauté, R. The heritability of behavior: A meta-analysis. *J. Hered.* **2019**, *110*, 403–410. [CrossRef]
42. Wilkins, A.S.; Wrangham, R.W.; Fitch, W.T. The "domestication syndrome" in mammals: A unified explanation based on neural crest cell behavior and genetics. *Genetics* **2014**, *197*, 795–808. [CrossRef] [PubMed]
43. Wilkins, A.S. A striking example of developmental bias in an evolutionary process: The "domestication syndrome". *Evol. Dev.* **2020**, *22*, 143–153. [CrossRef] [PubMed]
44. Brown, T.A.; Jones, M.K.; Powell, W.; Allaby, R.G. The complex origins of domesticated crops in the Fertile Crescent. *Trends Ecol. Evol.* **2009**, *24*, 103–109. [CrossRef] [PubMed]
45. Sánchez-Villagra, M.R.; Geiger, M.; Schneider, R.A. The taming of the neural crest: A developmental perspective on the origins of morphological covariation in domesticated mammals. *R. Soc. Open Sci.* **2016**, *3*, 160107. [CrossRef]
46. Harlan, J.R.; de Wet, J.M.J.; Price, E.G. Comparative evolution of cereals. *Evolution* **1973**, *27*, 311–325. [CrossRef] [PubMed]
47. Kendler, K.S.; Greenspan, R.J. The nature of genetic influences on behavior: Lessons from "simpler" organisms. *Am. J. Psychiatry* **2006**, *163*, 1683–1694. [CrossRef]
48. Jensen, P. Behavior genetics and the domestication of animals. *Annu. Rev. Anim. Biosci.* **2014**, *2*, 85–104. [CrossRef] [PubMed]
49. Price, E.O. *Animal Domestication and Behavior*; CABI Publishers: Wallingford, UK, 2002.
50. York, R.A. Assessing the genetic landscape of animal behavior. *Genetics* **2018**, *209*, 223–232. [CrossRef]
51. Cesarani, A.; Sechi, T.; Gaspa, G.; Usai, M.G.; Sorbolini, S.; Macciotta, N.P.P.; Carta, A. Investigation of genetic diversity and selection signatures between Sarda and Sardinian Ancestral black, two related sheep breeds with evident morphological differences. *Small Rumin. Res.* **2019**, *177*, 68–75. [CrossRef]
52. Luquetti, D.V.; Heike, C.L.; Hing, A.V.; Cunningham, M.L.; Cox, T.C. Microtia: Epidemiology and genetics. *Am. J. Med. Genet.* **2012**, *158A*, 124–139. [CrossRef]
53. Bartel-Friedrich, S. Congenital auricular malformations: Description of anomalies and syndromes. *Facial Plast. Surg.* **2015**, *31*, 567–580. [CrossRef]
54. Soundararajan, C.; Nagarajan, K.; Prakash, M.A. Occurrence of Microtia in Madras red Sheep-A study of 12 flocks. *Intas. Polivet.* **2016**, *17*, 255–257.
55. Kijas, J.W.; Lenstra, J.A.; Hayes, B.; Boitard, S.; Neto, L.R.P.; San Cristobal, M.; Servin, B.; McCulloch, R.; Whan, V.; Gietzen, K.; et al. Genome-wide analysis of the world's sheep breeds reveals high levels of historic mixture and strong recent selection. *PLoS Biol.* **2012**, *10*, e1001258. [CrossRef] [PubMed]
56. Saetre, P.; Lindberg, J.; Leonard, J.A.; Olsson, K.; Pettersson, U.; Ellegren, H.; Bergström, T.F.; Vilà, C.; Jazin, E. From wild wolf to domestic dog: Gene expression changes in the brain. *Brain Res. Mol. Brain Res.* **2004**, *126*, 198–206. [CrossRef] [PubMed]
57. Albert, F.W.; Somel, M.; Carneiro, M.; Aximu-Petri, A.; Halbwax, M.; Thalmann, O.; Blanco-Aguiar, J.A.; Plyusnina, I.Z.; Trut, L.; Villafuerte, R.; et al. A comparison of brain gene expression levels in domesticated and wild animals. *PLoS Genet.* **2012**, *8*, e1002962. [CrossRef]
58. Nätt, D.; Rubin, C.J.; Wright, D.; Johnsson, M.; Beltéky, J.; Andersson, L.; Jensen, P. Heritable genome-wide variation of gene expression and promoter methylation between wild and domesticated chickens. *BMC Genom.* **2012**, *13*, 1–12. [CrossRef]
59. Carneiro, M.; Piorno, V.; Rubin, C.J.; Alves, J.M.; Ferrand, N.; Alves, P.C.; Andersson, L. Candidate genes underlying heritable differences in reproductive seasonality between wild and domestic rabbits. *Anim. Genet.* **2015**, *46*, 418–425. [CrossRef]
60. Christie, M.; Marine, M.; Fox, S.; French, R.A.; Blouin, M.S. A single generation of domestication heritably alters the expression of hundreds of genes. *Nat. Commun.* **2016**, *7*, 10676. [CrossRef]
61. Axelsson, E.; Ratnakumar, A.; Arendt, M.L.; Maqbool, K.; Webster, M.T.; Perloski, M.; Liberg, O.; Arnemo, J.M.; Hedhammar, A.; Lindblad-Toh, K. The genomic signature of dog domestication reveals adaptation to a starch-rich diet. *Nature* **2013**, *495*, 360–364. [CrossRef]
62. Carneiro, M.; Rubin, C.J.; Di Palma, F.; Albert, F.W.; Alföldi, J.; Barrio, A.M.; Pielberg, G.; Rafati, N.; Sayyab, S.; Turner-Maier, J.; et al. Rabbit genome analysis reveals a polygenic basis for phenotypic change during domestication. *Science* **2014**, *345*, 1074–1079. [CrossRef]
63. Montague, M.J.; Li, G.; Gandolfi, B.; Khan, R.; Aken, B.L.; Searle, S.M.J.; Minx, P.; Hillier, L.W.; Koboldt, D.C.; Davis, B.W.; et al. Comparative analysis of the domestic cat genome reveals genetic signatures underlying feline biology and domestication. *Proc. Natl. Acad. Sci. USA* **2014**, *111*, 17230–17235. [CrossRef]

64. Mäkinen, H.; Vasemägi, A.; McGinnity, P.; Cross, T.F.; Primmer, C.R. Population genomic analyses of early-phase Atlantic Salmon (Salmo salar) domestication/captive breeding. *Evol. Appl.* **2015**, *8*, 93–107. [CrossRef] [PubMed]
65. Nunney, L. The cost of natural selection revisited. *Ann. Zool. Fenn.* **2003**, *40*, 185–194.
66. Roberge, C.; Normandeau, É.; Einum, S.; Guderley, H.; Bernatchez, L. Genetic consequences of interbreeding between farmed and wild Atlantic salmon: Insights from the transcriptome. *Mol. Ecol.* **2008**, *17*, 314–324. [CrossRef] [PubMed]
67. Christie, M.R.; Ford, M.J.; Blouin, M.S. On the reproductive success of early-generation hatchery fish in the wild. *Evol. Appl.* **2014**, *7*, 883–896. [CrossRef]
68. Floridi, L. *The Logic of Information. A Theory of Philosophy as Conceptual Design*; Oxford University Press: Oxford, UK, 2019; p. 240.
69. Khun, T.S. *The Structure of Scientific Revolutions*; University of Chicago Press: Chicago, IL, USA, 1962; p. 264.
70. Geffroy, B.; Sadoul, B.; Putman, B.J.; Berger-Tal, O.; Garamszegi, L.Z.; Møller, A.P.; Blumstein, D.T. Evolutionary dynamics in the Anthropocene: Life history and intensity of human contac shape antipredator responses. *PLoS Biol.* **2020**, *18*, e3000818. [CrossRef] [PubMed]
71. Food and Agriculture Organization of the United Nations. Available online: http://www.fao.org/3/i2414e/i2414e05.pdf (accessed on 2 March 2021).
72. Leonard, S.M.; Xin, H.; Brown-Brandl, T.M.; Ramirez, B.C.; Dutta, S.; Rohrer, G.A. Effects of Farrowing Stall Layout and Number of Heat Lamps on Sow and Piglet Production Performance. *Animals* **2020**, *10*, 348. [CrossRef]
73. Rollin, B. *The Frankenstein Syndrome: Ethical and Social Issues in the Genetic Engineering of Animals*; Cambridge University Press: Cambridge, UK; New York, NY, USA, 1995; p. 245.
74. Shriver, A. Prioritizing the protection of welfare in gene-edited livestock. *Anim. Front.* **2020**, *10*, 39–44. [CrossRef] [PubMed]
75. Turner, S.P.; White, I.M.S.; Brotherstone, S.; Farnworth, M.J.; Knap, P.W.; Penny, P.; Mendl, M.; Lawrence, A.B. Heritability of post-mixing aggressiveness in grower-stage pigs and its relationship with production traits. *Anim. Sci.* **2006**, *82*, 615–620. [CrossRef]
76. Meffert, L.M.; Hicks, S.K.; Regan, J.L. Nonadditive genetic effects in animal behavior. *Am. Nat.* **2002**, *160* (Suppl. 6), S198–S213. [CrossRef]
77. Angvall, B.; Jöngren, M.; Strandberg, E.; Jensen, P. Heritability and genetic correlations of fear-related behaviour in red junglefowl—possible implications for early domestication. *PLoS ONE* **2012**, *7*, e35162. [CrossRef]
78. Kjaer, J.B.; Sørensen, P. Feather pecking behaviour in White Leghorn chickens, a genetic study. *Br. Poult. Sci.* **1997**, *38*, 333–341. [CrossRef] [PubMed]
79. Riber, A.B.; Hinrichsen, L.K. Welfare consequences of omitting beak trimming in barn layers. *Front. Vet. Sci.* **2017**, *4*, 222. [CrossRef]
80. Ding, Y.; Berrocal, A.; Morita, T.; Longden, K.D.; Stern, D.L. Natural courtship song variation caused by an intronic retroelement in an ion channel gene. *Nature* **2016**, *536*, 329–332. [CrossRef]
81. Boissy, A.; Bouix, J.; Orgeur, P.; Poindron, P.; Bibé, B.; Le Neindre, P. Genetic analysis of emotional reactivity in sheep: Effects of the genotypes of the lambs and of their dams. *Genet. Sel. Evol.* **2005**, *37*, 381–401. [CrossRef] [PubMed]
82. Turner, S.P.; Roehe, R.; Mekkawy, W.; Farnworth, M.J.; Knap, P.W.; Lawrence, A.B. Bayesian analysis of genetic associations of skin lesions and behavioural traits to identify genetic components of individual aggressiveness in pigs. *Behav. Genet.* **2008**, *38*, 67–75. [CrossRef] [PubMed]
83. Drews, C. The concept and definition of dominance in animal behaviour. *Behaviour* **1993**, *125*, 283–313. [CrossRef]
84. Phillips, C.J.C. *Cattle Behaviour*; Farming Press Books: Ipswich, UK, 1993.
85. Knap, P.W.; Merks, J.W.M. A note on the genetics of aggressiveness of primiparous sows towards their piglets. *Livest. Prod. Sci.* **1987**, *17*, 161–167. [CrossRef]
86. Hofmeyr, I. The fighting bulls of Portugal: Genetics & handling. *Stockfarm* **2016**, *6*, 34–35.
87. Ulloa-Arvizu, R.; Gayosso-Vázquez, A.; Ramos-Kuri, M.; Estrada, F.J.; Montaño, M.; Alonso, R. A Genetic analysis of Mexican Criollo cattle populations. *J. Anim. Breed. Genet.* **2008**, *125*, 351–359. [CrossRef]
88. Buchanan, D.S.; Lenstra, J.A. Breeds of cattle. In *The Genetics of Cattle*, 2nd ed.; Garrick, D.J., Ruvinsky, A., Eds.; CABI Publishers: Wallingford, UK, 2015; p. 33.
89. Sartori, C.; Mantovani, R. Effects of inbreeding on fighting ability measured in Aosta Chestnut and Aosta Black Pied cattle. *J. Anim. Sci.* **2012**, *90*, 2907–2915. [CrossRef]
90. Sartori, C.; Manser, M.B.; Mantovani, R. Relationship between number and intensity of fighting: Evidence from cow fighting tournaments in Valdostana cattle. *Ital. J. Anim. Sci.* **2014**, *13*, 2907–2915. [CrossRef]
91. Sartori, C.; Guzzo, N.; Mantovani, R. Genetic correlations of fighting ability with somatic cells and longevity in cattle. *Animal* **2020**, *14*, 13–21. [CrossRef]
92. Mastrangelo, S.; Ben Jemaa, S.; Ciani, E.; Sottile, G.; Moscarelli, A.; Boussaha, M.; Montedoro, M.; Pilia, F.; Cassandro, M. Genome-wide detection of signatures of selection in three Valdostana cattle populations. *J. Anim. Breed. Genet.* **2020**, *137*, 609–621. [CrossRef] [PubMed]
93. Misztal, I. Breeding and genetics symposium: Resilience and lessons from studies in genetics of heat stress. *J. Anim. Sci.* **2017**, *95*, 1780–1787. [CrossRef] [PubMed]
94. Blasco, A. Animal breeding methods and sustainability. *Am. Nat.* **2012**, *139*, 749–770.
95. Frankham, R. Genetic adaptation to captivity in species conservation programs. *Mol. Ecol.* **2008**, *17*, 325–333. [CrossRef] [PubMed]
96. Crowell-Davis, S.L. Behavior problems in pet rabbits. *J. Exot. Pet Med.* **2007**, *16*, 38–44. [CrossRef]
97. Scott, J.P.; Fuller, J.L. *Genetics and the Social Behavior of the Dog*; University of Chicago Press: Chicago, IL, USA, 2012; Volume 570.

98. Ilska, J.; Haskell, M.J.; Blott, S.C.; Sánchez-Molano, E.; Polgar, Z.; Lofgren, S.E.; Clements, D.N.; Wiener, P. Genetic characterization of dog personality traits. *Genetics* **2017**, *206*, 1101–1111. [CrossRef] [PubMed]

99. McDonnel, S. *The Equid Ethogram: A Practical Field Guide to Horse Behaviour*; The Blood Horse Inc.: Lexington, KY, USA, 2003.

100. Deesing, M.J.; Grandin, T. Behavior Genetics of the Horse (*Equus caballus*). In *Genetics and the Behavior of Domestic Animals*, 2nd ed.; Academic Press: Cambridge, MA, USA, 2014; pp. 237–290.

101. Pulina, G.; Francesconi, A.H.D.; Stefanon, B.; Sevi, A.; Calamari, L.; Lacetera, N.; Dell'Orto, V.; Pilla, F.; Marsan, P.A.; Mele, M.; et al. Sustainableruminant production to help feed the planet. *Ital. J. Anim. Sci.* **2017**, *16*, 140–171. [CrossRef]

102. Pulina, G.; Tondo, A.; Danieli, P.P.; Primi, R.; Crovetto, G.M.; Fantini, A.; Macciotta, N.P.P.; Atzori, A.S. How to manage cows yielding 20,000 kg of milk: Technical challenges and environmental implications. *Ital. J. Anim. Sci.* **2020**, *19*, 865–879. [CrossRef]

103. EU Commission, 2016. Commission recommendation (EU) 2016/336 of 8 March 2016 on the application of Council Directive 2008/120/EC laying down minimum standards for the protection of pigs as regards measures to reduce the need for tail-docking. *OJEU 9.3* **2016**, *62*, 20–22.

104. Van de Weerd, H.; Ison, S. Providing Effective Environmental Enrichment to Pigs: How Far Have We Come? *Animals* **2019**, *9*, 254. [CrossRef] [PubMed]

105. Norton, T.; Chen, C.; Larsen, M.L.V.; Berckmans, D. Review: Precision livestock farming: Building 'digital representations' to bring the animals closer to the farmer. *Animal* **2019**, *13*, 3009–3017. [CrossRef] [PubMed]

106. Neethirajan, S. The role of sensors, big data and machine learning in modern animal farming. *Sens. Bio-Sens. Res.* **2020**, *29*. [CrossRef]

107. Word Organization for Animal Health. 2019. Available online: https://www.oie.int/fileadmin/Home/eng/Health_standards/tahc/current/chapitre_aw_introduction.pdf (accessed on 2 March 2021).

MDPI
St. Alban-Anlage 66
4052 Basel
Switzerland
Tel. +41 61 683 77 34
Fax +41 61 302 89 18
www.mdpi.com

Animals Editorial Office
E-mail: animals@mdpi.com
www.mdpi.com/journal/animals

9 783036 555621